D0076966

The Facts On File

DICTIONARY of BOTANY

The Facts On File

DICTIONARY of BOTANY

Edited by
Jill Bailey

Facts On File, Inc.

The Facts On File Dictionary of Botany

Facts On File, Inc.
132 West 31st Street
New York NY 10001

Library of Congress Cataloging-in-Publication Data

The Facts on File dictionary of botany / edited by Jill Bailey.
p. cm.
Includes bibliographical references (p.).
ISBN 0-8160-4910-6
1. Botany—Dictionaries. I. Title: Dictionary of botany. II. Bailey, Jill.
III. Facts on File, Inc.

QK9.F33 2002
580'.3—dc21 2002035202

Compiled and typeset by Market House Books Ltd, Aylesbury, UK

Printed in the United States of America

MP 10 9 8 7 6 5 4 3 2 1

This book is printed on acid-free paper

PREFACE

This dictionary is one of a series covering the terminology and concepts used in important branches of science. *The Facts On File Dictionary of Botany* is planned as an additional source of information for students taking Advanced Placement (AP) Science courses in high schools, but will also be helpful to older students taking introductory college courses.

This volume covers the whole area of pure and applied plant science including anatomy and plant morphology, plant physiology, biochemistry, cell biology, genetics, evolution, and ecology. It also covers the taxonomy and classification of plants, with entries for the higher-ranking taxa. The definitions are intended to be clear and informative and, where possible, we have provided helpful diagrams and examples. The book also has a selection of short biographical entries for people who have made important contributions to the field. The appendices include lists of webpages and an informative bibliography.

The book will be a helpful additional source of information for anyone studying AP Biology, notably the sections on Organisms and Populations, Structure and Function of Plants and Animals, and Molecules and Cells. However, we have not restricted the content to this syllabus. Modern plant science is a subject of considerable importance and we hope that this book will be useful to anyone interested in the subject.

ACKNOWLEDGMENTS

Consultant

Andrew Lack B.Sc., Pd.D.

Contributors

Eve Daintith B.Sc.
Elizabeth Tootill B.Sc.

CONTENTS

A

ABA *See* abscisic acid.

abaxial In structures such as a leaves and petals, the side facing away from the main axis, i.e. the lower surface. In lateral organs such as leaves, abaxial is synonymous with the underside. *Compare* adaxial.

abiotic environment The nonliving factors of the environment that influence ecological systems. Abiotic factors include climate, chemical pollution, geographical features, etc.

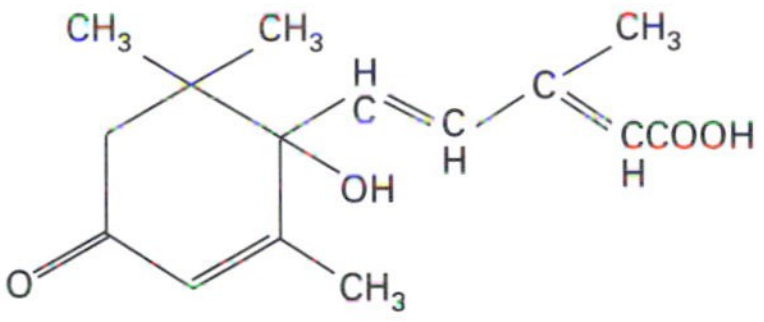

Abscisic acid

abscisic acid (**ABA**) A plant hormone, that functions chiefly as an inhibitor of growth and cell elongation. Abscisic acid has a variety of effects related to seed dormancy and stress responses: it regulates protein expression in seed development leading to dormancy and is one of the hormones involved in bud dormancy; it regulates stress responses by, for example, closing stomata in times of water shortage and increasing the ability of roots to carry water. It can also promote root growth and inhibit shoot growth. Despite its name it does not directly promote abscission, but only indirectly through increasing ethylene production. Formerly it was known as *abscisin II* or *dormin.*

abscission The organized loss of part of a plant, usually a leaf, fruit, or unfertilized flower. An *abscission zone* occurs at the base of the organ. Here a separation layer (*abscission layer*) is formed by breakdown or separation of cells and final severance occurs when the vascular bundles are broken mechanically, e.g. by wind or rain. The abscission layer is activated by increasing levels of ethylene and a decreasing concentration of auxin

absolute humidity *See* humidity.

absolute pollen frequency (**APF**) *See* pollen analysis.

absorption **1.** The uptake of liquid by cells and organs. In plants, water and mineral salts are absorbed mainly by the root hairs, just behind the root tips.
2. The capture of radiant energy by plant pigments. About 80% of the visible light falling on a leaf is absorbed, and about 10% of the infrared radiation.

absorption spectrum A plot of the absorbance by a substance of radiation at different wavelengths, usually of ultraviolet, visible, or infrared radiation. It can give information about the identity or quantity of a substance. Chlorophylls, for example, have absorption peaks in the red and blue (and therefore reflect green light). *Compare* action spectrum.

accessory cell *See* subsidiary cell.

accessory chromosome *See* B chromosome.

accessory pigment *See* photosynthetic pigments.

acellular Denoting relatively large tissues or organisms that are not composed of discrete cells and are, in effect, unicellular. In flowering plants, for instance, the early stages of endosperm development are often acellular. Other examples include aseptate fungal hyphae and certain green algae, such as *Acetabularia*. The term is used in preference to unicellular to distinguish such structures (which are often multinucleate) from conventional cells and show their equivalence to multicellular structures. *See also* coenocyte.

Aceraceae A family of temperate and tropical trees and shrubs that includes the maples and sycamores.

acetaldehyde (**ethanal**) An aldehyde, CH_3CHO, that is an intermediate in the conversion of pyruvic acid to ethanol during the final stage of glycolysis during anaerobic respiration in plants. It is involved in the synthesis and breakdown of the amino acid threonine.

acetic acid (**ethanoic acid**) A carboxylic acid, CH_3COOH, obtained by the oxidation of ethyl alcohol. Acetic acid is a component of vinegar (which is obtained by bacterial oxidation of wine waste). It is used as an alternative carbon source by certain green algae. When combined with COENZYME A to form ACETYL COA, it plays a key role in AEROBIC RESPIRATION.

acetocarmine A stain used to color chromosomes deep reddish-black for viewing with a light microscope. Tissues are fixed in acetic acid before applying acetocarmine. *See* fixation; staining.

acetyl CoA (**acetyl coenzyme A**) A compound made up of acetyl and coenzyme A linked by a sulfur bridge. Acetyl CoA plays a key role in metabolism, being a precursor of the KREBS CYCLE and GLYOXYLATE CYCLE, and the starting point for synthesis of fatty acids, terpenes, and some amino acids. The synthesis of acetyl CoA is a high-energy process, requiring energy from ATP, which is converted to AMP.

achene A dry indehiscent fruit formed from an ovary with a single carpel containing a single seed, e.g. oak (*Quercus*). Different types of achenes include the CARYOPSIS, CYPSELA, NUT, and SAMARA.

acicular Needle-shaped.

acid A substance that gives rise to hydrogen ions (or H_3O^+) when dissolved in water. An acid in aqueous solution will have a pH below 7. *Lowry–Brønsted theory* defines an acid as a substance that exhibits a tendency to release a proton, and a base as a substance that tends to accept a proton. Strong acids (e.g. HNO_3) react completely with water to give H_3O^+. Weak acids (e.g. CH_3COOH) are only partly dissociated because H_3O^+ is a stronger acid than the free acids.

acidic stain *See* staining.

acid rain The deposition of acids by natural precipitation, mainly by rain but also by snow and fog. Acids are formed by reaction of gaseous waste products, particularly sulfur dioxide, and also nitrogen oxides, with moisture in the air to form sulfuric and nitric acids. The subsequent precipitation has led to raised acidity in forests and lakes, especially in Scandinavia and some other parts of northern Europe, damaging the environment. Acid rain is often used more loosely for any atmospheric pollutant that dissolves in precipitation causing environmental damage, such as carbon monoxide or ozone, and these may interact with the acids.

The most serious pollutant is sulfur dioxide, which comes mainly from burning coal and is, consequently, less serious now than in the twentieth century. Sulfuric acid in soils may lead to the formation of ammonium sulfate, which causes the release of toxic aluminum and heavy metal ions that inhibit metabolic activity. In waterways these can damage the gills of fish. Unpolluted rain is normally slightly acidic with a pH of 5.0–5.6.

acid soil A soil with a pH less than 6.0. Such soils usually form in areas of heavy rainfall, which causes leaching of lime from the surface layers, or over acid substrata such as granite or sand. In the acidic conditions, decomposition of organic material in the soil is slow. Acid soils often contain substantial concentrations of iron and aluminum hydroxides. *See* brown earth; podsol.

Acrasiomycota (**cellular slime molds**) A class of protoctists in the phylum Rhizopoda. Formerly considered to be fungi, they are made up of independently living amebas that feed on bacteria, which they ingest by phagocytosis. Under conditions of food shortage or other stresses, they come together to form a slug-like aggregation of amebas called a *pseudoplasmodium*, which acts like an individual organism.

acrocarpous Describing mosses in which the reproductive organs are borne at the top of the main axis, which is usually erect, so that subsequent growth is sympodial.

acrocentric *See* centromere.

acropetal Describing a process (such as growth or development) that progresses from the base or point of attachment, so that the oldest parts are at the base and the youngest are at the tip. *Compare* basipetal.

actin A globular contractile protein (G-actin) that makes up 10–15% of the total cell protein in eukaryotic cells. It is a major component of the CYTOSKELETON. It also controls motility, cellular movements, and cyclosis. G-actin can polymerize into helical strands that coil together to form microfilaments called *fibrous actin* (F-actin).

Actinobacteria (**ray fungi; actinomycetes**) A phylum of Eubacteria that contains the true actinobacteria and the *coryneform bacteria*. The true actinobacteria form branching filaments that resemble small fungal mycelia, with hyphae usually less than 1.5 μm in diameter; the coryneform bacteria are rodlike structures. The Actinobacteria are distinguished by producing *actinospores*: entire cells encysted in thick walls to form resistant spores.

actinodromous (**palmate; digitate**) Describing a form of leaf VENATION in which three or more primary veins radiate from the base of the lamina toward the margin, as in sycamore (*Acer pseudoplatanus*), resulting in a leaf that has several large lobes or several leaflets all originating at the same point.

actinomorphy *See* radial symmetry.

actinomycetes *See* Actinobacteria.

actinostele A type of protostele in which the xylem is star shaped and the phloem lies between the points of the star. Protosteles are found in the primary roots of higher plants, and in some species of *Lycopodium* and *Psilotum*. *See* stele.

action spectrum A graph showing the effect of different wavelengths of radiation, usually light, on a given process. It is often similar to the absorption spectrum of the substance that absorbs the radiation and can therefore be helpful in identifying that substance. For example, the action spectrum of photosynthesis is similar to the absorption spectrum of chlorophyll. It shows the net assimilation plotted against wavelength. *See* absorption spectrum. See illustration overleaf.

activation energy The minimum extra energy that must be put into a system to enable a reaction to occur. It increases the energy levels of participating molecules, and hence raises their reactivity, often by raising electrons to an excited state. ENZYMES enhance reaction rates by lowering the activation energy. This is often achieved by the attraction of the reactants to the enzyme's active site, thus coming into closer proximity, or by the enzyme causing conformational or electrostatic changes in one or more of the reactants.

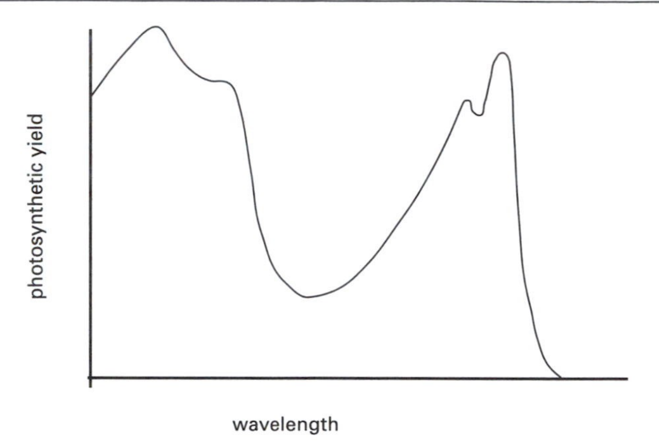

Action spectrum of photosynthesis

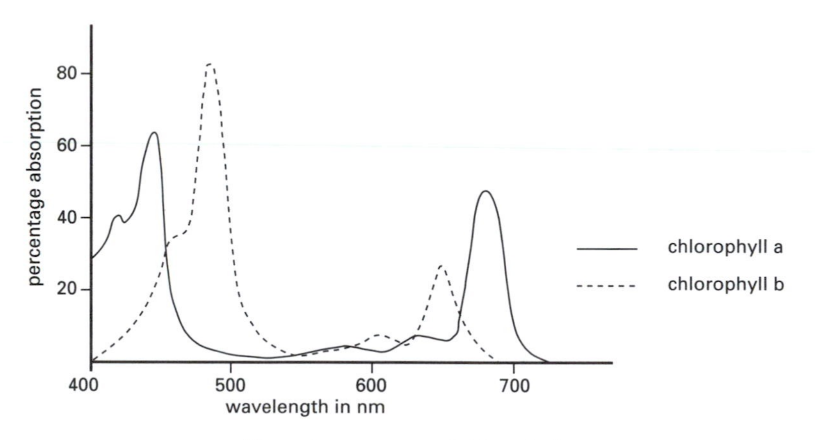

Absorption spectrum of chlorophylls

activator 1. A metal ion that acts together with an enzyme or its substrate to bring about a reaction.
2. A protein that positively regulates the transcription of a gene.

active site The particular part of an enzyme molecule that combines with and acts on the substrate. The active site consists of amino acids arranged in a configuration specific to a particular substrate or type of substrate. Binding of an inhibiting compound elsewhere on the enzyme molecule may change this configuration and hence the efficiency of the enzyme activity. *See* allosteric site.

active transport The transport of molecules or ions across a cell membrane against a concentration gradient, with the expenditure of energy, usually in the form of ATP. It is probably an attribute of all cells. Anything that interferes with the provision of energy will interfere with active transport. The mechanism typically involves a carrier protein that spans the cell membrane and transfers substances in or out of the cell by changing shape.

acuminate Describing a structure that gradually narrows to a point, such as certain leaves.

acylglycerol (**glyceride**) An ester of glycerol and one or more fatty acids. They may be mono-, di-, or triacylglycerols according to the number of –OH groups esterified. The fat stores of the body consist mainly of triacylglycerols (triglycerides). These can form a source of energy when carbohydrate levels are low, being broken down by lipases into fatty acids, which can enter metabolic pathways. Glycolipids (glycosyldiacylglycerols) are diacylglycerols with a sugar attached to the unesterified hydroxyl group. *See also* lipid.

adaptation The extent to which an organism, or a physiological or structural characteristic of an organism, is suited to a particular environment. Specialist organisms that have become highly adapted to one environment are then often not so adaptable as less specialized organisms and are at a disadvantage in a changing environment (adaptation versus adaptability).

adaptive enzyme *See* inducible enzyme.

adaptive radiation The formation through evolution of a number of different varieties or species from a common ancestor. It is often seen most clearly on isolated oceanic islands, e.g. the many forms of trees, shrubs, vines, and herbs of the genus *Hedyotis* on the Hawaiian Islands, but the term can apply to the range of species in any genus or family.

adaxial In structures such as a leaves and petals, the side facing toward the main axis, i.e. the upper surface. *Compare* abaxial.

adder's tongue ferns *See* Ophioglossales.

adenine A nitrogenous base (6-aminopurine) found in DNA and RNA. It is also a constituent of certain coenzymes, e.g. NAD and FAD, and when combined with the sugar ribose it forms the nucleoside ADENOSINE found in AMP, ADP, and ATP. Adenine has a purine ring structure.

adenosine (**adenine nucleoside**) A nucleoside formed from adenine linked to D-ribose with a β-glycosidic bond. Adenosine triphosphate (ATP) is a nucleotide derived from adenosine.

adenosine diphosphate *See* ADP.

adenosine monophosphate *See* AMP.

adenosine triphosphate *See* ATP.

Adiantaceae (**maidenhair ferns**) A family of ferns (Filicinophyta) found throughout the world, especially in the moist American tropics.

ADP (**adenosine diphosphate**) A nucleotide consisting of adenine and ribose with two phosphate groups attached. The

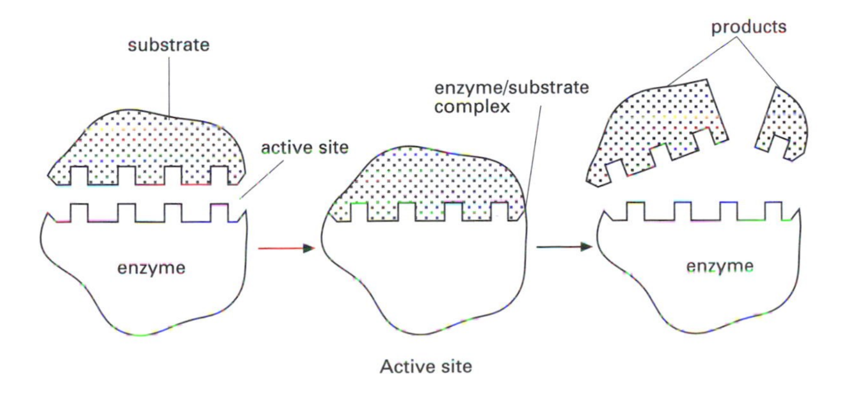

Active site

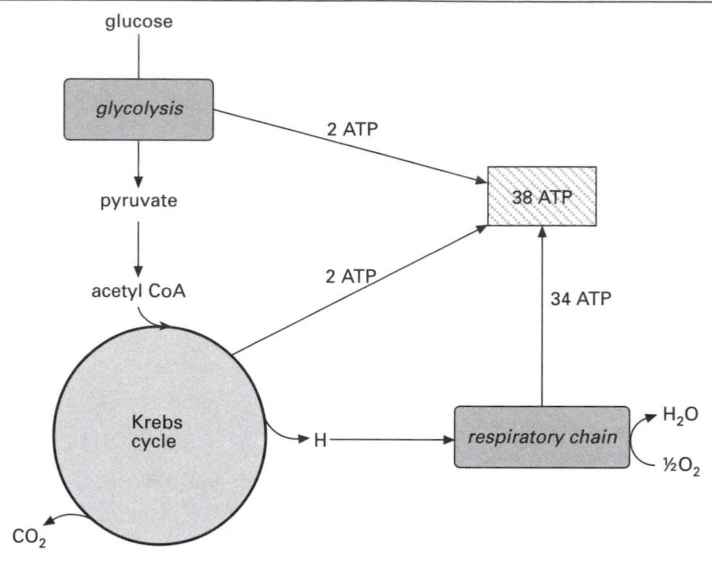

Aerobic respiration

second phosphate is attached to the first by a high-energy bond.

adventitious Describing plant organs that arise in unexpected places, for example the development of adventitious roots from stems, and adventitious buds from leaves.

aerenchyma A plant tissue containing large intercellular air spaces, usually formed as a consequence of the death of parenchyma cells. Aerenchyma is typical of the stems of many aquatic plants whose leaves float at the surface of ponds and lakes, giving added buoyancy to the tissues.

aerial root A root that arises above soil level. Examples include the roots of epiphytes and climbers, which hang down in the air or stick to a trunk or branch. The roots of many orchids and other epiphytes developed a sheath of dead cells, the *velamen*, which helps to absorb water from the atmosphere.

aerobe An organism that can live and grow only in the presence of free oxygen, i.e. it respires aerobically (*see* aerobic respiration). All plants and most bacteria and fungi are aerobes. *Compare* anaerobe.

aerobic respiration Respiration in which free oxygen is used to oxidize organic substrates to carbon dioxide and water, with a high yield of energy. The reaction overall is:

$$C_6H_{12}O_6 + 6O_2 = 6CO_2 + 6H_2O + \text{energy}$$

It occurs in a number of stages, the first of which (GLYCOLYSIS) also occurs in anaerobic respiration in the cell cytoplasm. With glucose as the substrate, a sequence of reactions results in the formation of pyruvate. The remaining stages, which do not occur in anaerobic respiration, take place in the mitochondria. Pyruvate is converted to ACETYL COA, which enters a cyclic series of reactions, the KREBS CYCLE, with the production of carbon dioxide and hydrogen atoms. These and other hydrogen atoms produced at earlier stages are now passed to the ELECTRON-TRANSPORT CHAIN (involving CYTOCHROMES and FLAVOPROTEINS). Here they combine with atoms of free oxygen to form water. Energy released at each stage of the chain is used to form ATP during a coupling process (*see* oxidative phosphorylation). There is a net production of 38 ATPs (2930 kJ) per molecule of glucose during aerobic respiration, a yield of about 19 times that of anaerobic respiration (300 kJ), and the mechanism of the majority of organisms. *Compare* anaerobic respiration.

aerotaxis (**aerotactic movement**) A TAXIS in response to an oxygen concentration gradient. For instance, motile aerobic bacteria are positively aerotactic, whereas motile obligate anaerobic bacteria are negatively aerotactic. Some photosynthetic eukaryotic unicells are also aerotactic.

aerotropism A TROPISM in which the orientating stimulus is oxygen.

afforestation The establishment of forest on land not previously forested, either by natural succession or by planting.

aflatoxin One of a group of carcinogenic toxins produced by fungi of the genus *Aspergillus*, especially *A. flavus*. This species is a common contaminant of crops, such as peanuts. Contaminated feed causes serious outbreaks of disease among livestock.

after-ripening The collective name for processes that are necessary before germination can take place in certain seeds, even though external conditions may be suitable. For example, a period of dormancy may be imposed on the seed, preventing premature germination before an unfavorable season such as winter. After-ripening is common in areas with marked seasonality.

agamospermy *See* apomixis.

agar A gelling agent prepared from seaweed (normally a red alga, *Gelidium*), used to set liquid nutrients. It can withstand sterilization at high temperatures and is resistant to attack by most bacteria. Agar gels are extensively used for growing microorganisms and tissue cultures.

Agaricales (**agarics**) An order of basidiomycetes that contains the mushrooms and toadstools. The cap of the fleshy fruiting body – the mushroom or toadstool – bears a series of parallel gills on its lower surface that greatly increase its surface area. These are covered in the spore-bearing layer (hymenium). Agarics are common in leaf litter and rotting wood, where they live as saprobes.

aggregate fruit (**compound fruit**) A fruit-like structure that is made up of several individual fruits derived from the carpels of a single flower, e.g. strawberry (*Fragaria*), blackberry (*Rubus fruticosus*). These may be achenes, follicles, berries, or drupelets. An aggregate fruit is sometimes called an *etaerio*, but this term may also be restricted to an aggregate of drupelets. A fruit that develops from a group of flowers is termed a *multiple fruit*, e.g. pineapple (*Ananas comosus)* and hop (*Humulus lupulus*).

Agrobacterium A genus of soil bacteria, the species *A. tumefaciens* being the causative agent of CROWN GALL, a type of tumor in plants. A segment of DNA (transferred DNA, T-DNA) from a PLASMID in the bacterium is transferred into the host DNA and induces tumor formation. Since the plasmid is capable of independent replication in host cells of many dicotyledonous plants, it has been used as a cloning vector in GENETIC ENGINEERING.

agroforestry A system of cultivation common in many parts of the tropics, especially in rainforest regions, in which forestry and arable farming are mixed. This reduces soil degradation, as freshly tilled soils are shaded and protected from heavy rains and run-off by the trees. The trees help to draw nutrients up from the deeper soil layers into the zone occupied by the roots of the crop plants.

agronomy A branch of agriculture dealing with soil management and crop production.

air bladder 1. A structure on the pollen grains of some conifers, such as *Pinus*, that aids the wind dispersal of the pollen. It consists of an air-filled protuberance on each side of the pollen grain developed by

the separation of the exine layers, giving the pollen a characteristic winged appearance.
2. An air-filled chamber in the thallus of many Phaeophyta (brown algae) that increases the buoyancy of the thallus, enabling it to float toward the surface when underwater to increase light absorption and gaseous exchange for photosynthesis. Air bladders are highly developed in certain wracks, such as the bladder wrack (*Fucus vesiculosus*).

air plant *See* epiphyte.

alanine A simple AMINO ACID, one of the early products of photosynthesis. Alanine is formed by transamination when an amino group is donated by gutamine to pyruvic acid. It may be deaminated back to pyruvate for use in the KREBS CYCLE.

albumin One of a group of simple, low-molecular-weight proteins found in plants, for example in the endosperm of barley and wheat seeds. Albumins are water-soluble and coagulate when heated.

albuminous cell **1.** A vertically elongated parenchyma cell, found in groups in the rays of the secondary phloem in gymnosperms, where they are associated with sieve cells. Unlike the companion cells of angiosperms, they are not derived from the same mother cell as the sieve cell.
2. An albumin-containing cell found in certain seeds.

alcohol A type of organic compound of the general formula ROH, where R is a hydrocarbon group. Examples of simple alcohols are methanol (CH_3OH) and ethanol (C_2H_5OH). Ethanol is a product of anaerobic metabolism.

alcohol dehydrogenase An enzyme that converts ethanol into acetaldehyde (ethanal). It is important in a plant's ability to withstand waterlogging. Ethanol is a product of anaerobic respiration in plants, and waterlogged roots suffer from a shortage of oxygen. Accumulation of ethanol is harmful to a plant.

alcoholic fermentation A form of ANAEROBIC RESPIRATION in which glucose is broken down to form ethanol and carbon dioxide. It is carried out by yeasts and some other fungi and certain bacteria. Alcoholic fermentation is catalyzed by enzymes of the zymase complex, which are secreted by the cells or released after the cells die. The process is self-limiting, since the organisms usually die once the alcohol concentration in the medium exceeds 15%. Alcoholic fermentation is the basis of the preparation of alcoholic beverages, and of bread making (in which the carbon dioxide released causes the dough to rise). During alcoholic fermentation acetaldehyde acts as a hydrogen acceptor instead of oxygen. The pyruvic acid formed by GLYCOLYSIS is broken down into acetaldehyde and carbon dioxide. The acetaldehyde is further reduced by NADH to form ethanal, releasing only about a tenth of the energy that would be released by aerobic respiration.

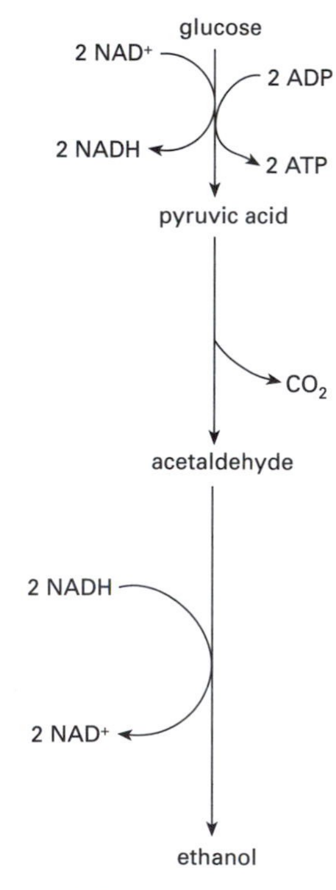

Alcoholic fermentation

aldehyde A type of organic compound with the general formula RCHO, where the –CHO group (the aldehyde group) consists of a carbonyl group attached to a hydrogen atom. Simple examples of aldehydes are methanal (formaldehyde, HCHO) and ethanal (acetaldehyde, CH_3CHO).

aldose A SUGAR containing an aldehyde (CHO) or potential aldehyde group. Examples include the sugars ribose (with five carbon atoms) and glucose (with six carbon atoms).

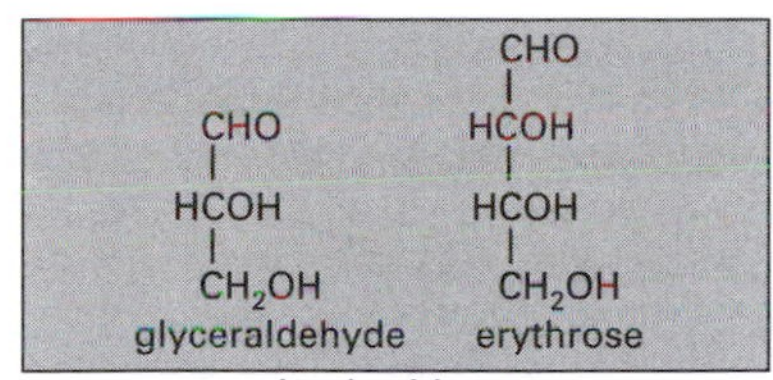

simple aldoses

ribose	arabinose	xylose
CHO	CHO	CHO
HCOH	HOCH	HCOH
HCOH	HCOH	HOCH
HCOH	HCOH	HCOH
CH_2OH	CH_2OH	CH_2OH

aldopentoses

glucose	mannose	galactose
CHO	CHO	CHO
HCOH	HOCH	HCOH
HOCH	HOCH	HOCH
HCOH	HCOH	HOCH
HCOH	HCOH	HCOH
CH_2OH	CH_2OH	CH_2OH

aldohexoses

Aldose

aleurone grain (**aleurone body**) A modified vacuole found in the embryo and endosperm of seeds and containing mostly reserve proteins, but also phytic acid and various enzymes associated with mobilization (digestion) of these reserves. The proteins and phytic acid are present in crystalline form in the dormant seed.

aleurone layer The outermost protein-rich layer of the endosperm of grass fruits (e.g. cereal grains). At germination, the embryo produces gibberellin, which stimulates the aleurone layer to synthesize enzymes, especially amylase. The latter causes hydrolysis of the starch in the endosperm. The enzymes are synthesized from the amino acids supplied by breakdown of ALEURONE GRAINS.

algae (*sing.* **alga**) A large mixed group of photosynthesizing eukaryotic organisms, now usually placed in the kingdom Protoctista. They often resemble plants and are found mainly in marine or fresh-water habitats, although some algae are terrestrial. Algae differ from plants in lacking any real differentiation of leaves, stems, and roots, in having no layer of sterile cells around the reproductive organs,and in not having an embryo stage in their life cycle. Algae can be unicellular (e.g. *Chlamydomonas*), colonial (e.g. *Volvox*), filamentous (e.g. *Spirogyra*), or thalloid (e.g. *Fucus*). All algae contain chlorophyll but this may be masked by various accessory pigments, these being one of the major characteristics used to divide the algae into their various phyla. Other characters used to classify the algae are the nature of storage products, the type of cell wall, the form and number of undulipodia (flagella), ultrastructural cell details, type of food reserves, and reproductive processes.

alkaline soil *See* calcareous soil.

alkaloid One of a group of organic compounds found in plants, which are poisonous insoluble crystalline compounds. They contain nitrogen and usually occur as salts of acids such as citric, malic, and succinic acids. Their function in plants remains obscure, but it is suggested that they may be nitrogenous end-products of metabolism, or they may have a protective function against herbivores, since they taste bitter. Important examples in human use are quinine, nicotine, atropine, opium, morphine, codeine, and strychnine. They occur mainly in the poppy family, the buttercup family, and the nightshade family of plants.

allele (**allelomorph**) One of the possible forms of a given gene. The alleles of a particular gene occupy the same positions (*loci*) on homologous chromosomes. A gene is said to be *homozygous* if the two loci have identical alleles and *heterozygous* when the alleles are different. When two different alleles are present, one (the *dominant* allele) usually masks the effect of the other (the *recessive* allele). The allele determining the normal form of the gene is usually dominant while mutant alleles are usually recessive. Thus most mutations show in the phenotype only when they are homozygous. In some cases one allele is not completely dominant or recessive to another allele (INCOMPLETE DOMINANCE). Thus an intermediate phenotype will be produced in the heterozygote. *See also* incomplete dominance; multiple allelism.

allelomorph *See* allele.

allelopathy Inhibition of the germination, growth, or reproduction of an organism effected by a chemical substance released from another organism. It is probably an anticompetition mechanism in plants; for example, barley secretes an alkaloid substance from its roots to inhibit competing weeds, but it is not known how important it is generally.

allogamy Cross-fertilization in plants. This promotes genetic variation in the population especially in plants that are DIOECIOUS and MONOECIOUS. Most plants are hermaphrodite, but have mechanisms such as self-INCOMPATIBILITY that promote allogamy. *Compare* autogamy. *See also* heterostyly; chasmogamy; dichogamy.

allopatric species *See* species.

allopolyploidy A type of polyploidy involving the combination of chromosomes from two or more different species. *Allopolyploids* usually arise from the doubling of chromosomes of a hybrid between species, the doubling often making the hybrid fertile. The properties of the hybrid, such as greater vigor and adaptability, are retained in the allopolyploid in subsequent generations and such organisms are often highly successful. Many plant species have been derived originally from allopolyploidy, e.g. cultivated wheat. *Compare* autopolyploidy.

allosteric enzyme An enzyme whose catalytic activity can be modified by the noncovalent binding of a particular metabolite (a modulator) at a site (the ALLOSTERIC SITE) other than the active site. It may either inhibit or enhance the enzyme activity. Allosteric enzymes may have more than one modulator.

allosteric site A part of an enzyme to which a specific effector or modulator can be attached. This attachment is reversible. Allosteric enzymes possess an allosteric site in addition to their ACTIVE SITE.

allotetraploid (**amphidiploid**) An allopolyploid whose chromosomes are derived from two different species and which therefore has four times the haploid number of chromosomes, e.g. *Spartina anglica*, derived from the diploids *S. alterniflora* and *S. maritima*. *See* allopolyploidy.

alluvial soil A type of soil formed on river floodplains and deltas, where new sediment is deposited on the land during floods.

alpha helix A highly stable structure in which peptide chains are coiled to form a spiral. Each turn of the spiral contains approximately 3.6 amino acid residues. The R group of these amino acids extends outward from the helix. Hydrogen bonding between successive coils holds the helix together. If the alpha helix is stretched the hydrogen bonds are broken but reform on relaxation. The alpha helix is found in muscle protein and keratin.

alpha-naphthol test (**Molisch's test**) A test for detecting the presence of carbohydrates in solution. The test solution is placed in a test tube, and a small amount of alpha-naphthol added; concentrated sulfuric acid is then trickled slowly down the side of the tube. If carbohydrate is present, a violet ring will form at the junction of the liquids.

alpine Describing a BIOME (regional community) of plants above the treeline and below the snowline on high mountains. The lower limit of the alpine zone varies in different mountain regions, according to the rainfall and other climatic and topographic factors, from 100 meters above sea level in parts of Scotland to 3700 meters in the western Himalayas. There are often considerable differences between the nature and distribution of vegetation on north- and south-facing slopes and between windward and leeward slopes. Alpine vegetation is often similar to TUNDRA vegetation, being adapted to harsh climate conditions, including high wind speeds.

alternate Describing a leaf arrangement in which there is only one leaf at each node, as in hazel. This is the commonest form of leaf arrangement. *See* phyllotaxis.

alternate host Any host other than the main (most common) one. For example, many rusts overwinter on alternate hosts. Some parasites and pests are able to live on more than one host, and some need a second host to complete their life cycle.

alternation of generations The occurrence of two, or occasionally more, generations during the life cycle of an organism. In all plants and some algae there is an alternation between sexual haploid and asexual diploid stages. They usually differ markedly in morphology. The haploid plant produces gametes mitotically and is thus termed the *gametophyte* while the diploid plant produces spores meiotically and is called the *sporophyte*. The gametes fuse to form a zygote, which develops into the sporophyte, and the spores germinate and produce the gametophyte, so forming a cycle. In bryophytes (mosses, liverworts and hornworts) the haploid gametophyte is the dominant phase of the life cycle and the sporophyte is represented only by the capsule, seta, and foot. In vascular plants the diploid sporophyte is the dominant phase and in the ferns, for example, the gametophyte is a small prothallus. The concept of an alternation of generations can be extended to the flowering plants, in which the embryo sac and pollen represent the much reduced female and male gametophyte generations respectively.

ameboid Describing an organism that resembles an ameba in shape and movement.

amensalism An association between two different species, at either the level of the individual organism or the population level, in which one is harmed and the other is unaffected. *Compare* commensalism; mutualism.

amino acids Compounds containing both carboxylic acid and amino groups in their molecules. The amino acids have the general formula $RCHNH_2–COOH$. Here, the group R ranges from a simple hydrogen atom to complex ring structures. All amino acids are white, crystalline, soluble in water, and with the sole exception of the simplest member, glycine, all are optically active.

Amino acids are the basic components of proteins, which consist of chains of amino acids. Plants also contain some 200 amino acids that do not occur in proteins, some of which have protective or storage roles, while others are intermediates in the synthesis of commoner amino acids. Amino acids contain both acids (carboxylic acids) and bases (amino group), so they react with both acids and bases. Thus the charge on the amino acid depends on the pH. This feature is used to separate amino acids for analysis by chromatography and other methods. Each amino acid has its own characteristic isoelectric point, where the net charge on the molecule is zero. See illustration overleaf.

amino acid	symbol	R group
alanine	Ala	$-CH_3$
valine	Val	$-CH<\begin{smallmatrix}CH_3\\CH_3\end{smallmatrix}$
leucine	Leu	$-CH_2-CH<\begin{smallmatrix}CH_3\\CH_3\end{smallmatrix}$
isoleucine	Ile	$-\underset{\underset{CH_3}{\mid}}{CH}-CH_2-CH_3$
proline	Pro	COO^-, C, H; $\overset{H_2}{C}-CH_2$; $\underset{H}{N}-CH_2$
methionine	Met	$-CH_2-CH_2-S-CH_3$
phenylalanine	Phe	$-CH_2-$(benzene ring)
tryptophan	Trp	$-CH_2-C$, HC, $\underset{H}{N}$ (indole ring)

amino acids with nonpolar R groups

The **amino acids** commonly found in proteins are alpha amino acids. In all except proline, where the complete structure is given, the remainder of the molecule has the structure

$$HOOC-\underset{NH_2}{\underset{|}{\overset{\overset{H}{|}}{C}}}-$$

Amino acids

amino acid sequencing The determination of the amino acid sequence of a protein or peptide. This sequence is the primary structure of a protein; it influences secondary, tertiary, and quaternary structure.

amino sugar A monosaccharaide sugar in which a hydroxyl group (OH) has been replaced by an amino group (NH_2). Glucosamine (from glucose) is one of the commonest amino sugars. Galactosamine

amino acid	symbol	R group
glycine	Gly	—H
serine	Ser	$-CH_2-OH$
threonine	Thr	$-C(OH)(H)-CH_3$
cysteine	Cys	$-CH_2-SH$
asparagine	Asn	$-CH_2-C(=O)NH_2$
glutamine	Gln	$-CH_2-CH_2-C(=O)NH_2$
tyrosine	Tyr	$-CH_2-C_6H_4-OH$

amino acids with uncharged polar R groups

amino acid	symbol	R group	
aspartic acid	Asp	$-CH_2-C(=O)O^-$	acidic amino acids
glutamic acid	Glu	$CH_2-CH_2-C(=O)O^-$	acidic amino acids
lysine	Lys	$-CH_2-CH_2-CH_2-CH_2-\overset{+}{N}H_3$	basic amino acids
arginine	Arg	$-CH_2-CH_2-CH_2-NH-C(=\overset{+}{N}H_2)-NH_2$	basic amino acids
histidine	His	$-CH_2-C-CH$ (imidazolium ring: HN, NH, C=, H, +)	basic amino acids

amino acids with charged polar R groups

Amino acids

(from galactose) is a major component of glycolipids. Amino sugars are important components of bacterial cell walls.

amitosis Nuclear division characterized by the absence of a nuclear spindle and leading to the production of daughter nuclei with unequal sets of chromosomes. The ordered process of division, duplication of chromosomes, dissolution of nuclear membrane, and production of a spindle as in mitosis is apparently absent. Cells produced amitotically inherit variable numbers of chromosomes. The chances of a daughter cell lacking essential genes are less than may be expected since many cells that characteristically divide amitotically are polyploid, e.g. the endosperm nucleus in angiosperms. *Compare* endomitosis; mitosis.

AMP (**adenosine monophosphate**) A nucleotide consisting of adenine, ribose, and phosphate. AMP has an important role in the regulation of glycolysis, promoting the production of fructose bisphosphate from fructose 6-phosphate. *See* ATP; cyclic AMP.

amphidiploid *See* allotetraploid.

amphimixis True sexual reproduction by fusion of gametes. *Compare* apomixis.

amylase An enzyme, found widely in plants, animals, and microorganisms, that hydrolyzes starch or glycogen to the sugars maltose, glucose, or dextrin. Amylase hydrolyzes the α(1-4)glycosidic bonds in starch. It occurs in two forms, α and β. β-amylase attacks the nonreducing ends of the starch molecule, while α-amylase can attack bonds within the starch molecule. Both α- and β-amylases occur in plants, the latter particularly in malt (being used in the brewing industry), but only β-amylase is found in animals, having an important role in digestion. During seed germination α-amylase breaks down the starch of the endosperm, the transcription of the α-amylase gene being greatly enhanced by gibberellic acid produced by the embryo. α-amylase is also involved in the release of stored starch from plastids.

amylopectin The water-insoluble fraction of STARCH. It is a branching polymer of glucose units.

amyloplast A plastid that synthesizes and stores starch grains. Amyloplasts are common in storage organs, e.g. the potato tuber. They have a physiological role in the root cap and elsewhere, where the starch grains act as statoliths.

amylose The water-soluble fraction of STARCH. It is an unbranched polymer of glucose units.

anabolism Metabolic reactions in which molecules are linked together to form more complex compounds. Thus, anabolic reactions are concerned with building up structures, storage compounds, and complex metabolites in the cell. Starch, fats, and proteins are all products of anabolic pathways. Anabolic reactions generally require an input of energy, usually provided by ATP produced by catabolism. An example is the Calvin cycle. *See also* metabolism. *Compare* catabolism.

anaerobe An organism that can live and grow in the absence of free oxygen, i.e. it respires anaerobically (*see* anaerobic respiration). Anaerobes can be *facultative*, in that they usually respire aerobically but can switch to anaerobic respiration when free oxygen is in short supply, as are most yeasts; or *obligate*, in that they never respire aerobically and may even be poisoned by free oxygen, as are denitrifying bacteria and lactic acid bacteria. *Compare* aerobe.

anaerobic respiration Respiration in which oxygen is not involved, found in almost all organisms. The organic substrate is not completely oxidized and the energy yield is low. The glycolytic pathway, whereby glucose is degraded to pyruvate, with the production of a small amount of energy, is the same as in aerobic respiration, but after that, in plants and fungi, ethanol and carbon dioxide are produced as the end products of the process, normally known as FERMENTATION. In animals lactic acid is produced and normally remetabolized with oxygen at a later stage. Only two molecules of ATP are produced by this process. *See* anaerobe; glycolysis. *Compare* aerobic respiration.

analogous Describing structures that are apparently similar (structurally or functionally) but have a different evolutionary origin, and thus a different embryological origin and structure. Phyllodes are analogous to leaf blades, but are derived from petioles (leaf stalks). *See also* homologous.

anaphase The stage in mitosis or meiosis when chromatids are pulled toward opposite poles of the nuclear spindle. In mitosis each of the chromatids moving toward the poles represents a single complete chromosome. During anaphase I of meiosis pairs of chromatids still connected at their centromere move to the spindle poles. During anaphase II the centromeres divide and single chromatids are drawn toward the poles, thus sister chromatids are separated.

anatomy The organization of the parts of the body and the structural relationships between them, including the arrangement of tissues and their component cells.

anatropous *See* ovule.

androdioecious Describing species in which there are two forms, one bearing only male flowers, the other bearing only hermaphrodite flowers (or occasionally bearing male and female flowers, i.e. MONOECIOUS). *Compare* gynodioecious.

androecium The collective name in higher plants for the male parts of a plant, i.e. the STAMENS. It is denoted in the floral formula by a letter *A*.

andromonoecious Describing species in which both male and hermaphrodite flowers are borne on the same plant. *Compare* gynomonoecious.

anemophily Pollination by wind. Plants pollinated in this manner (e.g. grasses) usually have insignificant unscented flowers with large, often feathery stigmas.

aneuploidy The condition, resulting from nondisjunction of homologous chromosomes at meiosis, in which one or more chromosomes are missing from or added to the normal somatic chromosome number. If both of a pair of homologous chromosomes are missing, *nullisomy* results. *Monosomy* and *trisomy* are the conditions in which one or three homologs occur respectively, instead of the normal two. *Polysomy*, which includes *trisomy*, is the condition in which one or more chromosomes are represented more than twice in the cell. *See* nondisjunction.

aneuspory The production of an unusual number of spores during meiosis of a spore mother cell. In dandelion (*Taraxacum*), for example, the chromosomes of the megaspore mother cell remain in one cell after the first meiotic division, forming a *restitution nucleus*. The second meiotic division thus gives rise to two cells with the diploid number of chromosomes, one of which divides parthogenetically to form an embryo. This is a form of APOMIXIS. Since crossing over can occur during the first meiotic division, it does allow for some genetic variation in the offspring. *See* parthenogenesis.

angiosperms The flowering plants constituting by far the largest phylum (Anthophyta or Angiospermophyta) of vascular seed plants. They differ from conifers and other gymnosperms by having the ovule enclosed within an ovary, which after fertilization develops into a fruit. The female gametophyte is represented by the embryo sac, the archegonia being absent. The contents of the pollen grain represent the male gametophyte. The pollen grain germinates on a special extension of the carpel, the stigma, and double fertilization results in a diploid zygote and triploid (usually) endosperm nucleus. Angiosperms are divided into two major classes, the monocotyledons (MONOCOTYLEDONAE) and dicotyledons (DICOTYLEDONAE), with a small basal group known as primitive dicotyledons.

angstrom Symbol: Å A former unit of length equal to 10^{-10} meter (one thousandth of a micrometer, one tenth of a nanometer). It is still used occasionally for measurements of wavelength or interatomic distance.

aniline stains A group of stains with various properties that are derived from aniline. For example, aniline hydrocholoride and aniline sulfate will stain lignin yellow. *See* staining.

anion A negatively charged ion, formed by addition of electrons to atoms or molecules. In electrolysis anions are attracted to the positive electrode (the anode). *Compare* cation.

anisogamy (heterogamy) The sexual fusion of nonidentical gametes. Anisogamy grades from situations in which the gametes differ only in size to the extreme of oogamy, in which one gamete is a large nonmotile ovum and the other a small motile sperm. *Compare* isogamy. *See* oogamy.

annual A plant that completes its life cycle within a year. Examples are the common field poppy (*Papaver rhoeas*) and the sunflower (*Helianthus*). *Compare* biennial; ephemeral; perennial.

annual ring (growth ring) The annual increase in girth of the stems or roots of woody plants, as a result of cambial activity. The annual rings of plants growing in temperate climates can be seen in cross-section as two consecutive rings of light- and dark-colored xylem tissue. These are formed from a zone containing larger vessel elements produced by the cambium in the spring (lighter layer), followed by a zone containing smaller vessel elements (darker layer) produced during the late summer. This process is repeated annually so that the number of light or dark rings indicate the age of that part of the plant.

annular thickening Rings of thickening laid down on the inner wall of protoxylem vessels and tracheids. Such thickening allows extension of the XYLEM between the rings so that it is not ruptured as the surrounding tissues grow.

annulus **1.** The ring of tissue surrounding the stalk of the mature fruiting body (mushroom or toadstool) of the basidiomycete fungi. The annulus is all that remains of the veil, which joined the rim of the pileus (cap) to the stalk in the immature mushroom or toadstool. It is sometimes termed the *velum*.
2. A special arc or ring of cells in the sporangia of ferns that constitutes the mechanism for spore dispersal. The cells of the annulus are thickened except on the outer wall, so that they contract on drying out. This causes the stomium to rupture, exposing the spores within the capsule. The capsule wall gradually bends back as the water in the annulus cells continues to evaporate until a point when the remaining water in the cells suddenly turns to vapor. This results in the wall springing back to its original position, the sudden movement dispersing any remaining spores.
3. A ring of large cells separating the epidermis from the operculum in certain bryophytes (e.g. *Funaria*).

anterior Designating the part of a flower or axillary bud facing away from the inflorescence axis or stem, respectively.

anther The part of the stamen that produces the pollen. The anther is usually joined to the tip of the filament (stalk) and is made up of two lobes. Each lobe contains two pollen sacs that produce very large quantities of small pollen grains. The pollen is released when the lobes split open longitudinally.

The anther is made up of an outer epidermis, a middle fibrous layer, and an inner nutritive layer, the *tapetum*. The haploid pollen cells develop in the tapetal zone from spore mother cells.

anther culture (pollen culture) The generation of haploid plants from immature pollen grains or intact excised anthers. The resultant plants are generally smaller than their diploid counterparts.

antheridial cell The cell from which the antheridium develops. In seed plants it is the generative cell in the pollen grain, which divides in the pollen tube to produce two sperm cells.

antheridium The male sex organ of the algae, mosses, liverworts, hornworts, ferns, horsetails, clubmosses, and fungi. It may be made up of one cell, or one or many layers of cells. It produces gametes that are usually motile. *Compare* archegonium.

antherozoid (**spermatozoid**) The male gamete of algae, some gymnosperms, and the non-seed-bearing plants. It is motile and is produced in an antheridium, except in certain gymnosperms (e.g. *Cycas*, *Ginkgo*) in which antherozoids develop from the generative cells of the pollen tube.

Anthocerophyta (**hornworts**) The small phylum of nonvascular plants. Like thallose liverworts, the hornworts have an alternation of generations in which the gametophyte is a dorsiventral flattened green thallus and the sporophyte is an upright structure that has a foot embedded in the gametophyte, from which it draws nutrition.

anthocyanin One of a group of water-soluble pigments found dissolved in the sap of higher plant cell vacuoles. Anthocyanins are red, purple, and blue and are widely distributed, particularly in flowers and fruits where they are important in attracting insects, birds, etc. They also occur in buds and sometimes contribute to the autumn colors of leaves. They are natural pH indicators, often changing from red to blue as pH increases, i.e. acidity decreases. Color may also be modified by traces of iron and other metal salts and organic substances, for example cyanin is red in roses but blue in the cornflower. *See* flavonoid.

Anthophyta *See* angiosperms.

antibiotic One of a group of organic compounds, varying in structure, that are produced by microorganisms and can kill or inhibit the activities of other microorganisms. One of the best-known examples is penicillin, which is produced by the mold *Penicillium notatum*. Another example is streptomycin, from the actinobacterium *Streptomyces griseus*. Antibiotics are also commonly used in medicine to combat bacterial infections. Antibiotics are widely used in research to inhibit protein synthesis. Some antibiotics selectively inhibit protein synthesis by the ribosomes of prokaryotes, but not by those of eukaryotes (excepting the ribosomes of chloroplasts and mitochondria), lending support to the theory that these organelles evolved from symbiotic microorganisms. Antibiotic-resistant genes are used as markers in genetic-engineering experiments.

anticlinal Describing something that is at right angles to a surface. An anticlinal division results in the formation of anticlinal walls between daughter cells, allowing a tissue to increase its circumference. The anticlinal wall of a cell is perpendicular to the surface of the plant body. In cylindrical organs, such as stems and roots, the term *radial* is often used instead of anticlinal. *Compare* periclinal.

anticodon A nucleotide triplet on transfer RNA that is complementary to and bonds with the corresponding codon (triplet) of messenger RNA in the ribosomes during protein synthesis, thus ensuring the correct sequence of amino acids is bonded together in the final peptide. *See* transfer RNA.

antioxidant A substance that slows or inhibits oxidation reactions, especially in biological materials or within cells, thereby reducing spoilage or preventing damage. Natural antioxidants include vitamin E and β-carotene.

antipodal cells The three haploid cells found in the embryo sac of seed-bearing plants that migrate to the chalazal end of the sac farthest from the micropyle. These nuclei arise as a result of the three meiotic divisions that produce the egg cell, synergid cells, and polar nuclei, but they do not themselves take part in the fertilization process. They are eventually absorbed by the developing embryo, and their function is uncertain. They do not occur in all flowering plants.

antiport A membrane transport protein in which the energy liberated by the passive movement of H^+ ions across the membrane is coupled to the active transport of another solute, e.g. Na^+, in the opposite direction but against its own electrochemical potential gradient.

antisense DNA The DNA strand that is not transcribed. In transcription the DNA double helix unwinds and only one of the strands acts as the template for messenger RNA synthesis.

antitranspirant A chemical that reduces transpiration. Antitranspirants may work by depositing a waterproof film over the stomata or by inducing stomatal closure.

ant plant A plant that is adapted to provide a home for ants. Such plants usually have one or more *domatia*, cavities in stems, leaves or roots that house ants. These may be elaborate systems of cavities with many chambers, e.g. *Myrmecodia*, or simple hollows in internodes, leaf veins (*Coffea arabica*), petioles (*Tococa guyanensis*) or stipular spines, e.g. whistling thorns (*Acacia sphaerocephala*).

APF *See* pollen analysis.

aphid (**greenfly; blackfly; plant louse**) A bug of the family Aphididae of the insect order Hemiptera. Aphids feed by sucking plant juices. Many occur in such numbers that they are serious pests of crops and greenhouse plants. Others carry plant viruses.

aphotic zone The lower part of a body of water where light intensity is insufficient to support photosynthesis. *Compare* photic zone.

Aphyllophorales *See* bracket fungi.

aphyllous Without leaves.

Apiaceae (**umbellifers; carrot family**) A family of dicotyledons distinguished by its INFLORESCENCES, which are umbels. Most species are herbaceous, but a few are shrubs and trees. The plants have a characteristic odor and some species, such as parsley, fennel, and dill, are used as herbs; the seeds of many others are popular spices, e.g. aniseed, caraway, coriander, cumin. Many species develop swollen taproots, and the family includes several important root vegetables, e.g. carrot and parsnip; the leaf stalks of celery are also eaten.

apical dominance The phenomenon in which the presence of a growing apical bud on a plant inhibits the growth of lateral buds. It is controlled by the interactions of plant hormones, particularly AUXIN (produced by the shoot tip) and ABSCISIC ACID.

apical meristem The actively dividing cells constituting the growing point at the tip of the root or stem in vascular plants. New cells are cut off on the lower side to form new stem tissue at the stem apex, and on both sides in the root apex to form root tissue and a protective root cap. The apical meristems in the lower plants consist of one cell only, as in the ferns, but become more complex and consist of groups of cells in the higher plants. *See* histogen theory; tunica–corpus theory; meristem.

aplanospore A nonmotile spore, characteristic of the pin molds and certain green algae of the phyla Clorophyta and Chrysomonada. It is an asexual spore formed in a sporangium and is usually thick walled.

apocarpous *See* carpel.

apocarpy An ovary made up of unfused carpels, as in the buttercup (*Ranunuculus*). *Compare* syncarpy.

apoenzyme An ENZYME whose cofactor or *prosthetic* group has been removed (e.g. via dialysis) rendering it catalytically inactive. It is the protein part of a conjugate enzyme. When combined with its prosthetic group (*coenzyme*) it forms a complete enzyme (*holoenzyme*). Examples include the cytochromes, which require ferrous or ferric ions for activity.

apogamy In certain ferns, algae, and fungi, the development of the sporophyte directly from a cell of the gametophyte, so fusion of gametes is bypassed. It frequently occurs in gametophytes that have been produced aposporously and are thus diploid. The term also describes the development of an unfertilized female gamete into the sporophyte, a phenomenon described as PARTHENOGENESIS. *See* apospory; apomixis.

apomixis Any form of asexual reproduction. The term is sometimes used in a restricted sense to describe a modified form of reproduction by plants in which seeds are formed without fusion of gametes, a process also termed *agamospermy*. A plant that reproduces in this way is called an apomict, e.g. many species of *Rubus*. Apomicts are often polyploids that would have difficulty reproducing sexually. Apomixis is comparable to the conditions of apogamy and apospory, which are seen in many ferns. It includes the process whereby a diploid cell of the nucellus develops into an embryo giving a diploid seed with a genetic constitution identical to the parent. Another form of apomixis in which seeds develop from unfertilized gametes can also be termed parthenogenesis. Seeds produced in this way may be either haploid or diploid depending on whether or not the megaspore mother cell undergoes meiosis. Often, in the process termed *pseudogamy*, entry of the male gamete is required to stimulate the development of the female gamete, even though nuclear fusion does not occur. Such cases of apomixis are difficult to distinguish from true sexual reproduction. *Compare* amphimixis. *See also* apogamy; apospory; parthenogenesis; sexual reproduction.

apoplast The system of cell walls and intercellular material extending through a plant body and along which water containing mineral salts, etc. can move passively. It is an important pathway for movement of these substances outside the xylem, for example across the root cortex. *Compare* symplast.

apospory The development of the gametophyte directly from the cell of a sporophyte, thus bypassing meiosis and spore production. Gametophytes produced in this manner are thus diploid instead of haploid. If such gametophytes produce fertile gametes, the resulting sporophyte is then tetraploid, and large polyploid series may subsequently be developed. Apospory is found in some mosses and ferns. In angiosperms, a diploid embryo may develop from a cell of the nucellus or chalaza; this is also apospory. *See also* apogamy; apomixis.

apothecium *See* ascoma.

apposition The deposition of successive layers of cellulose on the inner surface of a plant cell wall after elongation is completed, resulting in an increase in thickness of the wall. These extra layers are called the *secondary cell wall*. *Compare* intussusception.

appressorium (*pl.* **appressoria**) In many parasitic fungi, a flattened hyphal structure that presses closely against the host epidermis. From its undersurface a narrow infection hypha or penetration tube pushes through the cell wall or intercellular spaces and develops into a series of hyphae or a HAUSTORIUM.

aquaporin (**water pore**) A protein channel in the plasma membrane that facilitates water movement across the membrane. Aquaporins can open or close to increase or decrease water flow.

Archaea (**archaebacteria**) In the Five Kingdoms classification scheme, a subkingdom of the BACTERIA, but in most modern classifications, a DOMAIN in its own right. It contains bacteria-like organisms with distinctive ribosomal RNA whose nucleotide sequences differ significantly from those of other prokaryotes. Because RNA is considered to evolve very slowly, these differences are considered important in assigning the Archaea the status of domain. The shape and structure of their ribosomes are more similar to those of eukaryotes than to bacterial ribosomes. Whether they are more closely related to the eukaryotes than to the bacteria is hotly debated. They appear to have a mixture of prokaryotic and eukaryotic genes. Like eukaryotes, but unlike bacteria, they have introns in their transfer RNA, but like bacteria they have polycistronic operons (gene regulators). Archaean cell walls lack murein, a common component of bacterial cell walls, and their lipids have ether links rather than ester links.

The Archaea contain many organisms of extreme environments, such as hydrothermal vents, hot springs, saline sediments, volcanic craters and boiling muds. They include methane-generating organisms (methanogens), sulfate reducers, and extremophiles.

archaebacteria *See* Archaea.

Archean (**Archeozoic**) *See* Precambrian.

archegonium The female sex organ of the mosses, ferns, clubmosses, horsetails, most gymnosperms, and some red and brown algae. It is a multicellular flask-shaped structure made up of a narrow neck and a swollen base (venter) that contains the female gamete. *Compare* antheridium.

archesporium The single cell or group of plant cells in the sporophyte from which spore mother cells may eventually develop in a sporangium.

Arecaceae (**Palmae; palms**) A family of monocotyledonous plants found mainly in tropical regions. The typical growth form is an evergreen tree with unbranched trunk and a crown of large spirally arranged leaves at the apex. The trunk is often sheathed in old leaf bases. Some palms, e.g. the doum palms (*Hyphaene*), branch dichotomously. Many palms grow in the understorey of rainforests in Asia and the Americas.

arginine An AMINO ACID found especially in the histone proteins that surround DNA in chromosomes.

aril A brightly colored fleshy or hairy outgrowth from the funicle at the base of the ovule that may partly or completely cover the seed. The mace-yielding outgrowth around the fruit of the nutmeg is an example, as is the red fleshy cup surrounding a yew seed (*Taxus baccata*). *See also* caruncle.

artificial parthenogenesis *See* parthenogenesis.

ascocarp *See* ascoma.

ascogonium (*pl.* **ascogonia**) The female gametangium of certain fungi of the phylum Ascomycota (e.g. *Erysiphe*, *Eurotium*).

ascoma (**ascocarp**) The fruiting body of most fungi of the phylum Ascomycota, in which the asci are borne.

ascomycete *See* Ascomycota.

Ascomycota (**sac fungi; ascomycetes**) A large phylum of fungi characterized by their distinctive reproductive structure, the ASCUS. In some species the asci are cylindrical, and the ascospores are discharged violently as a result of hydrostatic pressure building up inside the ascus. In others, the asci are globular, and the ascospores are released passively. The tip of the ascus may open through a simple pore, or it may have a cap. Most ascomycotes form branching septate hyphae, but some are yeastlike.

Ascomycotes are found in a wide range of habitats. Some are saprobes, such as blue and green molds (e.g. *Neurospora*), morels, and truffles, which are important decomposers. There are also parasites, such as ergot of rye (*Claviceps purpurea*), peach leaf curl (*Taphrina*), the powdery mildews (Erysiphales), and *Fusarium* wilt. A few are symbionts in lichens and mycorrhizas. The yeasts are economically important in fermentation processes such as brewing (*Saccharomyces cerevisiae*), wine making, and bread making. Some yeasts are parasites of humans, e.g. *Candida albicans*; others, such as *Penicillium*, are sources of antibiotics.

ascorbic acid *See* vitamins.

ascospore One of (usually) eight haploid spores produced inside an ASCUS by members of the phylum Ascomycota.

ascus (*pl.* **asci**) The spore-producing cell of fungi of the phylum ASCOMYCOTA. It is a saclike structure that is formed either singly or in large numbers in ascomata. After meiosis the ascus contains four or eight haploid ASCOSPORES that are liberated through a pore at the end of the sac.

asexual reproduction The formation of new individuals from a single parent without the production of gametes or special reproductive structures. It occurs in many plants, usually by vegetative propagation or spore formation; in unicellular organisms usually by fission or budding; and in some algae by fragmentation. Such 'offspring' are genetically identical clones of their parents. *See* apomixis.

asparagine An AMINO ACID formed by the addition of ammonia to aspartic acid. This is a way of removing ammonia, which is highly toxic, from living cells. Asparagine can later be broken down and the nitrogen used in amino acid synthesis.

aspartic acid An AMINO ACID formed by transamination, in which the amino group of glutamic acid is transferred to oxaloacetic acid. Aspartic acid can be broken down in the KREBS CYCLE. It is an important precursor of several amino acids, and also acts as a source of the amino group in various metabolic reactions.

association A climax plant community named according to the dominant type of species. Examples are heath associations and coniferous forest associations. *See also* consociation.

assortative mating Sexual reproduction in which the pairing of male and female is not random. If similar phenotypes (e.g. plants with similar flower shapes and color that attract the same insects) breed together, this a positive assortative mating; it may lead to INBREEDING. This is likely in insect-pollinated plants, where the insect looks for similar colors and guide markings. Negative assortative mating occurs when dissimilar phenotypes breed together, leading to OUTBREEDING.

aster *See* centriole.

Asteraceae (**Compositae**) A large family of flowering plants, with some 21 000 species found worldwide and including almost every kind of growth form, but mainly herbs. The leaves, which may be simple or compound, are usually alternate, occasionally opposite or whorled, and often in basal rosettes. Latex is present in some genera, e.g. lettuce (*Lactuca*). Taproots are common. Cultivated composites include sunflower (*Helianthus annuus*), a source of oil and seeds; Jerusalem artichoke (*H. tuberosus*), grown for food; globe artichoke (*Cynara scolymus*), whose young flower heads are eaten; salsify (*Tragopogon porrifolius*), whose roots are eaten; chicory (*Cichorium intybus*), used as a substitute for coffee; lettuce (*Lactuca sativa*); and *Chrysanthemum cinerariifolium* and *C. coccineum*, which are sources of the insecticide pyrethrum.

astrosclereid An irregularly branched SCLEREID found in the mesophyll of leaves of certain dicotyledons.

ATP (**adenosine triphosphate**) The universal energy carrier of living cells. Energy from respiration or, in photosynthesis, from sunlight is used to make ATP from ADP. It is then reconverted to ADP in various parts of the cell, the energy released being used to drive cell reactions. ATP is a nucleotide consisting of adenine and ribose with three phosphate groups attached. Hydrolysis of the terminal phosphate bond releases energy (30.6 kJ mol^{-1}) and is coupled to an energy-requiring process. Further hydrolysis of ADP to AMP sometimes occurs, releasing more energy. The pool of ATP is small, but the faster it is used, the faster it is replenished.

ATPase (**adenosine triphosphatase**) An enzyme that catalyzes the hydrolysis of ATP to ADP and inorganic phosphate. Found in every cell, it is responsible for the release of energy to power metabolic reactions and active transport. In oxidative phosphorylation, it forms part of the enzyme complex ATP synthetase, catalyzing the synthesis of ATP from ADP and inorganic phosphate using energy from the proton gradient generated by the electron-transport chain.

atropous (**orthotropous**) *See* ovule.

autecology The study of the interactions of an individual organism or a single species with the living and nonliving components of its environment. *Compare* synecology.

autodiploid **1.** A diploid plant produced by doubling the chromosomes of a haploid plant. Such a plant is homozygous for every allele.
2. A gamete containing a diploid number of chromosomes, rather than the usual haploid number. Such a gamete may be produced by tetraploid tissues, or it may be the result of faulty meiosis. Two diploid gametes may fuse to give a tetraploid zygote, or gamete may develop directly by parthenogenesis.

autoecious Denoting rust fungi that require only one host species to complete the various stages of their life cycle. Examples are *Puccinium antirrhini* found on antirrhinum and the mint rust (*Puccinia menthae*). *Compare* heteroecious.

autogamy Self-fertilization in plants. This restricts genetic variation but allows isolated individuals to reproduce. It is found particularly in pioneer weed species and in ecosystems such as the tundra, where insect vectors are rare. Very few plants rely exclusively on autogamy. *Compare* allogamy.

autolysis The self-destruction of cells by digestive enzyme activity. It is the final stage of cell senescence resulting in complete digestion of all cell components. *See* lysosome.

autonomic movements (**autogenic movements; spontaneous movements**) Movements of plants in response to internal rather than external stimuli. Examples are cytoplasmic streaming, chromosome movement during nuclear division, and growth itself. *Compare* paratonic movements.

autophagy The process by which worn-out cell organelles are enclosed in vacuoles, then broken down by hydrolytic enzymes secreted into vacuole, the breakdown products being reabsorbed afterward. *See also* lysosome. *Compare* autolysis.

autopolyploidy A type of polyploidy involving the multiplication of chromosome sets from only one species. *Autopolyploids* may arise from the fusion of diploid gametes that have resulted from the NONDISJUNCTION of chromosomes at meiosis. Alternatively, like allopolyploids, they may arise by the nondisjunction of chromatids during the mitotic division of a zygote. Autopolyploids often have larger cells than the parent plants but are morphologically very similar. They may, however, be at least partially isolated from each other reproductively and occupy different ecological niches. *Compare* allopolyploidy.

autoradiography A technique whereby a thin slice of tissue containing a radioactive isotope is placed in contact with a photographic plate. The image obtained on development shows the distribution of the isotope in the tissue.

autosomes Paired somatic chromosomes that play no part in sex determination. *Compare* sex chromosomes.

autotetraploid An autopolyploid that has four times the haploid number of chromosomes. *See* autopolyploidy.

autotrophism A type of nutrition in which the principal source of carbon is inorganic (carbon dioxide or carbonate).

organic materials are synthesized from inorganic starting materials. The process may occur by use of light energy (*photoautotrophism*) or chemical energy (*chemoautotrophism*). Autotrophic organisms (autotrophs) are important ecologically as primary producers, their activities ultimately supplying the carbon requirements of all heterotrophic organisms. *Compare* heterotrophism. *See* chemotrophism; phototrophism.

auxanometer An instrument designed to measure the increase in length of a plant part. A growing plant is attached by thread to the end of a lever that magnifies any growth movement. The opposite end of the lever is used to record a trace on a slowly rotating drum.

auxin Any of a group of plant hormones, the most common naturally occurring one being indole acetic acid, IAA. Auxins are made continually in growing shoot and root tips. They are actively moved away from the tip, having various effects along their route before being inactivated. They regulate the rate of extension of cells in the growing region behind the shoot tip and are involved in phototropic and gravitropic curvature responses of shoot tips (but probably not root tips) moving laterally away from light and toward gravity.

Auxins stimulate cell enlargement, probably by stimulating excretion of protons leading to acid-induced wall loosening and thus wall extension. They help maintain apical dominance by inhibiting lateral bud development. Root initiation may be stimulated by auxin from the shoot, and auxins have been shown to move toward the root tips. Pollen tube growth is stimulated by auxin and its production by developing seeds stimulates fruit set and pericarp growth in fleshy fruits. It interacts synergistically with gibberellins and cytokinins in stimulating cell division and differentiation in the cambium. A high auxin:cytokinin ratio stimulates root growth but inhibits regeneration of buds in tobacco pith callus. It is antagonistic to abscisic acid in abscission.

awn In grass flowers (Poaceae), a stiff bristlelike structure formed from a continuation of the central nerves of the LEMMAS or GLUMES.

axenic culture A culture that consists of only one type of organism or cell, i.e. it is not contaminated by any other organisms or cells.

axial system All the cells derived from the fusiform initials in the vascular cambium, which are elongated parallel to the long axis of an organ. They compromise the secondary xylem and phloem. *See also* secondary growth.

axil The angle between the upper side of a leaf or branch and the shoot bearing it. Leaf axils are the site of lateral (axillary) buds.

axillary bud *See* bud.

axoneme The central '9+2' core of microtubules found in cilia and flagella (undulipodia) of eukaryotes, consisting of nine pairs of outer microtubules surrounding two single central microtubules. *See* undulipodium.

Azollaceae (**mosquito ferns**) A family of free-floating aquatic heterosporous ferns found in tropical and warm temperate parts of America and Asia and introduced elsewhere.

Azotobacter A genus of free-living aerobic nitrogen-fixing bacteria found in limestone soils and water. The cells are plump rods or cocci, surrounded by slime. A multilayered wall may be synthesized around the cell to produce a microcyst resistant to desiccation.

B

Bacillariophyta *See* Diatoms.

bacillus (*pl.* **bacilli**) Any rod-shaped bacterium. Bacilli may occur singly (e.g. *Pseudomonas*), in pairs, or in chains (e.g. *Lactobacillus*). Some are motile.

backcross A cross between an individual and an individual of the original parent generation. If a homozygous dominant AA is crossed with a homozygous recessive aa, the F_2 generation obtained by selfing the F_1 would be 25% AA, 50% Aa, and 25% aa. To distinguish between the phenotypically identical AA and Aa, these can be backcrossed to the homozygous recessive parent aa (a *test cross*). The offspring from AA × aa will all be identical (Aa) whereas in the cross Aa × aa, 50% will have the dominant phenotype (Aa) and 50% will show the recessive character (aa).

Bacteria A kingdom containing all the prokaryotic organisms, formerly called Monera or Prokaryotae. This is a large and diverse group of organisms, which, in terms of numbers and variety of habitats, includes the most successful life forms. In nature, bacteria are important in the nitrogen and carbon cycles, and some are useful to humans in various industrial processes, especially in the food industry, and in techniques of genetic engineering (*see also* biotechnology). However, there are also many harmful parasitic bacteria that cause diseases such as botulism and tetanus. Others are sources of antibiotics such as streptomycin and erythromycin. Some bacteria require oxygen for respiration, while others can survive without oxygen. There may be at least ten times more bacteria living in cracks in the rocks deep inside the earth than in the surface layers of the planet, including the soil. Some derive their carbon from inorganic sources, and their hydrogen from hydrogen sulfide or other inorganic compounds, while other bacteria may be photosynthetic or saprobic.

Bacterial cells are simpler than those of animals and plants. They lack well-defined membrane-bound nuclei and histone-coated chromosomes, and do not contain complex organelles such as chloroplasts and mitochondria. Their DNA is usually located in a specific part of the cell, called the nucleoid. Bacteria may divide every 20 minutes and can thus reproduce very rapidly. They also form resistant spores. The Bacteria are divided into two main groups or subkingdoms: the ARCHAEA, or archaebacteria, which often occur in extreme conditions, such as hot springs; and the EUBACTERIA, which include the vast majority of bacteria. Some taxonomists consider these groups to be full kingdoms in their own right. *See* Actinobacteria; Cyanobacteria; myxobacteria; sulfur bacteria. *See also* prokaryote.

bacterial chromosome The naked DNA of bacteria and Archaea. Bacteria do not have true CHROMOSOMES – linear pieces of DNA complexed with histone proteins. Their DNA is circular and is not complexed with proteins. It lacks the centromere, telomere, and other structures of the eukaryote chromosome.

bacterial transformation *See* transformation.

bactericidal Used to describe a compound that has a lethal effect on bacteria. A bactericidal compound may act by interfering with a vital biochemical pathway, or by destroying the molecular structure of the cell.

bacteriochlorophyll Any of several types of chlorophyll found in photosynthetic bacteria, such as the purple bacteria. There are seven forms, designated bacteriochlorophylls *a–g*. All are structurally similar to chlorophyll *a* of plants. The bacteriochlorophylls absorb light at longer wavelengths than chlorophyll *a* enabling far-red and infrared light to be used in photosynthesis. *See* photosynthetic pigments.

bacteriophage (**phage**) A virus that infects bacteria. Phages usually have complex *capsids* (protein coats) composed of a polyhedral head, containing the nucleic acid (DNA or RNA), and a helical tail, through which nucleic acid is injected into the host. Phages are usually specific to particular species or even strains of bacteria. Infection by a phage may or may not lead to the death of the bacterium, depending on the phage and on the conditions. All RNA phages cause the death of the host cell. After reproduction of the viral nucleic acid the host cell usually undergoes lysis (LYSOGENY). Some DNA phages, called TEMPERATE PHAGES, may become integrated into the host DNA and are replicated when it replicates. Some of these phages are capable of transferring genes between bacteria in a process called TRANSDUCTION. In GENETIC ENGINEERING, nonviral DNA can be inserted into a phage, which is then used as a cloning vector.

bacteriostatic Used to describe a compound that prevents reproduction of bacteria, but does not kill them.

bacteroid A modified bacterial cell in a root nodule, typically a cell of the bacterium *Rhizobium* in the root nodule of a leguminous plant (family Fabaceae).

Banks, Sir Joseph (1743–1820) British botanist. Banks used his considerable fortune to finance several important expeditions, including his most famous one to the South Pacific in 1768 with Captain James Cook on the *Endeavour*. This voyage lasted for three years and resulted in the discovery of the unique flora and fauna of Australia. On his return with his enormous collection of specimens, he received an invitation to meet King George III, who had expressed a great interest in the expedition. His continued friendship with the king increased his influence in scientific circles and he became president of the Royal Society for life in 1778. Following his appointment to the post of honorary director of Kew Gardens, he helped to establish a major collection of living plant species and to make Kew famous as a center of excellence for plant cultivation. He is also famous for introducing the tea plant to India (from China) and breadfruit to the West Indies (from Tahiti). His extensive library and herbarium are housed in the British Natural History Museum.

bark The outermost tissue of the stem and roots in woody plants, on the outside of the *phellogen* (cork cambium). If the same phellogen functions from year to year, as in beech and oak, then the bark is smooth and consists only of tissue external to the phloem, mainly cork. If new cork cambia arise every few years, as in most species of woody plants, then the bark includes dead phloem and cortex as well as cork, and is termed *rhytidome*. As bark increases in thickness, the outer layers may split in distinctive patterns. The texture and patterning of the bark is often characteristic of the species. The bark is punctuated by *lenticels*, small patches of loose cells, which allow air to penetrate the living tissues below.

basal body *See* kinetosome.

base 1. (*Chemistry*) A compound that reacts with an acid to produce water plus a salt. In solution, it forms ions that can react with hydrogen ions.
2. (*Biochemistry*) A nitrogenous molecule, either a PYRIMIDINE or a PURINE, that combines with a pentose sugar and phosphoric acid to form a nucleotide, the fundamental unit of nucleic acids. The most abundant bases are cytosine, thymine, and uracil

(pyrimidines) and adenine and guanine (purines).

base pairing The bonding relationship between purine and pyrimidine bases in the nucleotides of DNA and RNA. The pairing of these bases, sustained by hydrogen bonding, is highly specific: adenine (A) on one strand of the nucleic acid pairs only with thymine (T) in the complementary strand of DNA, or with uracil (U) in RNA; guanine (G) pairs only with cytosine (C). The large numbers of hydrogen bonds that form between the base pairs along a DNA or RNA molecule stabilize the double helix. The specificity of the pairing ensures that during REPLICATION and TRANSCRIPTION accurate copies are made of the genetic material. The number of base pairs is a measure of the length of the DNA or RNA molecule, which is often expressed in units of 100 bases (kilobases), meaning 100 base pairs.

base ratio The ratio of the molar quantities of two bases in DNA, or of one base pair with the other base pair. The base ratio of 1:1 for the bases adenine and thymine and the bases cytosine and guanine was the first clue to the specificity of BASE PAIRING in the nucleic acids. The base pair ratio, AT:GC, is constant within a species but differs between species.

basic stain *See* staining.

basidioma (*pl.* **basidiomata**) The fruiting body of the Basidiomycota, excepting the rusts and smuts, formerly called the *basidiocarp*. It may be of a fleshy, corky, or spongy nature. The basidioma is composed mainly of sterile tightly packed hyphae; in most species there is a distinct fertile layer, the *hymenium*, on the surface of the basidioma that bears the basidial hyphae which, after meiosis, give rise to the basidiospores (*see* basidium).

basidiomycete *See* Basidiomycota.

Basidiomycota (**basidiomycetes**) A phylum of the Fungi that includes the mushrooms and toadstools. Basidiomycotes are characterized by their spore-bearing structures (*see* basidium), each of which usually bears four sexually produced spores (*basidiospores*) on its outer surface. In most basidiomycotes the basidia are borne on a special fertile layer, the hymenium, on the surface of a fruiting body, the BASIDIOMA.

basidiospore *See* Basidiomycota; basidium.

basidium (*pl.* **basidia**) The cell of the BASIDIOMYCOTA that produces the sexual spores (*basidiospores*). In most basidiomycotes the basidia develop in the BASIDIOMA from hyphal tips in the hymenium and may be club-shaped or cylindrical.

basifixed Stamens in which the anther lobes are attached at their base to the filament, and are not capable of independent movement. *Compare* dorsifixed; versatile.

basipetal Developing from the apex downward so that the youngest structures are furthest from the apex. Basipetal differentiation is seen in the formation of proto- and metaxylem in the stem. The term may also be applied to the movement of substances toward the base, for example the movement of auxin in shoot tissues. *Compare* acropetal. *See also* centrifugal; centripetal.

B chromosome (**accessory chromosome**) An extra chromosome found in addition to the normal number of chromosomes characteristic of a species (the A chromosomes). B chromosomes are known from more than 20 families of plants and are also found in some fungi.

beet sugar *See* sucrose.

Benedict's test A test for the presence of a dissolved reducing sugar. Benedict's solution is an alkaline solution of copper(II) sulfate, which is reduced to insoluble copper(I) oxide by reducing sugars, giving a red precipitate.

Bentham, George (1800–84) British botanist. Having developed an interest in plant taxonomy from the age of 17, Bentham abandoned (1833) his career in law to devote himself entirely to his botanical collection and library. In 1854 he presented his collection to the Royal Botanic Gardens at Kew, where he worked for the rest of his life. His famous *Handbook of British Flora* was published in 1858. Between 1862 and 1883 he collaborated with Joseph HOOKER to publish *Genera Plantarum*, a world flora describing 7569 genera and 97 000 species, which is considered his greatest achievement.

benthic Of, relating to, or living in the zone at the bottom of a sea, lake, or other body of water. *See* benthos. *Compare* pelagic.

benthos The organisms that live on the bottom sediments of freshwater and marine ecosystems. They include plants and algae (*phytobenthos*) that grow below low tide mark.

berry A succulent fruit, usually containing more than one seed, that does not burst open when ripe. The ovary wall remains fleshy after fertilization of the ovule except for the development of a thin outer skin. A berry with a hard rind, such as a cucumber, is called a *pepo*. A berry with a leathery rind in which the seeds are separated by segments, as in the citrus fruits, is called a *hesperidium*. The epicarp and mesocarp may be brightly colored thus attracting animals, which disperse the fruit, as in the tomato. *Compare* drupe. *See also* pericarp.

beta-oxidation *See* oxidation.

beta-pleated sheet (**β-pleated sheet**) A type of protein secondary structure in which polypeptide chains run close to each other and are held together by hydrogen bonds at right angles to the main chain between the NH and C=O groups on one chain and the C=O and NH groups on the other chain. The structure is folded in regular 'pleats'. Fibers having this type of structure are usually composed of amino acids with short side chains. The chains may run in the same direction (parallel) or opposite directions (antiparallel). *See* conformation; secondary structure.

bicollateral bundle The arrangement of tissues in the vascular bundles in which the phloem is situated on both sides of the xylem. This is thought to aid transport of nutrients in plants with long trailing stems, such as the marrow. *Compare* concentric bundle; collateral bundle.

biennial A plant that completes its life cycle within two years. In the first year it produces foliage only and photosynthesizes. The food is stored during the winter in a swollen underground root or stem. In the second year, the stored food is used to produce flowers, fruits, and seeds. Many important crops, such as carrot and parsnip, are biennials. Some biennials can be induced to act as annuals and flower in the first year by appropriate cold or hormone treatments. *Compare* annual; ephemeral; perennial.

bilateral symmetry The arrangement of parts in an organism in such a way that the structure can only be divided into similar halves (mirror images) along one plane. Bilateral symmetry is characteristic of most free-moving animals, where one end constantly leads during movement. In plants, bilateral symmetry is seen particularly in flowers (e.g. snapdragon), the condition being termed *zygomorphy*. *See also* radial symmetry.

binomial distribution (**normal distribution; Gaussian distribution**) The idealized distribution of a population of statistical values centered around a mean, when departures from the mean are due the random occurrence of a large number of small independent effects. It represents the probability of each value occurring and is represented on a graph by a bell-shaped curve or histogram.

binomial nomenclature A system of classification introduced by Linnaeus, the Swedish botanist, in which each species is given two names. The first is the generic name, written with a capital letter, which designates the genus to which the species belongs. The second is the specific name, indicating the species. The generic and specific names are in Latin and are printed in italic type. For example, the sunflower belongs to the species *Helianthus annuus*. *Helianthus* is the generic name and *annuus* is the specific name. *See* International Code of Botanical Nomenclature.

bioassay An experimental technique for measuring quantitatively the strength of a biologically active chemical by its effect on a living organism. For example, the vitamin activity of certain substances can be measured using bacterial cultures. The increase in bacterial numbers is compared against that achieved with known standards for vitamins. Plant growth hormones can be estimated by their effect in causing curvature of oat coleoptiles.

Biochemical Oxygen Demand (BOD) The standard measurement for determining the level of organic pollution in a sample of water. It is the amount of oxygen used by microorganisms feeding on the organic material over a given period of time. Sewage effluent must be diluted to comply with the statutory BOD before it can be disposed of into clean rivers.

biochronology The measurement of geological time using biological events. Particular strata of rocks can be approximately dated according to the presence or absence of certain fossils. The first and last appearances of particular fossil species represent specific moments in geological time.

biodegradable Describing organic compounds that can be decomposed by bacteria and other microorganisms, such as the constituents of sewage, as compared with nonbiodegradable compounds, such as most plastics. *See also* decomposer; pollution.

biodiversity The number and variety of organisms in a given locality, *community*, or *ecosystem*. High biodiversity is typical of complex and highly productive ecosystems, such as tropical rainforests, where a small area can contain many different species of animals, plants, and other organisms. Biodiversity is often used as an indicator of the health of such ecosystems. *See also* Convention on Biological Diversity.

bioenergetics The study of energy transfer in living organisms.

biogenesis The theory that living things originate only from other living things as opposed to nonliving matter. The theory became accepted as a result of the work of Redi and Pasteur, who showed that dirt, for example, does not itself produce bacteria or maggots, but that bacteria and maggots come only from spores or eggs already existing in the dirt. This theory satisfactorily explains the occurrence of existing organisms, but not the origins of the first organisms.

biogenic Describing structures produced as a result of the activity of living organisms, in particular rocks (e.g. chalk, limestone, diatomaceous earth) and fossils.

biogeochemical cycle The movement of chemical elements from the physical environment into organisms and back to the environment in a more or less circular pathway. If the elements concerned are essential to life, this cycle is called a nutrient cycle, e.g. the NITROGEN CYCLE. Some of these cycles are regulated mainly by physical factors. For example, the main constraint on the phosphorus cycle is the slow release of phosphates from rocks by weathering. Others, such as the nitrogen cycle, involve biological feedback mechanisms, as nitrates are incorporated into proteins and released by decomposition. In the CARBON CYCLE, photosynthetic organisms are thought to be important in balancing the cycle by carbon fixation; carbon compounds are slowly released as dead organisms decompose.

biogeographical barrier A barrier to the migration of plants and animals that prevents different populations mixing. Such barriers may be physical, such as mountain ranges, large rivers, lakes, or seas; or they may be climatic.

biogeographical region A subdivision of the Earth's surface based on the particular species of plants and animals present. There are several different schemes of biogeographical regions, but the most widely recognized are: Boreal (northern Eurasia and North America), Paleotropical (Africa and southeast Asia as far as New Guinea), Neotropical (South America, except the far south), Australasian, Cape (South Africa), and Antarctic (New Zealand and southern South America). These biogeographical regions are subdivided into *biogeographical provinces*.

biogeography The study of the distribution of plants and animals, both past and present, and its interpretation in terms of ecology, in particular the distribution of world vegetation types and the interactions between humans and their environment.

bioinformatics The creation and maintenance of databases of biological information. With the rapid expansion of biological knowledge and research, sophisticated computerized information-retrieval systems are necessary to speed up access to biological information.

biological clock The internal mechanism of an organism that regulates CIRCADIAN RHYTHMS and various other periodic cycles.

biological conservation The preservation of the world's natural biological diversity and natural habitats and the maintenance, as far as is possible, of the abiotic factors that influence these. This may involve human management of a habitat. Modern biological conservation may also include sustainable use of biological resources, which sometimes conflicts with the conservation of maximum diversity. *See* World Conservation Union.

biological control The use of natural predators or parasites, instead of chemicals, to control pests. The most famous successful example was the introduction of the gray moth, *Cactoblastis cactorum*, into Australia to control the prickly pear, *Opuntia inermis*, which was overrunning vast tracts of land. The moth's caterpillars eat the shoots of the plant.

biology The study of living organisms, including their structure, function, evolution, interrelationships, behavior, and distribution.

biomass The weight or volume of living organisms of one particular species per unit area (*species biomass*), or of all the species in the community (*community biomass*). *See also* pyramid of biomass.

biome A major regional community of plants and animals with similar life forms and environmental conditions. It is the largest geographical biotic unit, and is named after the dominant type of vegetation, such as tropical rain forest, grassland, or coral reef.

biometry The application of statistics to the analysis of biological phenomena.

biorhythm A periodic physiological or behavioral change that is controlled by a biological clock. Circadian rhythms are an example.

biosphere The part of the earth and its atmosphere that is inhabited by living organisms. The Earth is made up of three concentric zones: the lithosphere (solid earth), the hydrosphere (water layer), and the atmosphere. The biosphere includes parts of all three of these zones. It is sometimes considered to represent a single ecosystem, the *ecosphere*.

biosynthesis Chemical reactions in which a living cell builds up its necessary molecules from other molecules present. *See* anabolism.

biosystematics The area of systematics in which experimental taxonomic techniques are applied to investigate the relationships between taxa. Such techniques include serological methods, biochemical analysis, breeding experiments, comparative morphology, cytological examination, and RNA analysis, in addition to the more established procedures of comparative anatomy. Evidence from ecological studies may also be brought to bear. *See also* molecular systematics.

biota The plants, animals, and other living organisms that live together in a particular place.

biotechnology The application of technology to biological processes for industrial, agricultural, and medical purposes. For example, bacteria such as *Penicillium* and *Streptomycin* are used to produce antibiotics and fermenting yeasts produce alcohol in beer and wine manufacture. Recent developments in genetic engineering have enabled the large-scale production of hormones, blood serum proteins, and other medically important products. Genetic modification of farm crops, and even livestock, offers the prospect of improved protection against pests, or products with novel characteristics, such as new flavors or extended storage properties. *See also* genetic engineering.

biotic environment The biological factors acting on an organism, which arise from the activities of other living organisms, as distinct from physical factors.

biotin A water-soluble vitamin generally found, together with VITAMINS in the B group, in the vitamin B complex. It is widely distributed in natural foods, egg yolk, kidney, liver, and yeast being good sources. Biotin is required as a coenzyme for carboxylation reactions in cellular metabolism.

biotrophic Obtaining nutrients from the living tissues of another organism. *Compare* autotrophism.

biotype **1.** A naturally occurring group of individuals all with the same genetic composition, i.e. a clone of a pure line. *Compare* ecotype.
2. A physiological race or form within a species that is morphologically identical with it, but differs in genetic, physiological, biochemical, or pathogenic characteristics.

bipinnate Describing a pinnate leaf whose leaflets are pinnately subdivided, e.g. sensitive plant (*Mimosa pudica*).

bisexual *See* hermaphrodite.

Biuret test A test for the presence of proteins, in which sodium hydroxide is added to the test solution, followed by drops of 1% copper(II) sulfate solution. A violet color indicates the presence of proteins or peptides (excluding dipeptides). Amino acids have no effect.

bivalent Describing any pair of homologous chromosomes when they pair up during MEIOSIS. Pairing of homologous chromosomes (*synapsis*) commences at one or several points on the chromosome and is clearly seen during PACHYTENE of meiosis I.

black earth *See* chernozem.

bladder **1.** A modified leaf, found on the stems of members of the bladderwort family, that develops into a distended structure for trapping small invertebrates. The bladder is thin-walled and closed by a one-way valve.
2. An air-filled sac found in large numbers in the thalli of certain seaweeds (e.g. bladderwrack).

blending inheritance Inheritance in which the characters of the parents appear to blend into an intermediate form in the offspring. It is seen in characters controlled by a single gene whose alleles are codominant, as in pink flowers of *Antirrhinum*, and also in characters such as yield, which are determined by many genes. Until the rediscovery of Mendel's work, it was thought that all characteristics become blended in the offspring. *See* Mendel's laws.

blepharoplast *See* kinetosome.

blight A plant disease in which serious leaf damage appears suddenly. It is usually due to fungi or oomycetes, such as the potato blight (*Phytophthora infestans*), or occasionally to bacterial infections, as in fireblight of pears, which is caused by *Erwinia amylovora*.

bloom A sudden growth of microorganisms in a freshwater or marine ecosystem, often known as an *algal bloom*. This may occur naturally in spring, when growth of microorganisms is for a time unchecked by herbivores. It may also be brought on by nutrient enrichment of the water, often due to fertilizers being washed off agricultural land, or to seawater entering the marine environment.

blue-green bacteria *See* Cyanobacteria.

BOD *See* Biochemical Oxygen Demand.

bog A plant community found in wet acid areas where rainfall is high, the atmosphere is continually moist, and decomposition rates are slow, favoring PEAT formation. Bogs are common in uplands and western areas of north temperate and some southern temperate regions. Typical bog plants of north temperate regions include the bog mosses (*Sphagnum*), cotton grasses (*Eriophorum*), rushes (*Juncus*), sedges (*Carex*), and certain heaths (e.g. *Calluna vulgaris*, *Erica tetralix*).

Boletales (**boletes**) An order of basidiomycote fungi having fleshy mushroom-shaped fruiting bodies in which the hymenium (spore-bearing surface) lines the inner surface of vertical tubes (pores) on the underside of the cap (e.g. *Boletus*). *Compare* Agaricales.

bolting Premature flowering and seed production. Biennial plants will sometimes bolt in their first year; this seriously reduces their yield.

borax carmine A dye used to stain nuclei for light microscopy.

Boreal floral region (**Holarctic floral province**) A floral region that consists of northern Eurasia and North America. *See* floral province.

boron A MICRONUTRIENT found in very low concentrations in plant tissues but essential for plant growth. It is involved in phenol metabolism and membrane function. It is implicated in the uptake of calcium, the development of apical meristems, and pollen germination.

botanical **1.** Relating to the study of plants.
2. A cultivated plant of an unaltered wild species.
3. An insecticide derived from a plant, e.g. pyrethrum (from *Chrysanthemum* flowers).

botany The scientific study of plants.

bracket fungi A member of the basidiomycote order Aphyllophorales, whose fruiting bodies (basidiomata) typically project from the trunks of trees like shelves or domed brackets. These fruiting bodies are usually tough, sometimes woody. The spores are produced on small teeth or lamellae rather than gills.

bract A modified leaf that develops below a flower or an inflorescence on its axis. It may be reduced or highly colored, as in the scarlet bracts of *Poinsettia*. In the daisy family the numerous bracts which subtend the inflorescence are known as an *involucre*. In the grasses bracts are borne in pairs below the spikelets and are called *glumes*. Each individual floret of the spikelet is also subtended by two bracts, the lower being called the *lemma* and the upper the *palea*.

bracteole A small leaflike organ on the pedicel of a flower in an inflorescence that is itself subtended by a bract, as in the bluebell. *Compare* bract.

bract scale The structure found in large numbers in the female cone of gymnosperms, each bearing an ovuliferous scale in its axil.

branch A lateral stem arising from another stem of the same form.

Brassica A genus of plants of the cabbage family (Brassicaceae) that includes many vegetables and fodder plants.

Brassicaceae (**Cruciferae**) A family of annual, biennial, or perennial plants, mainly herbs, that includes many commercially important crops, especially the BRASSICA species; salad vegetables, such as cress (*Sinapis*), rocket (*Eruca*), and watercress (*Rorippa*); and ornamental plants, e.g. *Alyssum*, *Aubretia,* wallflower (*Erysimum*), stock (*Matthiola*), and honesty (*Lunaria annua*).

breeding system The way in which a species breeds, including the method of mate selection, genetic determination of sex and incompatibility, method of mate selection or pollination, and other factors that place limits on the genetic diversity of the offspring. *See* self-incompatibility; sex determination; sexual reproduction.

breeding true (**pure breeding**) The production of offspring whose genetic makeup for particular alleles is identical to that of their parents. This requires both parents to be homozygous for same alleles in question.

Brown, Robert (1773–1858) British botanist. Brown studied medicine and served as a military medical officer. During a visit to London he was introduced to Sir Joseph Banks, who encouraged him to apply for the post of naturalist on an expedition (1801–06) to survey the coast of Australia in the *Investigator*. Having returned with some 4000 plant specimens, he spent the next few years on their classification. His painstaking observations led him to discover many important features of plants. He was the first to observe and name the nucleus of cells and to recognize its importance. He also observed that gymnosperm ovules were not enclosed in an ovary wall, leading to the establishment of one of the basic differences between angiosperms and gymnosperms. In 1827, upon observing a suspension of pollen grains in water, he discovered BROWNIAN MOTION.

brown algae *See* Phaeophyta.

brown earth The type of soil found under deciduous forests in temperate climates. It is rich in humus and slightly acid, due to leaching of lime. When cleared, it provides good fertile agricultural soil.

Brownian motion The random motion of microscopic particles due to their continuous bombardment by the much smaller and invisible molecules in the surrounding liquid or gas. Particles in Brownian motion can often be seen in colloids under special conditions of illumination.

Bryophyta A phylum of simple, mainly terrestrial, plants called mosses, commonly found in moist habitats. They show a heteromorphic alternation of generations, the gametophyte being the dominant generation. When mature the gametophyte shows differentiation into stem and leaves but there are no roots or vascular tissues, but mosses do possess elongated conducting cells. The sporophyte, which is wholly dependent on the gametophyte, is simply a spore capsule borne on a stalk. The spores germinate to produce a slender, filamentous haploid protonema, which produces buds that develop into leafy moss plants. Formerly, the Bryophyta also included the liverworts and hornworts, which in the Five Kingdoms classification are now placed in separate phyla (HEPATOPHYTA and ANTHOCEROPHYTA). The collective name for mosses, liverworts, and hornworts is *bryophytes*.

bud **1.** A compacted undeveloped shoot consisting of a shortened stem and im-

mature leaves or floral parts. The young leaves are folded about the growing tip, and the outermost leaves may be scaly and reduced to protect the growing point. A bud has the potential to develop into a new shoot or flower. *Terminal buds* are formed at the stem or branch tip. *Axillary* or *lateral buds* develop in the leaf axils. Buds can develop adventitiously on other parts of a plant and are sometimes a means of asexual reproduction.
2. The swelling produced by a unicellular organism that is reproducing asexually by BUDDING.

budding **1.** The production of buds on plants.
2. A type of grafting in which the grafted part is a bud.
3. A type of asexual reproduction in which a new individual is produced as an outgrowth (bud) of the parent organism. It occurs in some unicellular fungi, such as the yeasts.

buffer A solution that resists any change in acidity or alkalinity. Buffers are important in living organisms because they guard against sudden changes in pH. They involve a chemical equilibrium between a weak acid and its salt or a weak base and its salt. In biochemistry, the main buffer systems are the phosphate ($H_2PO_4^-$/HPO_4^{2-}) and the carbonate (H_2CO_3/HCO_3^-) systems.

bulb A modified shoot that acts as an organ of perennation and often vegetative reproduction. The stem is reduced to a disklike structure, bearing concentric layers of fleshy leaf bases that comprise the food store. Each leaf has a bud at its base that is able to develop into a subsidiary bulb. The whole bulb is protected by scale leaves and adventitious roots arise from the base of the stem. In spring or summer one or more buds grow and produce leaves and flowers, exhausting the food supply. The new leaves photosynthesize and food is stored in their bases thus giving rise to a new bulb. If more than one bud develops, eventually more than one bulb develops. *See* perennating organ; asexual reproduction.

bulbil A small bulblike organ of asexual reproduction that may form in a leaf axil, as in some lilies, or may replace flowers in an inflorescence, as in the wild leek (*Allium amploprasum*). It may detach itself from the parent plant and become independent.

bundle sheath The ring of parenchymatous or sclerenchymatous tissue, usually one cell thick, that surrounds the vascular bundle in an angiosperm leaf. The individual cells are closely packed with no apparent intercellular spaces, and conduct water and solutes from the vascular bundle to the surrounding tissues. Chloroplasts may be present in the bundle sheath, and are thought to be connected with starch storage in the tropical grasses.

bunt (**stinking smut**) A seed-borne disease of wheat (*Avena*) and other grasses, caused by the basidiomycote *Tilletia*.

bur A kind of pseudocarp (false fruit) in which the fruit is encased in a barbed involucre. The barbs catch in the fur of passing animals and the fruits are thus dispersed. Examples include burdock (*Arctium*) and cocklebur (*Xanthium*).

butenedioic acid *See* succinic acid.

buttress root A kind of flattened prop root found in many large tropical trees, e.g. padauk (*Pterocarpus*); it starts as an adventitious root that grows down from the trunk and forms a planklike structure attached to the trunk all the way, becoming wider toward the ground. Some buttress roots extend for several meters outward from the trunk at ground level. These roots provide the tree with extra support, especially in shallow saturated soil.

C

Cactaceae (**cacti**) A large family of succulent xerophytic dicotyledonous plants, found mainly in warm dry parts of the Americas; some species, notably the prickly pears (*Opuntia*), have been introduced to the Mediterranean and Australia.

Caenozoic *See* Cenozoic.

Cainozoic *See* Cenozoic.

calcareous soil (**basic soil; alkaline soil**) A soil with a pH greater than 6.0. Calcareous soils occur where lime accumulates in the surface layers, usually as a result of a calcareous bedrock such as limestone, or a calcareous sand deposit near coasts. They are most common in regions of light rainfall and are usually rich in calcium ions and often other elements. *See* rendzina.

calcicole Describing plants that thrive on neutral to alkaline soils rich in calcium carbonate, i.e. chalk or limestone soils, such as marls. *Compare* calcifuge.

calcifuge Describing plants that grow on soils containing very little calcium carbonate, such as sandy or peaty soils and cannot grow on basic substrates such as most chalk and limestone soils. *Compare* calcicole.

calcium Symbol: Ca. An essential mineral for animal and plant growth. One of its most important functions in plants is as an intracellular messenger with steep gradients of calcium concentration forming across membranes with a system of calcium pumps. It is an important constituent of the middle lamella between plant cell walls as calcium pectates. The calcium ions (Ca^{2+}) cause cross-linking of adjacent chains of pectic acid, thus strengthening the cell wall.

callose An insoluble polysaccharide that is laid down around the perforations in sieve plates, at wound or infection sites, and by growing pollen tubes. As sieve tubes age, the callose layers become thicker, eventually blocking the SIEVE ELEMENT either seasonally or permanently. At wound and infection sites callose provides a physical barrier. When pollen germinates on a stigma the growing pollen tubes deposit callose as small plugs, but if the plant is self-incompatible, tubes from its own pollen will deposit more callose, which then blocks the tube, preventing fertilization. *See* self-incompatibility.

callus **1.** A mass of undifferentiated parenchyma. If a plant is injured, the surrounding uninjured parenchyma cells form a suberin-impregnated callus, sometimes, sealing off the wound.
2. In tissue cultures, cells are first dedifferentiated to form a callus that can then be induced to form adventitious shoots and roots. *See also* graft.

calmodulin A Ca^{2+}-binding protein found in high levels in eukaryote cells. It mediates many calcium-regulated processes in cells. When bound to calcium ions, calmodulin changes shape and binds to other proteins, such as kinases and phosphatases, that are involved in signaling pathways, activating or inactivating them. *See* cascade.

Calvin, Melvin (1911–97) American biochemist noted for his investigations of the DARK REACTIONS of photosynthesis (using radioactive carbon). He was awarded the 1951 Nobel Prize for Chemistry.

Calvin cycle *See* photosynthesis.

calyptra **1.** The root cap: a conical-shaped structure that is formed by the activity of the meristem at the root apex and forms a protective cap around the root tip. It is constantly replaced by newly formed cells as the older tissue is sloughed off with growth of the root through the soil.
2. A layer of cells derived from the venter of the archegonium that covers the developing sporophyte. In mosses and liverworts it ruptures as the seta elongates, being taken up as a hood over the capsule in mosses, and forming the sheath of tissue at the base of the seta in liverworts. The presence of the calyptra is necessary for the proper development of the capsule in mosses and the embryo in ferns.

calyptrogen A layer of meristematic cells covering the root apical meristem in some plants (e.g. grasses) that gives rise to and maintains the root cap.

calyx (*pl.* **calices**) The outermost part of a flower, enclosing the other floral parts during the bud stage. It consists of sepals, often green but sometimes brightly colored. The symbol *K* denotes the calyx in the floral formula. If the sepals are joined at their lateral margins a *calyx tube* is formed, the mouth of which may be extended into lobes or teeth.

CAM *See* crassulacean acid metabolism.

cambium The ring of dividing cells (meristem) responsible for lateral growth in plants. The primary cambium is found in the stem and root between the phloem and xylem cells, and by division gives rise to the secondary phloem and xylem in woody dicotyledonous plants. There are two zones of cambium: the VASCULAR CAMBIUM, which produces secondary xylem and phloem, and the PHELLEM (cork cambium), which produces secondary cortex and cork or bark. *See* secondary growth; wood. *See also* annual ring; meristem.

Cambrian The earliest period of the Paleozoic era, about 590–510 million years ago. It is characterized by the appearance of algae and a proliferation of marine invertebrate animal forms, but no true plants. In Britain, Cambrian rocks are found in Wales and North-West Scotland. *See also* geological time scale.

Camerarius, Rudolph Jacob (1665–1721) German botanist. Camerarius was professor of medicine and director of the botanic gardens at Tubingen. In his experiments on sexuality in dioecious plants (with male and female flowers on separate plants) and monoecious plants (with male and female flowers on the same plant), he demonstrated that if the female flowers were isolated from the male flowers or the male flowers were removed, the seeds produced by the female flowers were always sterile. His results were published in *De sexu plantarum* (On the Sex of Plants) in 1694. He also described the process of pollination and showed that the stamens were the male organs and that the stigma and style were the female organs of a plant.

campylotropous (**orthocampylotropous**) *See* ovule.

Canada balsam A yellowish resin with similar optical properties to glass, used as a mounting medium for microscope slides.

Candolle, Augustin Pyrame de (1778–1841) Swiss botanist. In 1806 the French government invited Candolle to carry out a botanical and agricultural survey of France that took him six years to complete. His *Théorie élémentaire de la botanique* (Elementary Theory of Botany; 1813) introduced the term 'taxonomy' for classification. Candolle based his classification on resemblances between the plan of symmetry of the sex organs in various plants, even though these were often disguised by loss, fusion, or degeneration. His idea was that even when these parts appeared to be different, they were in fact homologous and had a common ancestry. This classification replaced that of LINNAEUS, eventually being improved upon by BENTHAM and HOOKER. He is also noted for having recognized the influence of the soil on the distribution of vegetation.

cane sugar *See* sucrose.

canker A localized plant disease in which there is considerable NECROSIS of the cortex tissue, especially in woody plants, which becomes surrounded by layers of callus tissue. An example is apple canker, caused by the fungus *Nectria galligena*.

canopy In a woodland, forest, or shrub community, the uppermost layer of vegetation, formed from the branches of the tallest plants. *Compare* ground layer.

capillarity (**capillary action**) The rise or fall of liquids in narrow tubes as a result of the surface tension of the liquid, which causes the water to adhere to solid surfaces, such as soil particles or the walls of xylem vessels. When water in the soil or in the xylem is put under tension, due to the evaporation of water from a particular site (which creates a water potential gradient in the direction of the evaporation), capillarity enables water to overcome the cohesive forces within the water and move against gravity in response to the water potential gradient. It is a key feature in the movement of water through the xylem and through the finer pores of the soil.

capitulum (*pl.* **capitula**) *See* inflorescence.

capsid The protein coat of a VIRUS, which surrounds its nucleic acid. Capsid proteins are involved in recognition and infection of the host and can be used to identify the virus by serological means or amino acid sequencing.

capsule **1.** A dry dehiscent fruit that is formed from several fused carpels. The numerous seeds may be released through pores as in snapdragon (*Antirrhinum*), or by complete splitting of the capsule (e.g. *Iris*).
2. The structure within which spores of the sporophyte generation of mosses and liverworts are formed. It is borne at the end of a stalk (the *seta*), and ruptures to release the spores.
3. The mucilaginous covering often found around the cell membrane in bacteria.

carbohydrates A class of compounds occurring widely in nature and having the general formula type $C_x(H_2O)_y$. (Note that although the name suggests a hydrate of carbon these compounds are in no way hydrates and have no similarities to classes of hydrates.) Carbohydrates are generally divided into two main classes: SUGARS and POLYSACCHARIDES. Simple sugars are monosaccharides – single sugar units, usually with five (pentoses) or six (hexoses) carbon atoms. Ribose and glucose are examples. Oligosaccharides are polymers of two to ten monosaccharide units. Polysaccharides are long linear or branched chains of monosaccharide units, which have a structural or storage role in the plant, since they tend to be insoluble and relatively unreactive.

Carbohydrates are both stores of energy and structural elements in living systems. Living cells are able to build up polysaccharides from simple units (anabolism) or break the larger units down to more simple units for releasing energy (catabolism).

carbon Symbol: C. An essential element in plant and animal nutrition that occurs in all organic compounds and thus forms the basis of all living matter. It enters plants as carbon dioxide and is assimilated into carbohydrates, proteins, and fats, forming the backbones of such molecules. The element carbon is particularly suited to such a role as it can form stable covalent bonds with other carbon atoms, and with hydrogen, oxygen, nitrogen, and sulfur atoms. It is also capable of forming double and triple bonds as well as single bonds and is thus a particularly versatile building block. Carbon, like hydrogen and nitrogen, is far more abundant in living materials than in the Earth's crust, indicating that it must be particularly suitable to fulfill the requirements of living processes. *See also* carbon cycle.

carbon cycle The circulation of carbon between living organisms and the environment. The carbon dioxide in the atmosphere is taken up by autotrophic organisms (mainly green plants) and incorporated into carbohydrates. The carbohydrates so produced are the food source of the heterotrophs (mainly animals). All organisms return carbon dioxide to the air as a product of respiration and of decay. The burning of fossil fuels also releases CO_2. In water, carbon, combined as carbonates and bicarbonates, is the source for photosynthesis.

carbon-14 dating *See* radiometric dating.

Carboniferous The second most recent period of the Paleozoic era, some 355–280 million years ago. It is characterized by the evolution on swampy land of amphibians, a few primitive early reptiles, and trees. Aquatic life included sharks and coelacanths. The period is named after the extensive coal deposits that formed from the remains of vast forests of swamp plants. These were dominated by tree-sized lycophytes (related to present-day clubmosses) and sphenophytes (horsetails), ferns, and some early seed plants (mainly pteridosperms or 'seed ferns') some of which when rapidly buried and compressed over many millions of years formed the coal measures. *See also* geological time scale.

carboxylase An enzyme that catalyzes the incorporation of carbon dioxide into a substrate molecule. Carboxylases are found in yeasts, bacteria, plants, and animals. The most important carboxylase in plants is ribulose bisphosphate carboxylase (*rubisco*), which catalyzes the carboxylation of ribulose bisphosphate, the 'carbon-fixing' first step in the Calvin cycle.

carboxylic acid An organic compound of general formula RCOOH, where R is an organic group and –COOH is the carboxylate group. Many carboxylic acids are of biochemical importance. Those of particular significance are: **1.** The lower carboxylic acids (such as citric, succinic, fumaric, and malic acids), which participate in the respiration in the KREBS CYCLE, and in CRASSULACEAN ACID METABOLISM. Some plants accumulate high concentrations of certain carboxylic acids: citrus fruits contain a lot of citric acid, and rhubarb contains oxalic acid.
2. The higher acids, which are bound in LIPIDS. These are also called *fatty acids*, although the term 'fatty acid' is often used to describe any carboxylic acid of moderate-to-long chain length. The fatty acids contain long hydrocarbon chains, which may be saturated (no double bonds) or unsaturated (C=C double bonds). Plant fatty acids are often unsaturated: oleic acid is the commonest example.

carcerulus (*pl.* **carceruli**) A type of capsular fruit that breaks up at maturity into one-seeded segments or nutlets. The carcerulus is typical of the Lamiaceae family, e.g. deadnettle (*Lamium*).

carnivorous plant (**insectivorous plant**) A plant adapted to supplementing photosynthesis by obtaining nutrients, especially nitrates, from insects and other small animals. The plants have various means of trapping and killing the insects, which are then digested by enzymes secreted by the plant. Examples include butterworts (*Pinguicula*), with slippery inrolling leaves; sundews (*Drosera*), with sticky glandular hairs that close over the trapped insects; Venus' fly trap (*Dionaea muscipula*), with hinged leaves that snap shut over the victim, triggered by sensitive hairs; and pitcher plants, with leaves modified to form pitchers into which small animals fall and drown.

carotene A carotenoid pigment, examples being lycopene, and α- and β-carotene. The latter compounds are important in animal diets as a precursor of vitamin A. Red, yellow, or orange carotene pigments occur in all photosynthetic cells, where they function as accessory pigments by screening other photosynthetic pigments from photo-oxidation. Some carotenes give bright colors to animal-dispersed fruits, e.g. lycopene in tomatoes.

carotenoids A group of yellow, orange, or red pigments comprising the carotenes and xanthophylls. They are found in all photosynthetic organisms, where they function mainly as accessory pigments in photosynthesis. They contribute, with anthocyanins, to the autumn colors of leaves since the green pigment chlorophyll, which normally masks the carotenoids, breaks down first. They are also found in some flowers and fruits, e.g. tomato. Carotenoids have three absorption peaks in the blue–violet region of the spectrum.

carpel The female reproductive organ of a flowering plant. It consists of an ovary, containing one or more ovules, a stalk, or style (absent in some plant groups), and a terminal receptive surface, the stigma. Each flower may have one or more carpels borne singly giving an apocarpous gynoecium or fused together giving a syncarpous gynoecium. The carpel is homologous with the megasporophylls of certain spore-bearing vascular plants (pteridophytes) and the ovuliferous scales of gymnosperms. It has evolved by fusion of the two edges of the megasporophyll. This development can be most clearly seen in simple carpels such as those of the Fabaceae.

carrier 1. An organism that carries a recessive, often harmful, gene masked in the phenotype by a normal dominant gene.
2. An individual infected with pathogenic microorganisms without showing symptoms of disease. Such carriers can transmit the infection to others.

caruncle A fleshy outgrowth from a seed, similar to, but smaller than, an aril. Caruncles are often waxy or oily. They may arise from the placenta, micropyle, or funicle and are seen in castor oil seeds (*Ricinus*) at the micropyle and in violet seeds (*Viola*) at the hilum. The oily caruncles of violet seeds attract ants, which help to disperse them.

caryopsis (**grain**) (*pl.* **caryopses**) A dry indehiscent fruit, typical of the grasses. It is similar to an ACHENE except that the ovary wall is fused with the seed coat.

cascade A sequence of biochemical reactions, each triggered by the previous one. Cascades are important in signaling pathways, where extracellular signals (such as hormones or ions) bind with receptors on the cell membrane, triggering changes in membrane proteins that activate proteins on the cytoplasmic side of the membrane, setting off a chain of reactions.

Casparian strip An impervious band of thickening on the radial and transverse walls of the endodermis. It consists of deposits of suberin and cutin and ensures that all the water and solutes entering the stele pass through the cytoplasm of the endodermal cells, thus allowing for selective filtering of solutes and control of flow rate.

catabolism Metabolic reactions involved in the breakdown of complex molecules to simpler compounds as, for example, in respiration. The main function of catabolic reactions is to provide energy, which is stored as ATP, for use in the synthesis of new structures, cell-to-cell signalling, or the movement of motile cells. *See also* metabolism. *Compare* anabolism.

catalase An enzyme present in both plant and animal tissues that catalyzes the breakdown of hydrogen peroxide, a toxic compound produced during metabolism, into oxygen and water. It also catalyzes the oxidation of substrates by hydrogen peroxide. Catalase is found in the cytoplasm in small organelles called peroxisomes.

catalyst A substance that increases the rate of a chemical reaction without being used up in the reaction. Enzymes are highly efficient and specific biochemical catalysts. They lower the activation energy needed to initiate reaction by combining temporarily with the reactants.

cation A positively charged ion, formed by removal of electrons from atoms or molecules. In electrolysis, cations are attracted to the negatively charged electrode (the cathode). *Compare* anion.

catkin *See* inflorescence.

C_2 cycle *See* photorespiration.

cdl *See* critical day length.

cell The basic unit of structure of all living organisms, excluding viruses. Cells were discovered by Robert HOOKE in 1665, but SCHLEIDEN and SCHWANN in 1839 were the first to put forward a clear *cell theory*, stating that all living things were cellular. Prokaryotic cells (typical diameter 1 μm) are significantly smaller than eukaryotic cells (typical diameter 20 μm). The largest plant cells are the fibers of false nettles (*Boehmeria nivea*), up to 550 mm in length and stinging nettles (*Urtica dioica*) up to 75 mm long; the smallest are the epidermal cells of roses (0.04 mm in diameter). Among algae, the internode cells of stoneworts (*Chara* spp.) reach lengths up to 80 mm, while the individual independent cells of *Chlamydomonas* are only 0.092 mm in diameter. All cells contain, or once contained genetic material in the form of DNA, which controls the cell's growth and some of its activities; in eukaryotes this is enclosed in the membrane-bounded nucleus. All contain cytoplasm, containing various organelles and are surrounded by a plasma membrane. This controls entry and exit of substances. Plant cells and most prokaryotic cells are surrounded by rigid cell walls. In multicellular organisms cells become specialized for different functions; this is called differentiation. Within the cell, further division of labor occurs between the organelles.

cell cycle The ordered sequence of phases through which a cell passes leading up to and including cell division (mitosis). It is divided into four phases G_1, S, and G_2 (collectively representing interphase), and M-phase, during which mitosis takes place. Synthesis of messenger RNA, transfer RNA, ribosomes, and protein and general growth of the cell occurs in G_1, and replication of DNA and synthesis of histones occurs during the S phase, as chromosomes divide to form chromatids. The materials required for spindle formation are formed in G_2, when there is a further burst of growth before prophase begins. The time taken to complete the cell cycle varies in different tissues.

cell division *See* meiosis; mitosis; amitosis.

cell fractionation The separation of the different constituents of the cell into homogenous fractions. This is achieved by breaking up the cells in a mincer or grinder and then centrifuging the resultant liquid. The various components settle out at different rates in a centrifuge and are thus separated by appropriately altering the speed and/or time of centrifugation. The components are thus separated according to their size and weight.

cell lineage The theory stating that cells arise only from pre-existing cells. The cell lineage of a structure traces the successive stages that the cells pass through from the time of their formation in the zygote to their appearance in the mature functional structure.

cell membrane *See* plasma membrane.

cellobiose *See* cellulase.

cell plate A structure that appears in late anaphase in dividing plant cells and is involved in formation of a new cell wall at the telophase stage of mitosis. It is formed by fusion of vesicles from the Golgi apparatus, resulting in a flattened membrane-bounded sac spreading across and effectively dividing the cell. Cell-wall polysaccharides and pectic substances contained in the vesicles contribute to growth of the new wall and middle lamella inside the sac. The process begins at the *phragmoplast*, a barrel-shaped region at the former site of the spindle equator, where microtubules remain and the vesicles cluster, possibly oriented by the microtubules. Endoplasmic reticulum and ribosomes are also present. The cell-plate membranes eventually form the two new plasma membranes of the daughter cells. *See* cytokinesis.

cell theory The theory that all organisms are composed of CELLS and cell products and that growth and development results from the division and differentiation of cells.

cellular slime molds *See* Acrasiomycota.

cellulase An enzyme that hydrolyzes 1,4-glycosidic linkages in cellulose, yielding *cellobiose* (a disaccharide) and glucose. It is important in the degradation of plant cell walls in living plants (e.g. in leaf abscission).

cellulose A polysaccharide forming the framework of the CELL WALLS of all plants, many algae and other protoctists, and some fungi. Cellulose molecules are unbranched chains that together form a rigid structure of high tensile strength. Bundles of molecules form microfibrils, which may be aligned in the primary cell wall either transversely or longitudinally. Cellulose forms an important source of carbohydrate in the diets of herbivores, and is a major constituent of dietary fiber in human diets. The individual units in cellulose are β-1,4 linked D-glucose molecules.

cell wall A rigid wall surrounding the cells of plants, fungi, bacteria, and algae. Plant cell walls are made of cellulose fibers in a cementing matrix of other polysaccharides. The walls of some algae differ, e.g. the silica boxes enclosing diatoms, and the calcareous layer on the cell walls of calcareous algae. Fungi differ, with their walls usually containing chitin. The walls of some fungal spores contain the pigment melanin, which protects against the damaging effects of ultraviolet radiation. Bacterial walls are more complex, containing peptidoglycans – complex polymers of amino acids and polysaccharides. Cell walls are freely permeable to gases, water, and solutes. They have a mechanical function, allowing the cell to become turgid by osmosis, but preventing bursting. This contributes to the support of herbaceous plants. Plant cell walls can be strengthened for extra support by addition of lignin (as in xylem and sclerenchyma) or extra cellulose (as in collenchyma). They are an important route for movement of water and mineral salts. Other modifications include the uneven thickening of guard cells, the sieve plates in phloem, the suberization of endodermal cells, the waterproof coverings of epidermal and cork cells.

The cellulose in cell walls takes the form of microfibrils. At cell division in plants the *primary wall* is laid down on the middle lamella of the cell plate as a loose mesh of cellulose microfibrils. This gives an elastic structure that allows cell expansion during growth. Later the *secondary wall* grows and acquires greater rigidity and tensile strength. New cellulose fibers are laid down in layers, parallel within each layer, but orientated differently in different layers. The Golgi apparatus provides polysaccharide-filled vesicles that deposit wall material by exocytosis, guided by microtubules.

Cenozoic (**Caenozoic; Cainozoic**) The present geological era, beginning some 65 million years ago, and divided into two periods, the Tertiary and the Quaternary. It is characterized by the rise of modern organisms, especially mammals and flowering plants. During this era, the world's tallest mountain ranges such as the Rockies, Andes, European Alps and Himalayas were formed, with dramatic effects on climate and plant distribution. *See also* geological time scale.

centrarch Denoting a protostele in which protoxylem is at the center of the axis. *Compare* endarch; exarch; mesarch.

centric bundle *See* concentric bundle.

centrifugal Developing from the center outwards so the youngest structures are at the outer edge. Centrifugal xylem differentiates from the center outwards, and a centrifugal inflorescence (e.g. the dichasial cyme) is one in which the progression of flower opening is from the center to the periphery. *Compare* centripetal. *See also* acropetal; basipetal.

centriole A cell organelle consisting of two short tubular structures orientated at right angles to each other. Each tubular structure is about 500 nm long and 150 nm in diameter. The centriole lies outside the nucleus of animal cells and many fungal and protoctist cells, but is absent in cells of most higher plants. Prior to cell division it replicates, and the sister centrioles move to opposite ends of the cell to lie within the spindle-organizing structure, the centrosome. At the start of cell division, fibrils appear to radiate from the centrioles, forming starlike ASTERS. These form the poles of the developing spindle. However, the centriole is not essential for spindle formation, although an analogous structure, the basal body, is responsible for organizing the microtubules of UNDULIPODIA.

centripetal Developing from the outside inwards so the youngest structures are at the center. Centripetal xylem differentiates from the outside toward the center, and a centripetal inflorescence (e.g. the capitulum) is one in which the outer flowers open before those in the center. *Compare* centrifugal. *See also* acropetal; basipetal.

centromere The region of the chromosome that contains the kinetochore, the structure that becomes attached to the nuclear spindle during mitosis and meiosis. It contains DNA of no known significance. Following the replication of chromosomes, resultant chromatids remain attached at the centromere. The centromere is a specific genetic locus and remains relatively uncoiled during prophase, appearing as a primary constriction. It does not stain with basic dyes.

centrosome A structure found in all eukaryotic cells, except fungi, that forms the SPINDLE during cell division. It lies close to the nucleus in nondividing cells. Microtubules extend across the cell from the centrosome, which controls the distribution and assembly of the microtubules that make up the cytoskeleton, thus controlling the cell's shape and, in some cells, movement. The centrosome probably plays an important role in the reorganization of the cell that occurs during differentiation. At the start of cell division the centrosome divides, and the sister centrosomes move to opposite ends of the cell, organizing the spindle. In animal and protoctist cells the centrosome contains two short barrel-shaped structures, the CENTRIOLES, but these are not directly involved in spindle formation. In fungi, the function of the centrosome is served instead by the spindle pole body.

chalaza (*pl.* **chalazae** or **chalazas**) The region of an angiosperm ovule where the nucellus and integuments merge. When ovule orientation is orthotropous the chalaza corresponds to the point where the funicle is attached but in anatropous and campylotropous ovules the chalaza is some distance from the funicle.

chalazogamy A method of fertilization in angiosperms in which the pollen tube enters the ovule by the chalaza instead of through the micropyle, penetrating the placenta en route. Chalazogamy is seen in certain trees and shrubs, e.g. beech. *Compare* porogamy.

chamaephyte A perennial plant that is able to produce new growth from resting buds near (within 0.25 m of) the soil surface. Chamaephytes are usually small bushes (e.g. heather (*Calluna vulgaris*)). *See also* Raunkiaer's plant classification.

chaparral Sclerophyllous scrub vegetation of coastal and inland mountain regions of southwestern North America, maintained by burning and grazing and, in some areas, possibly from past land clearance. It consists of dense thickets of broad-leaved evergreen bushes and shrubs and twisted dwarf trees up to 2.5 meters (8 ft) tall. The dominant plants are sages (Lamiaceae) and evergreen oaks and, in drier areas, chamise (*Adenostoma fasciculatum*), manzanita (*Arctostaphylos pungens*), and species of *Ceanothus*. *See* maquis. *Compare* garrigue.

Charophyceae A class of the Chlorophyta (green algae) that includes the stoneworts, formerly placed in a separate phylum *Charophyta*. They resemble true plants in many ways and are thought to be the most closely related algal group to plants.

Charophyta *See* Charophyceae.

chasmogamy The production of flowers that open their petals so that cross pollination is possible. *Compare* cleistogamy. *See also* anemophily; entomophily.

chelation The process of forming a stable chemical complex (*chelate*) with a given substance, usually a metal ion, that involves the formation of at least two chemical bonds. Chelates are usually organic ring compounds, the ion being chelated to form the closure of the ring. Chelating agents can be used to remove toxic metal ions from soil or water. Some plants achieve tolerance to toxic ions by chelating them, usually by combination with an organic compound, such as citrate or malate; the chelate is then deposited in the vacuole. Chlorophyll is also a chelate, the porphyrin part of the molecule being bound to iron.

chemical fossils Particularly resistant organic chemicals present in geological strata that are thought to indicate the existence of life in the period when the rocks were formed. Chemical fossils (e.g. alkanes and porphyrins) are often the only evidence for life in rocks of PRECAMBRIAN age.

chemiosmotic theory The theory proposed by Peter Mitchell in 1961 to explain how electron transport and ATP synthesis are coupled during photosynthesis and respiration. It requires a membrane that permits the movement of protons from inside to outside, but not vice versa. An energy-requiring 'proton pump' (the ELECTRON-TRANSPORT CHAIN) causes movement of protons (H^+ ions) from the internal matrix of the organelle to the outside, creating a pH gradient across the membrane. As the H^+ ions diffuse back across the membrane. This gradient of potential energy drives reactions at specific sites on the membrane, where the enzyme ATPase uses the energy to drive ATP synthesis, so the energy can be transferred to reaction sites elsewhere.

chemoautotrophism (**chemosynthesis**) *See* autotrophism; chemotrophism.

chemoheterotrophism *See* chemotrophism; heterotrophism.

chemoreceptor An organ or structure that responds to a particular chemical stimulant.

chemosynthesis *See* autotrophism; chemotrophism.

chemotaxis (**chemotactic movement**) A TAXIS in response to a chemical concentration gradient. The spermatozoids of primitive plants are often positively chemotactic, swimming toward the female organs in response to a chemical secreted by the latter. For example, the archegonium (female organ) of the moss *Funaria* secretes sucrose. Motile organisms may also be negatively chemotactic, swimming away from concentrations of toxic chemicals.

chemotaxonomy (**biochemical systematics; biochemical taxonomy**) A system of classification of plants based on the nature of their metabolic products (metabolites). Techniques involved include chromatography and electrophoresis, and the sequencing of nucleic acids and proteins.

chemotrophism A type of nutrition in which the source of energy for the synthesis of organic requirements is chemical. Most chemotrophic organisms are heterotrophic (i.e. *chemoheterotrophic*) and their energy source is always an organic compound; animals, fungi, and most bacteria are chemoheterotrophs. If autotrophic (i.e. *chemoautotrophic* or *chem-*

osynthetic) the energy is obtained by oxidation of an inorganic compound; for example, by oxidation of ammonia to nitrite or a nitrite to nitrate (by nitrifying bacteria), or oxidation of hydrogen sulfide to sulfur (by colorless sulfur bacteria). Only certain specialized bacteria are chemoautotrophic. *Compare* phototrophism.

chemotropism (**chemotropic movement**) A TROPISM in which the stimulus is chemical. The hyphae of certain fungi (e.g. *Mucor*) are positively chemotropic, growing toward a particular source of food. Pollen-tube growth down the style is both positively and negatively chemotropic: the tube grows toward chemicals secreted by the pistil, and away from atmospheric oxygen.

chernozem (**black earth**) A freely draining CALCAREOUS SOIL found in regions with high summer rainfall, characterized by humus distributed fairly evenly throughout the profile. The upper horizon of the soil is very thick and black with humus; it merges into a brown lime-rich horizon below. Chernozems are rich in plant nutrients and have a good crumb structure, allowing drainage and aeration. They are associated with grasslands in temperate climates such as the prairies of North America, the pampas of South America, and the Russian steppes, where they support highly productive agricultural systems.

chiasma (*pl.* **chiasmata**) A connection between homologous chromosomes seen during the prophase stage of meiosis. Chiasmata represent a mutual exchange of material between homologous, nonsister chromatids (CROSSING OVER) and provide one mechanism by which RECOMBINATION occurs, through the splitting of linkage groups. Chiasmata also help to hold homologs together until the start of anaphase, thus helping to ensure the correct distribution of chromosomes into the daughter nuclei.

chimera An individual or part of an individual in which the tissues are a mixture of two genetically different tissues. It may arise naturally due to mutation in a cell of a developing embryo, producing a line of cells with the mutant gene, and hence different characteristics compared to surrounding cells. It may also be induced experimentally. In plants, chimaeras produced from two different species are known as *graft hybrids*. For example, a bud may develop at the junction between the scion and stock with a mixture of tissues from both.

chi-squared test A statistical test used to compare observed frequencies of experimental sampling results with expected (hypothetical) frequencies, usually calculated from the results assuming no effect of the experimental variable. It is suitable for observations that fall into discrete classes. The value:

$$(\text{observed} - \text{expected})^2/\text{expected}$$

is calculated for each class, then the values are added together. The sum is checked against a table of χ^2 values at various degrees of freedom to determine the degree of significance of the difference between observed and expected values is significant.

chitin A nitrogen-containing polysaccharide found in the cell walls of most fungi and in some animals. Chitin is a polymer of *N*-acetylglucosamine. It consists of many glucose units, in each of which one of the hydroxyl groups has been replaced by an acetylamine group (CH_3CONH).

chlamydospore A thick-walled resting spore produced by certain fungi (e.g. smuts) and funguslike protoctists (e.g. *Mucor*).

Chlorella *See* Chlorococcales.

chlorenchyma A form of parenchyma in which the cells contain many chloroplasts and relatively large intercellular spaces, as in the mesophyll of the leaf.

chlorinated hydrocarbon A hydrocarbon in which one or more hydrogen atoms are replaced by chlorine atoms. Some chlorinated hydrocarbons are used as insecticides, e.g. aldrin, dieldrin, and DDT.

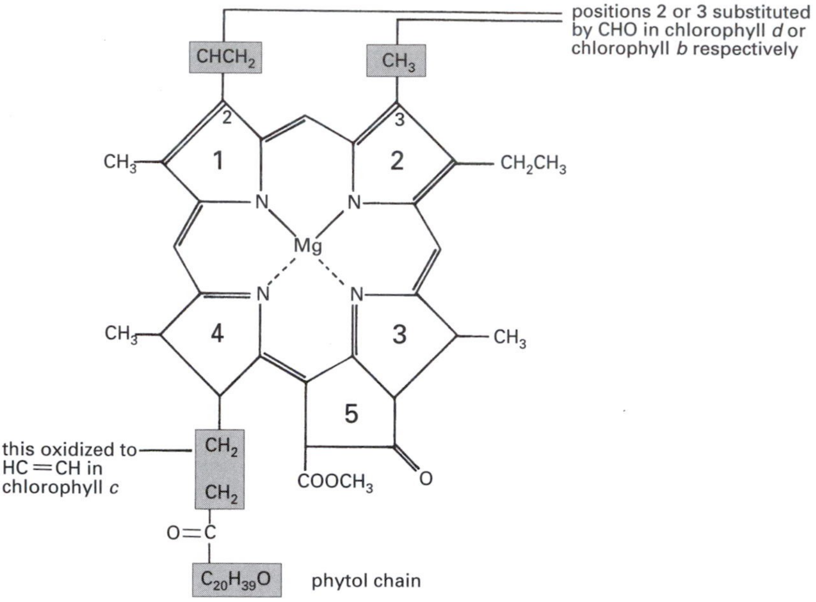

positions 2 or 3 substituted by CHO in chlorophyll *d* or chlorophyll *b* respectively

this oxidized to $HC{=}CH$ in chlorophyll *c*

phytol chain

Chlorophyll: the structural formula of chlorophyll a

chlorine Symbol: Cl. An element found in trace amounts in plants. The chloride ion (Cl-) can pass easily through cell membranes, and is involved in the maintenance of osmotic and ionic balance in the plant.

Chlorococcales An order of Chlorophyta (green algae) that contains single-celled and colonial algae that lack undulipodia, e.g. *Chlorella*, *Volvox*, found in freshwater plankton and most terrestrial habitats. Only the gametes and zoospores are motile. They reproduce asexually by means of zoospores or nonmotile spores and sexually by fusion of similar-looking gametes (isogamy).

chlorophylls A group of photosynthetic pigments. They absorb blue-violet and red light and hence reflect green light, imparting the green color to green plants. The molecule consists of a hydrophilic (water-loving) head, containing magnesium at the center of a porphyrin ring, and a long hydrophobic (water-hating) hydrocarbon tail (the *phytol chain*), which anchors the molecule in the lipid of the membrane. Different chlorophylls have different chemical groups attached to the head. Chlorophylls are located in the chloroplasts, and are involved in the light reactions of photosynthesis, absorbing light energy and initiating electron transport. There are four kinds of chlorophyll, *a*, *b*, *c*, and *d*, which differ in their molecular structure. Chlorophyll *a* is universal in all photosynthesizing eukaryotes. It converts the absorbed light energy to chemical energy. Chlorophylls *b*, *c*, and *d* are accessory pigments. Chlorophyll *b* occurs in plants and in the Chlorophyta; the other chlorophylls occur in certain groups of algae. *See* absorption spectrum; photosynthetic pigments. *See also* bacteriochlorophyll.

Chlorophyta (**green algae**) A phylum of protoctists comprising mainly freshwater algae with some marine and terrestrial forms. They contain the pigments chlorophyll *a* and *b* together with carotenes and xanthophylls. The Chlorophyta store food as starch and fat and have cell walls containing cellulose and hemicellulose. Some of the commoner orders include: Volvocales, unicellular and colonial plants, e.g. *Chlamydomonas*, *Volvox*; Chlorococcales, including unicellular and colonial plants, e.g. *Chlorella*, *Pediastrum*; Ulotrichales, filamentous, and thallose plants, or coenocytic tubes, e.g. *Ulothrix*, *Ulva*, *Enteromorpha*; and Oedogoniales, e.g. the filamentous *Oedogonium*.

chloroplast A photosynthetic PLASTID containing chlorophyll and other photosynthetic pigments. It is found in all photosynthetic cells of plants and protoctists but not in photosynthetic prokaryotes. It has a membrane system containing the pigments and on which the light reactions of PHOTOSYNTHESIS occur. The surrounding gel-like ground substance, or *stroma*, is where the dark reactions occur. The typical higher-plant chloroplast is lens-shaped and about 5 μm in length. Various other forms exist in the algae, e.g. spiral in *Spirogyra*, stellate in *Zygnema*, and cup-shaped in *Chlamydomonas*. The number per cell varies, e.g. one in *Chlorella* and *Chlamydomonas* and in most hornworts (Antherocyta), two in *Zygnema*, and about one hundred in palisade mesophyll cells of leaves.

Chloroplast membranes form elongated flattened fluid-filled sacs called *thylakoids*. The sheetlike layers of the thylakoids are called *lamellae*. In all plants except algae, the thylakoids overlap at intervals to form stacks, like piles of coins, called *grana*. In this way the efficiency of the light reactions seems to be improved. On the thylakoid membranes are spherical structures partially embedded in the membrane. These are the photosynthetic units (formerly called *quantasomes*), which contain the pigments and associated lipids and proteins of photosystems I and II. Photosystem I units are about 17.5 nm in diameter, and photosystem II units about 11 nm. Other, smaller particles on the thylakoid membranes are ATPases. The stroma may contain storage products of photosynthesis, e.g. starch grains. Apart from enzymes of the dark reactions, the stroma also contains typical prokaryotic protein-synthesizing machinery including circular DNA (*see* chloroplast DNA) and smaller ribosomes. There is now strong evidence that chloroplasts and other cell organelles, such as mitochondria, represent prokaryotic organisms that invaded heterotrophic eukaryotic cells early in evolution and are now part of an indispensable symbiotic union (*see* endosymbiont theory). In C_4 plants there are two types of chloroplast.

chloroplast DNA (cp-DNA) The DNA found in a chloroplast. It is in circular form, rather like that of prokaryotes. It codes for only some of the proteins needed by the chloroplast.

chlorosis The loss of chlorophyll from plants resulting in yellow (chlorotic) leaves. It may be the result of the normal process of senescence, lack of key minerals for chlorophyll synthesis (particularly iron and magnesium), or disease.

chlor-zinc-iodide *See* Schultze's solution.

chromatid One of a pair of replicated chromosomes found during the prophase and metaphase stages of mitosis and meiosis. During mitosis, sister chromatids remain joined by their centromere until anaphase. In meiosis it is not until anaphase II that the centromere divides, the chromatids being termed daughter chromosomes after separation.

chromatin The loose network of threads seen in nondividing nuclei that represents the chromosomal material, consisting of DNA, protein (mainly histone), and small amounts of RNA. It is classified as *euchromatin* or *heterochromatin* on the basis of its staining properties, the latter staining much more intensely with basic stains because it is more coiled and compact. Euchromatin is thought to be actively involved in transcription and therefore protein synthesis, while heterochromatin is inactive. Euchromatin stains more intensively than heterochromatin during nuclear division. *See also* chromosome.

chromatin loop A region of uncoiled DNA, usually about 200 nm long, that extends out in a loop from the chromosome. This uncoiled condition is thought to facilitate TRANSCRIPTION and probably indicates an area where genes are active.

chromatography A method of analyzing materials involving the separation by selective absorption of the various compounds as identifiable bands. For instance, a mixture of substances in solution is passed slowly down a long column packed with alumina. The different compounds move at different rates and separate into bands. The medium, together with the bands of separated material, are collectively called a *chromatogram*.

In general chromatography involves a test material being carried by a *moving phase* (liquid or gas) through a *stationary phase* (solid or liquid). Different substances move at different rates (depending on their absorption–desorption) and are therefore spatially separable, the least readily absorbed being carried the farthest. Colorless materials can be used if some means of detecting them is used (electronic detection, radioactive labeling, or ninhydrin developer). *See also* paper chromatography; thin-layer chromatography; gas chromatography; gel filtration.

chromatophore A name generally applied to a pigment-bearing structure, usually a membrane lamella or vesicle.

chromomere A small beadlike deeply staining structure seen in chromosomes during prophase of mitosis and meiosis. It is thought to represent an area where the chromosome is relatively tightly coiled. Homologous chromosomes have chromomeres in identical places along their length.

chromoplast (**chromatophore**) A colored plastid, i.e. one containing pigment. They include chloroplasts, which contain the green pigment chlorophyll and are therefore photosynthetic, and nonphotosynthetic chromoplasts. The term is sometimes confined to the latter, which are best known in flower petals, fruits (e.g. tomato), and carrot roots. They are yellow, orange, or red owing to the presence of carotenoid pigments.

chromosome 1. One of a group of threadlike structures of different lengths and shapes in nuclei of eukaryotic cells. They consist of DNA with RNA and protein (mostly histones) and carry the GENES. During nuclear division the chromosomes are tightly coiled and are easily visible through the light microscope. After division, they uncoil and may be difficult to see, the long narrow threads of uncoiled DNA being only about 2 nm in diameter. Along the length of the uncoiled chromosomes are beadlike structures called *nucleosomes*, highly organized aggregations of DNA and histones.

The number of chromosomes per nucleus is characteristic of the species except in aneuploid plants. Normally one set (haploid) or two sets (diploid) of chromosomes are present in the nucleus. However, many or perhaps a majority of plants are polyploid in origin, i.e. they have multiple sets of chromosomes, though most act as diploids in their reproduction. In some species there are diploid and polyploid varieties that look almost identical, though sometimes they may be referred to as separate species. In bryophytes the main plant body is haploid, as is the case in many algae (*see* alternation of generations). In early prophase of mitosis and later prophase of meiosis, the chromosomes split lengthwise into two identical chromatids held together by the CENTROMERE. The length of the arms of the chromosome is determined by the position of the centromere. The centromere contains the *kinetochore*, a structure involved in the attachment of the chromosomes to the spindle during nuclear division. In diploid cells, there is a pair of *sex chromosomes*; the remainder are termed *autosomes*. Each chromosome contains one DNA molecule, which is folded and coiled. Histones play a major architectural role, holding the coiled (helical) DNA in supercoils or beads, about 10 nm in diameter and consisting of about 170 base pairs. These in turn are coiled to form a hollow fiber, 30 nm in diameter, which itself forms a hollow coil measuring about 240 nm in diameter. It is this that is visible as the typical looped threadlike chromatin material of the nondividing

chromosome. *See* chromomere; chromosome map.
2. *See* bacterial chromosome.

chromosome map (**genetic map**) A diagram showing the order of genes along a chromosome. Such maps have traditionally been constructed from information gained by linkage studies to give a *linkage map*. The techniques employed differ according to the type of organism being studied. Many plants and animals can be crossed experimentally to study inheritance patterns of particular genes. However, the advent of new molecular techniques has dramatically changed the nature of chromosome mapping in all organisms. The growing science of PROTEOMICS seeks to analyze the genome by starting with the protein products of genes and working back to the genome.

Mapping a genome involves several steps. The first is to assign each gene to a particular chromosome. This can be achieved by, for example, somatic-cell hybridization or using a *gene probe*. The next step is to determine the relative positions of the genes on a particular chromosome. This involves comparing RESTRICTION FRAGMENT LENGTH POLYMORPHISMS between individuals and constructing a linkage map of all restriction sites, i.e. sites that are cleaved by restriction enzymes. These restriction sites can then be used as markers for closely neighboring genes. The last step is to construct a physical map of the base sequence of the chromosomal DNA. One approach uses cloned DNA segments obtained from a gene library of the chromosome. These clones can be fitted together to form a series of overlapping segments (*contig*) that corresponds to a particular region of the chromosome. The base sequence of the contig is then determined, and hence the base sequence of the chromosomal DNA. *See* gene library; restriction map.

chromosome mutation A change in the number or arrangement of genes in a chromosome. If chromosome segments break away during nuclear division they may rejoin the chromosome the wrong way round, giving an *inversion*. Alternatively, they may rejoin a different part of the same chromosome, or another chromosome, giving a *translocation*. If the segment becomes lost, this is termed a *deficiency* or *deletion*; it is often fatal. A part of a chromosome may be duplicated and occur either twice on the same chromosome or on two different nonhomologous chromosomes: this is a *duplication*. Chromosome mutations can occur naturally but their frequency is increased by the effect of x-rays and chemical mutagens. Such aberrations may be visible by microscopy, especially with the use of certain stains. *See also* gene mutation.

chromosome number The number of chromosomes possessed by a particular species. It may be expressed as either the haploid (n) or diploid (2n) number. For many species, chromosome number is a useful taxonomic character. Exceptions include species exhibiting APOMIXIS. Chromosome number also gives an indication of polyploidy, where numbers vary by exact multiples between related species.

Chrysomonada A phylum of golden-brown algae, formerly called the *Chrysophyta*, common in fresh temperate waters, with a few species in marine waters. Chrysomonads are mainly motile unicellular algae, but there are some colonial forms and a few complex filamentous forms.

ciliate **1.** Describing a structure, such the margin of a leaf, fringed with fine hairs.
2. A unicellular protoctist of the phylum Ciliophora, whose members are usually covered in short undulipodia (cilia) and have two different kinds of nuclei, micronuclei and macronuclei.

cilium (*pl.* **cilia**) A whiplike extension of certain eukaryotic cells that beats rapidly, thereby causing locomotion or movement of fluid over the cell. Cilia are short UNDULIPODIA, 2–10 μm long.

The beat of each cilium comprises an effective downward stroke followed by a gradual straightening (limp recovery). Cilia beat in such a way that each is slightly out of phase with its neighbor (*metachronal rhythm*), thus producing a constant rather than a jerky movement or flow of fluid. The basal bodies of cilia are connected by threadlike strands (*neuronemes*), which coordinate the beating of neighboring cilia.

circadian rhythm (**diurnal rhythm**) A daily rhythm of various metabolic activities in animals and plants. Such rhythms persist even when the organism is not exposed to 24-hour cycles of light and dark, and are thought to be controlled by an endogenous biological clock. Circadian rhythms are found in the most primitive and the most advanced of organisms. Thus *Euglena* shows a diurnal rhythm in the speed at which it moves to a light source. Stomata open and close in a daily cycle, which persists for several days in total darkness. Experiments have failed to reveal the type and location of the control mechanisms involved. *See also* biorhythm.

circinate *See* vernation.

circumnutation The spontaneous spiral growth of the shoot tips of certain plants, particularly climbers. The direction of rotation is often constant in a species; for example, *Convolvulus* always rotates in an anticlockwise direction. For climbers nutation increases the likelihood of a solid support being found. It is also marked in many leaf tendrils and to a lesser extent in roots, flower stalks, and the sporangiophores of some fungi. Gravity has been shown to be involved in the response: if a plant is turned upside down, the tip and the last few spirals unwind and straighten out before winding up in the characteristic way in the new direction.

cisterna (*pl.* **cisternae**) A flattened membrane-bounded sac of the endoplasmic reticulum or the Golgi apparatus, being the basic structural unit of these organelles.

cis–trans **effect** (**complementation test**) The phenomenon resulting from recombination within a gene (cistron), in which a mutation is only expressed in the phenotype if the mutant pseudoalleles are on different homologous chromosomes.

cis–trans **test** A test that determines whether two mutations that have the same effect occur in the same gene or in different genes. The mutations may be in either the *cis* position (i.e. on the same chromosome) or the *trans* position (one on each homolog). If the mutations are in different genes then a normal phentotype results whether the mutations are in the *cis* or *trans* position, since they are masked by corresponding dominant genes on the other homolog. However, if the mutations are in the same gene then a normal phenotype will result only if the mutations are in the *cis* position. In the *trans* position the mutant phenotype is expressed, since both alleles of the locus are mutants.

cistron A unit of function, i.e. a segment of DNA that determines a single polypeptide chain of a protein molecule or a messenger RNA molecule. Its boundaries may be defined by the *cis–trans* test.

CITES The Convention on International Trade in Endangered Species of Wild Fauna and Flora: an agreement signed between some 160 countries, prohibiting commercial trade in endangered species and products derived from them. As more species have become endangered, the list of protected species has lengthened. CITES sets out degrees of protection for different species, allowing threatened but not endangered species to be exported and imported under license.

citric acid A six-carbon carboxylic acid, occurring in the juice of citrus fruits, particularly lemons, and present in many other fruits. Citric acid is biologically important because it participates in the KREBS CYCLE, where it is formed by the condensation of acetyl CoA and oxaloacetate, a rate-limiting step of the cycle. The systematic name is 2-hydroxypropane-1,2,3-tricarboxylic acid

citric acid cycle *See* Krebs cycle.

cladistics A method of classification in which the relationships between organisms are based on selected shared characteristics. These are generally assumed to have been derived from a common ancestor, in the evolutionary process of cladogenesis, although the 'transformed cladists' believe that shared characteristics alone provide a logical basis for classification without postulating evolutionary relationships. The patterns of these shared characteristics are demonstrated in a branching diagram called a *cladogram*. The branching points of the cladogram may be regarded either as an ancestral species (as in an evolutionary tree) or solely as representing shared characteristics.

cladode A modified internode of the stem that functions as a leaf, being flattened and highly photosynthetic. It may be a xerophytic adaptation and is seen in butcher's broom (*Ruscus aculeatus*) and prickly pear (*Opuntia*). The true leaves may be reduced to spines or scales. Unlike a leaf, a cladode has buds on its surface.

cladogram *See* cladistics.

Cladophorales An order of Chlorophyta (green algae), generally called siphonaceous algae. They are multicellular saclike forms or branched filaments with branching rhizoidlike holdfasts, e.g. *Cladophora*.

class A taxonomic rank that is subordinate to a phylum (or sometimes a division in plant taxonomy) and superior to an order. The Latin names of plant classes end in *–ae*, e.g. Dicotyledonae. Classes may be divided into subclasses, with the Latin ending *–idae*, e.g. Rosidae.

classification The grouping and arrangement of organisms into a hierarchical order. The aim is usually to aid in identification of organisms or represent their phylogenetic relationships, or, ideally, both. An important aspect of classifications is their predictive value. For example if a characteristic is found in one member of a group of plants, then it is also likely to be found in the other members of that group even though the characteristic in question was not used in the initial construction of the classification.

Artificial classification systems are designed to aid identification, and may not reflect true relationships. e.g. LINNAEUS's system, which was based on the number of stamens and carpels in the flower. *Natural classification systems* are based on more characteristics, and thus thus are better for predicting relationships. *Phenetic classification systems* are based on many characters, and may also be weighted in favor of particular characters. *Phylogenetic classification systems* attempt to reflect the evolutionary relationships of taxa. The validity of these systems is now being tested by mRNA studies. *Compare* cladistics. *See* taxon; phylum; class; order; family; genus; species.

clay Extremely fine-textured SOIL made up of small mineral particles, less than 0.002 mm in diameter, formed mainly from aluminum and magnesium silicates. Clay soils become very sticky and difficult to work when wet and can easily become waterlogged. Nutrient availability to plants can be a problem as the nutrients may become chemically bound to the surfaces of the particles.

clearing **1.** In the preparation of permanent microscope slides, the stage between dehydration and embedding, the purpose of which is to remove the dehydrating agent and replace it by a substance that is miscible with the embedding substance. Clearing also renders the tissues transparent. Clearing agents include benzene and xylene.
2. In the preparation of cell material for staining, the dissolving away of cell contents so that the distribution of tissues may be better observed, using reagents such as sodium hypochlorite.
3. A treeless, usually grassy area in a woodland or forest.

cleavage polyembryony The splitting of an embryo into several identical parts, each of which may develop into a mature embryo. It is a form of natural cloning. Cleavage polyembryony occurs in some gymnosperms, e.g. in *Pinus*, but usually only one embryo develops to maturity. *See* clone.

cleistogamy The production of closed flowers. It is a method of ensuring self-pollination and occurs toward the end of the flowering season in certain plants, e.g. wood sorrel (*Oxalis acetosella*) particularly when little or no seed has been set by the cross-pollinating flowers. *Compare* chasmogamy.

cleistothecium (*pl.* cleistothecia) *See* ascoma.

climacteric The rise in respiration rate, found in the fruits of some species, associated with fruit ripening and senescence. It may be induced artificially by treatment with ethylene (ethene).

climatic climax A plant community that is in equilibrium with a stable climate. It represents the CLIMAX of a succession.

climatic factors Aspects of climate, such as temperature, rainfall, humidity, air movement, and light, that affect living organisms. *Compare* biotic environment; edaphic factors.

climax The final community in a succession of natural plant communities in one area under a particular set of conditions. Mixed woodland with oak is the natural climax vegetation in much of lowland Britain, but on chalk it is often beechwood. A climax is self-perpetuating, at least for a time, although there are constant small changes, e.g. when a tree dies pioneering plants may invade or the tree may be replaced by a different species. Succession to a climax can be held at any stage by human intervention, such as grazing on chalk downland. Such an equilibrium is called a *subclimax*, *plagioclimax*, *disclimax*, or *biotic climax*. *See* clisere.

climbing ferns Ferns of the family Schizaeaceae, found in most parts of the world in warm and tropical regions, including North America. There are four genera: *Anemia*, *Lygodium*, *Mohria*, and *Schizaea*. Many climbing ferns have long creeping rhizomes.

cline A graded series of characters, usually morphological, exhibited by a species or other related group of organisms, usually along a line of environmental or geographical transition. The populations at each end of the cline may be substantially different from one another.

clinostat *See* klinostat.

clisere A succession of climax communities in a given area, each giving way to the next as a result of climatic changes. *See also* climax; sere.

clone A group of organisms or cells that are genetically identical. In nature, clones are derived from a single parental organism or cell by mitotic cell division, asexual reproduction, or parthenogenesis. In genetic engineering, multiple identical copies of a gene are produced in *cloning vectors*, such as plasmids and phages. *See* gene cloning. *See also* apomixis.

clubmoss *See* Lycophyta.

clubroot A serious disease of the cabbage family (Brassicaceae) caused by *Plasmodiophora brassicae* (phylum Plasmodiophora), in which the roots become swollen and misshapen, with characteristic gall-like swellings. Above-ground symptoms include wilting in bright sunlight and a reddish, bluish, or yellowish tinge to the leaves.

CoA *See* coenzyme A.

coadaptation The evolution and maintenance of certain genetic characteristics that increase the effectiveness of a relationship between two species. Flower/pollinator relationships are an example, in which adaptations of flowers for pollination by animals is to the mutual advantage of both species (coevolution). *See* mutualism.

cobalt *See* micronutrient.

coccus (*pl.* **cocci**) A spherical-shaped bacterium. Cocci may be found singly, in pairs (e.g. *Diplococcus*), or chains (e.g. *Streptococcus*), or in regularly or irregularly packed clusters.

co-dominance *See* incomplete dominance.

codon A group of three nucleotide bases (i.e. a nucleotide triplet) in a messenger RNA (mRNA) molecule that codes for a specific amino acid or signals the beginning or end of the message (start and stop codons). Since four different bases are found in nucleic acids there are 64 (4 × 4 × 4) possible triplet combinations. The arrangement of codons along the mRNA molecule constitutes the *genetic code*. When synthesis of a given protein is necessary the segment of DNA with the appropriate base sequences is transcribed into mRNA. When the mRNA migrates to the ribosomes, its string of codons is paired with the anticodons of transfer RNA molecules, each of which is carrying one of the amino acids necessary to make up the protein. Most amino acids are coded for by more than one codon. The genetic code is therefore said to be *degenerate*. The term codon is sometimes also used for the triplets on DNA, which are complementary to those on mRNA, but contain the base uracil instead of thymine. *See* transfer RNA.

coenocyte An area of cytoplasm containing many nuclei and enclosed by a cell wall, typically found in many fungi, certain green algae (e.g. *Cladophora*), and some members of the Xanthophyta. Such a cell is described as *coenocytic*.

coenzyme *See* cofactor.

coenzyme A (**CoA**) A coenzyme that is important in the synthesis and reactions of fatty acids. In the KREBS CYCLE it combines with pyruvic acid, leading to loss of carbon dioxide.

coenzyme Q (**ubiquinone**) A coenzyme that is an essential component of the respiratory electron-transport chain.

cofactor (**coenzyme**) A nonprotein substance that helps an enzyme to carry out its activity. Cofactors may be cations or organic molecules, known as coenzymes. Many enzymes require metal cofactors, e.g. ferredoxin requires ferrous and ferric ions. The metal ion may be the catalytic center of the cofactor, or it may serve to bind the enzyme and substrate together to hold the enzyme in a catalytically active conformation. Unlike enzymes cofactors are, in general, stable to heat. When a catalytically active enzyme forms a complex with a cofactor a *holoenzyme* is produced. An enzyme without its cofactor is termed an *apoenzyme*. *See also* prosthetic group.

cohesion theory A theory to explain how sap rises up the xylem. It suggests that the lowering of water potential in the leaf due to an increase in evaporation causes water to flow toward the leaf down a water potential gradient. Strong cohesive forces between the water molecules in the xylem prevent the water column breaking.

colchicine An alkaloid drug, obtained from the autumn crocus *Colchicum autumnale,* which is used experimentally to prevent spindle formation in mitosis or meiosis by preventing tubulin polymerizing into microtubules. It has the effect of halting cell division at metaphase, the stage at which the chromosomes have duplicated to give four homologs for each chromosome. *See* anther culture.

cold hardening (**cold acclimation**) Processes taking place in plants before the onset of winter that prepare them to withstand low temperatures. For example, plants may accumulate solutes, such as oligosaccharides and the amino acid proline, to make their tissues frost-resistant.

coleoptile A sheathlike structure that protects the developing plumule in grasses. Some think that, together with the SCUTELLUM, it represents the COTYLEDON, while others believe it is the first plumular leaf. The coleoptile contains very little chlorophyll and is usually light-sensitive.

coleorhiza (*pl.* **coleorhizae**) The protective sheath surrounding the radicle in grasses.

coliform bacteria Gram-negative rod-shaped bacteria able to obtain energy aerobically or by fermenting sugars to produce acid or acid and gas. Most are found in the vertebrate gut (e.g. *Escherichia coli*), but some are present in soil, water, or as plant pathogens. Many are pathogenic to humans (e.g. *Salmonella*).

collateral bundle A kind of vascular bundle in which the phloem is external to the xylem and on the same radius. *Compare* bicollateral bundle; concentric bundle.

collenchyma A specialized type of parenchyma, usually located just beneath the epidermis, that functions as supporting tissue. The cell walls are irregularly thickened with cellulose and pectin, the thickening giving distinct patterns to the cells in cross-section. The thickening is usually greatest in the corners of the cells or adjacent to the intercellular spaces. Collenchyma is the first strengthening tissue to be formed in young plants and is able to expand as the young tissues continue development. *Compare* sclerenchyma.

colloid A substance that is composed of particles dispersed through another substance. Such particles range from 10^{-7} to 10^{-3} cm; they are too small to see with a light microscope, yet cannot pass through a selectively permeable membrane. Proteins in the cytoplasm are an example.

colony A body form in which many cells with similar form and function are grouped together, as in many algae, e.g. *Volvox*. Such a colony is often surrounded by mucilage.

columella **1.** The structure present in sporangia of many zygomycete fungi (e.g. *Mucor*) produced by formation of a dome-shaped septum cutting off the sporangium from the sporangiophore.
2. The central column of sterile tissue in the sporangium of liverworts and mosses.

commensalism An association between two organisms in which one, the *commensal*, benefits and the other remains unaffected either way, e.g. an epiphyte on the branch of a tree. *Compare* amensalism; mutualism; parasitism; symbiosis.

community A general term covering any naturally occurring group of different organisms living together in a certain environment, and interacting with each other. *See* association; consociation.

community biomass *See* biomass.

companion cell An elongated thin-walled cell cut off longitudinally from the same meristematic cell as the SIEVE ELEMENT with which it is closely associated. It has a nucleus, dense cytoplasm, and many mitochondria, and is thought to provide, via PLASMODESMATA, the needs of the less metabolically active, enucleated sieve element and perhaps plays a role in regulating the flow of metabolites through the sieve tube.

compass plant A plant with its leaf edges permanently aligned due north and south. Such plants thus avoid receiving the strong midday rays of the sun directly on the leaf blades, but are positioned to use fully the weaker rays of the morning and evening sun from the east and west. The best-known example is the compass plant of the prairies (*Silphium laciniatum*).

compatibility The coexistence of different bacterial plasmids in the same host cell. Plasmids that cannot exist in the same host cell belong to the same *incompatibility group*.

compensation depth *See* photic zone.

compensation point The concentration of carbon dioxide at which, in a closed system, the rate of photosynthesis is exactly balanced by the combined rates of respiration and photorespiration, so that net exchange of oxygen and carbon dioxide is zero.

The point at which photosynthesis does not increase with increased light intensity is termed the *light saturation point*. This point occurs at much higher light intensities in C_4 plants than C_3 plants.

competence A characteristic of embryonic cells and meristematic cells whereby they have the potential to differentiate into any of several different types of cell. Once plant cells have differentiated, most lose competence, except under special conditions. *See* totipotency.

competition The utilization of the same resources by one or more organisms of the same or different species living together in a community, when the resources are not sufficient to fill the needs of all the organisms. Plants, particularly the dominant plants of a community, compete mainly as seedlings are establishing, but once this has been achieved, the adult plant often outcompetes any potential competitor unless affected by herbivores or pathogens.

competitive inhibition A form of enzyme inhibition in which the inhibitor competes with the substrate to bind with the active site of the enzyme. A competitive inhibitor is usually structurally similar to the substrate, e.g. malonate is a competitive inhibitor of the enzyme succinate dehydrogenase. It may bind to the site but will not react to produce products.

complementary DNA (**cDNA**) A form of DNA synthesized by genetic engineering techniques from a messenger RNA template using a reverse transcriptase enzyme. It is used in cloning to obtain gene sequences from mRNA isolated from the tissue to be cloned. It differs from the original DNA sequence in that it lacks intron and promoter sequences. Labeled single-stranded cDNA is used as a gene probe to identify common gene sequences in different tissues and species. *See* gene cloning.

complementary genes Genes that can only be expressed in the presence of other genes; for example if one gene controls the formation of a pigment precursor and another gene controls the transformation of that precursor into the pigment, then both genes must be present for the color to develop in the phenotype. Such interactions between genes lead to apparent deviations from the 9:3:3:1 dihybrid ratio in the F_2.

Compositae *See* Asteraceae.

composite fruit *See* multiple fruit.

compound fruit *See* aggregate fruit.

concentric bundle (**centric bundle**) A vascular bundle in which the xylem and phloem are arranged in rings, one completely surrounding the other. *Compare* bicollateral bundle; collateral bundle.

conceptacle A flask-shaped reproductive cavity that develops on the swollen tips (receptacles) of the thalli of certain brown algae, (e.g. bladder wrack). Female conceptacles are lined with unbranched sterile hairs (*paraphyses*) and the oogonia develop on short stalks projecting from the chamber wall. Male conceptacles contain branched paraphyses that bear the antheridia. Both female and male conceptacles open to the exterior via a pore, the *ostiole*.

condensation reaction A chemical reaction in which two molecules are joined together with the elimination of a simpler molecule or group. Many condensation reactions involve the elimination of water, e.g. the formation of peptide bonds between amino acids in the formation of peptides. *Compare* hydrolysis.

conduplicate *See* ptyxis.

cone *See* strobilus.

confocal microscopy A form of light microscopy involving the use of two microscopes, one on each side of the specimen. The first microscope focuses an illuminated aperture onto the specimen, while the other also focuses on the specimen, but receives the light transmitted through it. This produces an image of a single plane in the specimen. It can be repeated at different depths of focus to build up a three-dimensional picture.

conformation The spatial arrangement of atoms in a molecule. In affecting the molecule's shape and the distribution of chemical groups and electrostatic charges on its surface, the conformation determines the potential energy and reactivity of the molecule that results from the attractive and repulsive forces between the different parts of the structure. The conformation of PROTEINS is critical to their effectiveness as enzymes. Their conformation is affected by the sequence of amino acids and the type, location, and nature of the bonds between different parts of the molecule, or between the polypeptide chain and various cofactors and prosthetic groups.

conidiophore *See* conidium.

conidiospore *See* conidium.

conidium (**conidiospore**) (*pl.* **conidia**) An asexual spore of certain fungi, especially of the Ascomycota and Fungi Anamorphici e.g. *Pythium* and *Albugo*. Conidia are cut off externally in chains at the apex of a specialized hypha, the *conidiophore*.

Coniferophyta (**conifers**) The largest phylum of gymnosperms, comprising about 560 species of evergreen trees and shrubs, with many important species, e.g. *Pinus* (pine), *Picea* (spruce), *Taxus* (yew), and *Abies* (fir). They dominate the vast boreal forests of the northern hemisphere. Most are evergreen, but *Larix* (larches) and *Taxodium* (swamp cypresses) are deciduous.

conjugated protein A protein that on hydrolysis yields not only amino acids but also other organic and inorganic substances. They are simple proteins combined with nonprotein groups (prosthetic groups). *See also* glycoprotein; lipoprotein.

conjugation **1.** A type of sexual reproduction found in the conjugating green algae (phylum Gamophyta), and certain fungi, in which a conjugation tube forms between cells of two individuals and the gamete formed in one cell (the male gamete) moves through the tube and fuses with the gamete of the other cell (the female gamete) to form a zygote. *Spirogyra* and *Rhizopus* are examples.
2. In prokaryotes, a means of transferring genetic information from one bacterium to another across a bridgelike structure. In *Escherichia* and related genera, two individuals join by a conjugation bridge and part of the genetic material of one, the donor (or male) cell, is transferred to the recipient (or female) cell.

connective The parenchymatous tissue that joins the two lobes of the anther and contains the vascular strand. *See* stamen.

consociation A climax of natural vegetation dominated by one particular species, such as oakwood, dominated by the oak tree, or *Calluna* heathland dominated by the heather, *Calluna vulgaris*. Many consociations together may form an association, for example oakwood, beechwood, and ashwood consociations together make up a deciduous forest association. *See also* association.

conspecific Describing individuals that belong to the same species.

consumer An organism that feeds upon another organism, e.g. all animals and parasitic and insectivorous plants. *Compare* producer. *See also* trophic level.

continental drift The theory that present-day continents have arisen by the breaking up and drifting apart of a previously existing ancient land mass (Pangaea). There is much evidence to support the theory, and it serves to explain the distribution of contemporary and fossil plants and animals. Continental drift is now be-

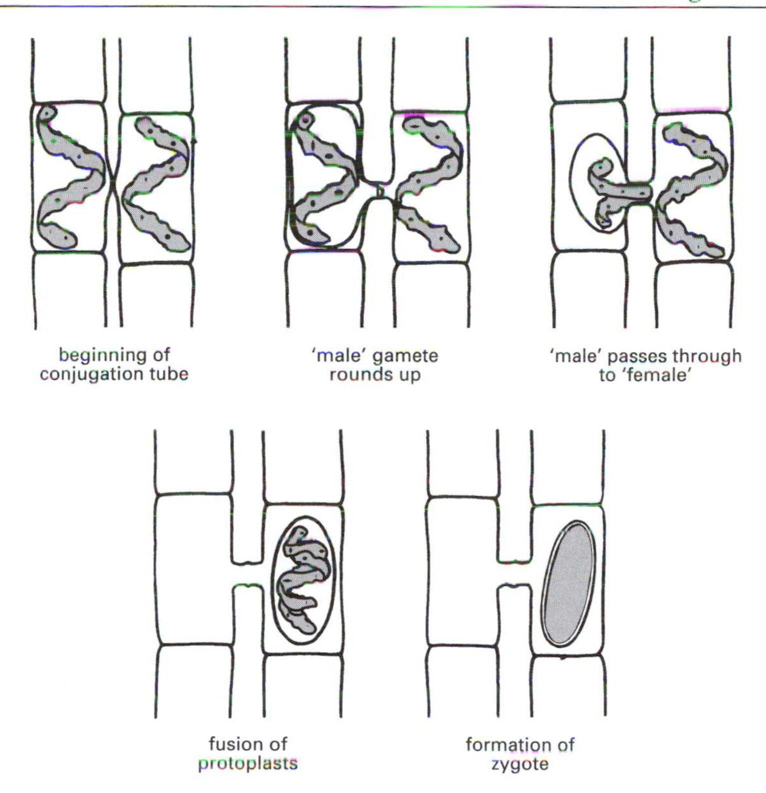

Conjugation: the stages occurring in *Spirogyra*

lieved to reflect the movement over geological time of underlying plates in the earth's crust – the theory of plate tectonics.

continuous variation (**quantitative variation; multifactorial inheritance**) A form of variation in which the characters show continuous variation within a certain range of values. Examples are grain yield in wheat and height in humans. Such traits are typically controlled by many different genes (polygenes). The identification and manipulation of genetic loci determining quantitative traits (*quantitative trait loci* (*QTLs*)) is important in plant and animal breeding.

contractile root A specialized root developed by certain bulb- and corm-forming plants that serve to pull the bulb or corm down to the appropriate depth in the soil, e.g. *Crocus*.

Convention on Biological Diversity An international convention, signed at the United Nations Rio Summit in 1992, that aims to conserve biodiversity, the sustainable use of natural resources, and the equitable sharing of the benefits arising from exploitation of genetic resources. It covers all aspects of biodiversity, from genetic resources to ecosystems, and aims to conserve resources while promoting sustainable development and to encourage the sharing of costs and benefits between developing and developed countries. The United Nations Environment Program (UNEP) responded by commissioning the Global Biodiversity Assessment, to collect and review data, theories, and opinion relating to current issues worldwide, which provided valuable information on the rapid rate at which environmental change is taking place and natural ecosystems are being modified.

convergent evolution (**convergence**) The development of similar (analogous) structures in unrelated organisms as a result of living in similar ecological conditions. An example is the evolution of similar succulent forms in the Cactaceae and the Euphorbiaceae, in which the leaves are reduced to pale spines. *See also* analogous.

copper Symbol Cu. A metal trace element (micronutrient), essential for plant growth. It is found in the enzyme cytochrome *c* oxidase, which is part of the respiratory electron-transport chain, and in the protein plastocyanin, which is involved in photosynthetic electron transport. Copper deficiency results in chlorosis, suggesting that copper may be involved in the synthesis of chlorophyll. *See also* micronutrient.

coppice (**copse**) A woodland managed for wood production by cutting trees back to ground level at regular intervals (usually 10–15 years) and allowing adventitious shoots to grow up from the base. The young shoots, which are often very straight, are used for fencing, charcoal burning, and firewood. Occasional trees may be left to mature: these are termed *standards*.

coralloid roots A type of root regularly produced by cycads, which contains symbiotic nitrogen-fixing cyanobacteria in root nodules, giving the roots a knobbly coral-like appearance.

cordate Heart-shaped, usually referring just to the base of leaves, e.g. the leaves of the common blue violet (*Viola papilionacea*).

cork *See* phellem.

cork cambium *See* phellogen.

corm An organ of perennation and vegetative reproduction, consisting of a short erect fleshy swollen underground stem, usually broader than high and covered with membranous scales. It stores food material in the stem and bears buds in the axils of the scalelike leaf remains of the previous years' growth. One or more new corms form from these buds, and lie above the shrivelled remains of the previous year's corm. The corm's position in the soil is often maintained with the aid of contractile roots. Examples of corms are crocus and gladiolus. *Compare* bulb.

corolla A collective term for the petals of a flower. The corolla is denoted in the floral formula by the symbol C. Where the margins of the petals are partly or completely fused, a *corolla tube* is formed.

corona **1.** In flowering plants, any type of outgrowth from the petals or sepals, such as the trumpet of the daffodil flower (*Narcissus*).
2. A group of cells at the tip of the oogonium in certain green algae (Chlorophyta), such as the stonewort *Chara* and the mermaid's cup (*Acetabularia*).

corpus The central region of the meristem below the tunica where cell divisions are in all directions giving both increased width and length to the apex. The tissues of the stele and cortex are derived from the corpus. *See* tunica–corpus theory.

cortex (*pl.* **cortices** or **cortexes**) A primary tissue in roots and stems of vascular plants derived from the corpus meristem, that extends inwards from the epidermis to the phloem. It usually consists of PARENCHYMA cells but other tissues (e.g. collenchyma and sclerenchyma) may be present.

corymb *See* inflorescence.

cothallus *See* thallus.

cotyledon (**seed leaf**) The first leaf of the embryo of seed plants, which is usually simpler in structure than later-formed leaves. Cotyledons play an important part in the early stages of seedling development. For example they act as storage organs in some seeds, such as peas and beans when mature. They form the first photosynthetic organ in seeds showing epigeal germination, e.g. sunflower (*Helianthus*) but in others they remain underground. Monocotyledons and dicotyledons are so termed because they normally contain one and two cotyledons respectively although there are exceptions.

covalent bond A chemical bond in which one or more pairs of electrons are shared between two atoms. It is often represented by a single line between the symbols of the two atoms that are bonded together. *Compare* hydrogen bond; ionic bond.

cp-DNA *See* chloroplast DNA.

C_3 plant A plant in which the first product of photosynthesis is a 3-carbon acid, glycerate 3-phosphate (phosphoglyceric acid). Most plants are C_3 plants. They are characterized by high carbon dioxide compensation points owing to PHOTORESPIRATION. *Compare* C_4 plant.

C_4 plant A plant in which the first product of photosynthesis is a 4-carbon dicarboxylic acid, oxaloacetic acid. C_4 plants have evolved from C_3 plants by a modification in carbon dioxide fixation, leading to more efficient carbon dioxide uptake. The modified pathway is called the *Hatch–Slack pathway* or the C_4 *dicarboxylic acid pathway*. In the leaves, the mesophyll cells surrounding the vascular bundles (bundle sheath cells) contain the carbon dioxide-fixing enzyme phosphoenolpyruvate carboxylase (PEP carboxylase) in their cytoplasm. This has a higher affinity for carbon dioxide than ribulose bisphosphate carboxylase (RUBP carboxylase). The product of carbon dioxide fixation is oxaloacetate, which is rapidly reduced to the C_4 acid malate or transaminated to the C_4 acid aspartate. These C_4 acids are transported to the bundle-sheath cells encircling the leaf veins, where they are decarboxylated, releasing CO_2, which is then refixed as in C_3 plants.

C_4 plants are mainly tropical or subtropical, including many tropical grasses (e.g. maize, sugar cane, and sorghum). They are more efficient than C_3 plants, producing more glucose per unit leaf area. *See also* bundle sheath; crassulacean acid metabolism.

crassulacean acid metabolism (CAM) A form of photosynthesis first discovered in the family Crassulacaeae and since found in many other succulent plants, such as cacti. It allows plants of arid environments to keep their stomata closed to conserve water during the day, opening them at night to exchange gases. Carbon dioxide taken in through the stomata at night is converted into an organic acid for storage. Instead of combining with ribulose bisphosphate, it combines with the 3-carbon phosphoenolpyruvate (PEP) to form oxaloacetic acid, which is then converted to malic acid for storage in the vacuole. The following day, when the stomata are closed, the malic acid is broken down and the carbon dioxide is released to take part in the light reactions of photosynthesis. When stressed by drought, plants with crassulacean metabolism (*CAM plants*) can last for days or even weeks without opening their stomata, conserving water and living on stored resources. *Compare* C_3 plant; C_4 plant.

cremocarp A dry fruit splitting into two one-seeded portions. The portions are termed *mericarps*.

Cretaceous The most recent period of the Mesozoic era, 145–66 million years ago. The dominant vegetation on land in the early Cretaceous was forests of cycads, conifers, gingkgoes and ferns. Most modern-type ferns, gymnosperms, and angiosperms arose during the Cretaceous, and by the end of the period the flowering plants had replaced the gymnosperms as the dominant terrestrial vegetation, forming vast broad-leaved forests with magnolias, figs, poplars, sycamores, willows, and herbaceous plants and a great variety of insect pollinators. The Cretaceous is named after the large amounts of chalk (fossilized plankton) found in rocks of the period. *See also* geological time scale.

crista (*pl.* **cristae**) The structure formed by folding of the inner mitochondrial membrane, giving a large surface area for electron transport.

critical day length (**cdl**) The amount of light per day that is the maximum a short-day plant may receive and still flower, and conversely, the minimum a long-day plant needs to flower. Cocklebur, a short-day plant, will not flower if given more than 15½ hours light per day; i.e. its cdl is 15½ hours. Henbane, a long-day plant, will only flower if given more than 11 hours light per day; i.e. the cdl is 11 hours. In reality, however, it is the length of the dark period that it critical, short-day plants requiring more than a critical length of dark period before they will flower, and vice versa. *See* photoperiodism.

cross **1.** The act of cross-fertilization. **2.** The organism resulting from cross-fertilization.

crossing over The exchange of material between homologous chromatids by the formation of chiasmata. This results in the recombination of linked genes. Such reassortment of genes is the main source of genetic variation during sexual reproduction. *Unequal crossing over*, in which one chromatid ends up with one copy of a chromosome segment while the other has three copies, results from crossing over after pairing between homologous chromosomes that are not properly aligned. *See also* chiasma; independent assortment; tetrad.

crossover frequency *See* chromosome map.

cross-pollination The transfer of pollen from the anthers of one individual to the stigma of another individual, usually of the same species, with subsequent germination of the pollen and growth of the pollen tube. This transfer is achieved by intermediary agents such as birds (ornithophily), bats, insects (entomophily), wind (anemophily), or water (hydrophily). *See* incompatibility. *Compare* self-pollination.

crown gall A disease of plants caused by the soil-borne bacterium *Agrobacterium tumefasciens*. It causes galls both and above and below ground on a wide range of plants, especially fruit trees.

Cruciferae *See* Brassicaceae.

crumb structure The texture of a soil in terms of the size of the soil particles and how they interact, often determined simply by rubbing the soil with fingers. Some soils, such as sandy soils, have a loose crumb structure with good drainage, whereas in clay soils the particles are small and stick together to give a dense crumb structure and poor drainage

cryptogam In early classifications, any plant that reproduces by spores or gametes rather than by seeds. Cryptogams were thus named because early botanists considered their method of reproduction to be hidden (cryptic). They included the algae, fungi, mosses, liverworts, hornworts, ferns, clubmosses, and horsetails, the vascular plant group often being termed *vascular cryptogams*. *Compare* phanerogam.

cryptophyte (**geophyte**) A plant in which the resting buds are below the soil surface or in water. *See also* Raunkiaer's plant classification.

cultivar Any agricultural or horticultural 'variety'. The term is derived from the words *culti*vated *var*iety.

culture A population of microorganisms or dissociated cells of a tissue grown on or within a solid or liquid medium for experimental purposes. This is done by inoculation and incubation of the nutrient medium. *See also* tissue culture.

culture medium A mixture of nutrients used, in liquid form or solidified with agar, to cultivate microorganisms, such as bacteria or fungi, or to support tissue cultures.

cushion plant A plant that forms a tight low hummock as an adaptation to cold and dry or windy situations. Cushion plants usually have small thick leaves to minimize water loss, or hairy leaves to reduce transpiration. Cushion plants are common in arctic–alpine habitats.

cuticle A waterproof, protective layer of wax secreted by the epidermis that covers the external surface of the plant, with the exception of the stomata and lenticels. Some seeds also have a cuticle. Its thickness varies with the species and environment. Xerophytes tend to have thick cuticles, and dry conditions often induce cuticular thickening. Its main function appears to be restriction of water loss from the plant. On average, only about 5% of the water lost from a plant is via the cuticle (cuticular transpiration).

cutin A group of substances chemically related to fatty acids forming a continuous layer called the cuticle on the epidermis of plants, interrupted only by stomata or lenticels. Being fatty in nature, cutin is water-repellent, therefore helping to reduce transpiration. It is also protective, for example preventing invasion by parasites. *See* suberin.

cutinization The impregnation of a plant cell wall with cutin.

cutting A part of a plant that is removed from the parent and encouraged to grow into another plant. Cuttings are a means of asexual propagation and may vary in size from buds, leaves or root segments to large shoots.

Cyanobacteria A phylum of Eubacteria containing the blue-green bacteria (formerly called blue-green algae) and the green bacteria (chloroxybacteria). Both groups convert carbon dioxide into organic compounds using photosynthesis, generally using water as a hydrogen donor to yield oxygen, like green plants. Cyanobacteria are an ancient group, and their fossils (*stromatolites*) have been dated at up to 3500 million years old. They are spherical (coccoid) or form long microscopic filaments of individual cells.

Cycadophyta (**cycads**) A phylum of cone-bearing gymnosperms (about 75 living species) with palmlike compound leaves and special CORALLOID ROOTS at or near the ground surface, which contain symbiotic nitrogen-fixing cyanobacteria.

cyclic AMP (**cAMP**; **adenosine-3′,5′-monophosphate**) A form of adenosine monophosphate (*see* AMP) formed from ATP in a reaction catalyzed by the enzyme adenyl cyclase. It has many functions, acting as an enzyme activator, genetic regulator, chemical attractant, secondary messenger, and as a mediator in the activity of many hormones, including epinephrine, norepinephrine, vasopressin, ACTH, and the prostaglandins.

cyclic photophosphorylation In PHOTOSYNTHESIS, the flow of electrons from Photosystem I, after excitation by absorbed light energy, along the ferredoxin/cytochrome *b*/*f*/plastocyanin electron-transport chain and back to Photosystem I to be reenergized. Cyclic photophosphorylation drives ATP synthesis, but does not produce NADPH.

cyclins Regulatory proteins whose concentration varies with certain stages of the CELL CYCLE. Cyclins help control progress through the various stages of the cell cycle by activating protein kinases (cyclin-dependent protein kinases).

cyclosis (**cytoplasmic streaming**) The streaming of cytoplasm in a circular motion around the cell observed in some plants, particularly young sieve tube elements.

cycyanocobalamin (**vitamin B_{12}**) One of the water-soluble B-group of VITAMINS.

cymose inflorescence (**cyme; definite inflorescence**) *See* inflorescence.

cypsela An achenelike fruit that develops from an inferior ovary and contains some tissues not derived form the carpel. It is common in the Asteraceae, e.g. dandelion (*Taraxacum*), whose fruits bear a ring of hairs (the pappus) derived from the calyx.

cyst A thick-walled resting spore.

cysteine A sulfur-containing AMINO ACID synthesized from methionine and serine. It is involved in the synthesis of biotin. It also acts as a store of sulfur for biosynthesis, and can be broken down to pyruvate.

cystine A compound formed by the joining of two cysteine AMINO ACIDS through a –S–S– linkage (a *cystine link*). Bonds of this type are important in the conformation of proteins, acting as cross-links between different parts of a polypeptide or between different polypeptide chains, thus determining the pattern of folding. It is also required to the synthesis of the coenzyme thiamin pyrophosphate, which is involved in many decarboxylation reactions.

cystolith A deposit of calcium carbonate arising internally on a stalk from the cell walls of large modified epidermal cells in some flowering plants, e.g. stinging nettle (*Urtica dioica*).

cytidine (**cytosine nucleoside**) A NUCLEOSIDE formed when cytosine is linked to D-ribose via a β-glycosidic bond.

cytochrome oxidase (**cytochrome** ***aa*****$_3$**) The enzyme that catalyzes the final step in the respiratory electron transport chain, in which water is formed from hydrogen ions and oxygen. *See* respiration.

cytochromes Conjugated proteins containing heme, that act as intermediates in the electron-transport chains of photosynthesis and respiration. The iron ion in the porphyrin prosthetic group of cytochromes changes its oxidation state from ferric (Fe^{3+}) to ferrous (Fe^{2+}) when the cytochrome is reduced. There are four main classes, designated *a*, *b*, *c*, and *d*, most bound to membranes.

cytogamy The fusion of gametes during fertilization. *Compare* karyogamy.

cytokinesis The division of the cytoplasm after nuclear division (mitosis or meiosis). In the cells of plants and most algae it involves formation of a new plant cell wall by means of a structure called the cell plate. In primitive protoctists, as in animals, the cell membrane ingrows under the control of a ring of actin microfilaments, and a furrow constricts the cell equator until it pinches the cell in two. *See* cell plate.

cytokinin One of a class of plant hormones concerned with the stimulation of cell division, nucleic acid metabolism, and root–shoot interactions. Cytokinins are often purine derivatives: e.g. *kinetin* (6-furfuryl aminopurine), an artificial cytokinin commonly used in experiments to promote cell division in tissue cultures; and zeatin, found in maize cobs. Natural cytokinins are all derivatives of adenine.

Cytokinins are produced in roots, where they stimulate cell division. They are also transported from roots to shoots in the transpiration stream, where they are essential for healthy leaf growth. Subsequent movement from the leaves to younger leaves, buds, and other parts may occur in the phloem and be important in sequential leaf senescence up the stem.

cytology The study of cells; cell biology.

cytoplasm The living contents of a cell, excluding the nucleus and large vacuoles, in which many metabolic activities occur. It is contained within the plasma membrane and comprises a colorless substance (*hyaloplasm*) containing organelles and various inclusions (e.g. crystals and insoluble food reserves). The cytoplasm is about 90% water. It is a true solution of ions (e.g. potassium, sodium, and chloride), small molecules (e.g. sugars, amino acids, and ATP), and a colloidal solution of large molecules (e.g. proteins, lipids, and nucleic acids). It can be gel-like, usually in its outer regions, or sol-like. A complex cytoskeleton of F-actin filaments and microtubules is often present, and appears to control the structure of the cytoplasm, cell movements, and movements of the cytoplasm and conformation of the membranes. *See* cytoskeleton; organelle; protoplasm.

cytoplasmic inheritance (**extrachromosomal inheritance**) The determination of certain characters by genetic material contained in plasmids or organelles other than the nucleus, e.g. mitochondria and chloroplasts. Characters controlled by the DNA of extranuclear organelles are not inherited according to Mendelian laws and are transmitted only through the female line, since only the female gametes have an appreciable amount of cytoplasm. Cytoplasmic inheritance is known in a wide variety of plants, animals, and unicellular organisms.

cytoplasmic streaming *See* cyclosis.

cytosine A nitrogenous base found in DNA and RNA. Cytosine has the PYRIMIDINE ring structure. In DNA and RNA it base-pairs with guanine.

cytoskeleton A network of fibers within the cytoplasm of a cell that maintains its shape, enables movement of the cell, and provides anchorage and movement of its organelles. It appears to be involved in the alignment of cellulose fibers during cell wall formation, and in changes in conformation of membranes. It comprises various elements, including microtubules, microfilaments of actin, and intermediate filaments. These consist of units that can assemble and disassemble rapidly by polymerization/depolymerization. *See* microfilament; microtubule; spindle.

cytosol The soluble fraction of cytoplasm remaining after all particles have been removed by centrifugation.

cytotaxonomy The use of chromosome number, size, and shape in the classification of organisms. The position of the centromere and the behavior of chromosomes during meiosis also important. Pairing of homologous parts of chromosomes provides clues to their taxonomic affinities. Increasingly, gene sequencing is assuming a greater importance in determining phylogeny. *See also* taxonomy.

D

2,4-D (**2,4-dichlorophenoxyacetic acid**) A synthetic AUXIN used as a potent selective weedkiller. Monocotyledenous species with narrow erect leaves (e.g. cereals and grasses) are generally resistant, while dicotyledonous plants are often very susceptible. The compound is thus particularly effective in controlling weeds in cereal crops and lawns.

dark-ground illumination A light microscopy technique used to study transparent material, such as living cells or microbes. The specimen is lighted from the side, so that only diffracted light from the specimen passes through into the objective. The specimen thus appears bright against a dark background. The difference in refractive index between organelles and the surrounding cytoplasm causes the boundaries between organelles and cytoplasm to reflect more light.

dark reactions (**light-independent reactions**) A group of reactions that follow the LIGHT REACTION in PHOTOSYNTHESIS and form glucose and other reduced products from carbon dioxide. They are not dependent on light, although they can take place in the light. The dark reactions use energy (as ATP) and reducing power (as NADPH) generated by the light reaction of photosynthesis to reduce carbon dioxide. In eukaryotes, the dark reactions take place in the stroma of the chloroplast by one of two pathways, according to whether the plant is a C_3 PLANT or a C_4 PLANT. The reduced carbon dioxide is incorporated into carbohydrates in the Calvin cycle. *See also* crassulacean acid metabolism.

Darwin, Charles Robert (1809–82) British naturalist noted for his theory of evolution by NATURAL SELECTION. Darwin first became interested in natural history at Cambridge University. He was inspired by the geologist Adam Sedgwick and encouraged by the professor of botany at Cambridge, John Henslow, to apply for the position of naturalist on a survey of the Pacific and South America on HMS *Beagle* in 1831. On this voyage, he observed many different animal species and was struck by the way in which they gradually changed from one location to another, in particular, the finches of the Galápagos Islands. He recorded 14 species of finches on these islands, each species dominating a particular part of the islands in different conditions. He speculated that it was probably unlikely that they could all have arisen spontaneously as separate species and thought that it was more probable that they had evolved from a mainland species of finch, possibly from Ecuador. It took 20 years for Darwin to collect enough evidence to publish his seminal work *On the Origin of Species by Means of Natural Selection* (1859), having been stimulated to publish by an unexpected letter in 1858 from the British naturalist Alfred Russel WALLACE, outlining theories similar to Darwin's own. Darwin and Wallace read a joint paper on their theories to the Linnaean Society in London in the summer of 1858. The publication of Darwin's book led to a storm of controversy among scientists and churchmen as his theory challenged the biblical account of the Creation. Darwin continued to publish more works, including his *The Descent of Man* (1871), in which he applied his theory of evolution to humans, implying that they had descended from apelike creatures

Darwinism Darwin's explanation of

the mechanism of evolutionary change, namely, that in any varied population of organisms only the best adapted to that environment will tend to survive and reproduce. Individuals that are less well adapted will tend to perish without reproducing. Hence the unfavorable characteristics, possessed by the less well-adapted individuals, will tend to disappear from a species, and the favorable characteristics will become more common. Over time the characteristics of a species will therefore change, eventually resulting in the formation of new species. Darwin called this process of selective birth and death NATURAL SELECTION. *See also* neo-Darwinism.

dating techniques Methods used to determine the age of rocks, fossils, or archaeological remains. There are two main methods. *Relative dating* assesses the age of a specimen in comparison to other specimens. *Absolute dating* involves assessing the actual age of a specimen by using some reliable measure of time. *See also* dendrochronology; radiometric dating.

day-neutral plant A plant that requires no particular photoperiod to flower. *See* photoperiodism.

deamination A type of chemical reaction in which an amino group (NH_2) is removed from a compound. Amino acids can be broken down by oxidative deamination. Transamination of many amino acids yields glutamate, which can be deaminated to α-ketoglutarate, which can then enter the Krebs cycle to be further broken down with the release of energy. *See* transamination.

De Bary, Heinrich Anton (1831–88) German botanist. De Bary abandoned a career in medicine in Frankfurt to devote the rest of his life to his botanical studies, becoming professor of botany at a number of German universities. His work on fungi represents a major landmark in the understanding of and fight against plant diseases. His observations of fungi in their natural habitat and his methods of culturing them enabled him to work out their development and life cycles, thus laying down the foundations of modern mycology. In his *Researches on Fungal Blights* (1853) outlining certain rust and smut diseases, he explained that fungi are the cause and not the effect of diseases. He also showed that a lichen was an association between an alga and a fungus and coined the term ‘symbiosis’ for dissimilar plants living together to their mutual benefit.

decarboxylase An enzyme that catalyzes the decarboxylation of carboxylic acids, including the conversion of amino acids to amines.

deciduous Denoting plants that seasonally shed all their leaves, for example before the winter, or dry season. It is an adaptation to prevent excessive water loss by transpiration when water is scarce. *Compare* evergreen.

decomposer An organism that feeds upon dead organisms, breaking them down into simpler substances. Decomposers recycle nutrients, making them available to producer organisms. Bacteria and fungi are important decomposers in most ecosystems. *See also* food chain; trophic level.

decussate Describing leaves in opposite pairs and with alternating pairs at right angles to each other up the stem.

dedifferentiation The reversion of a differentiated cell to the meristematic (undifferentiated) state, which is capable of differentiating again. It occurs naturally just before the production of secondary meristems or in response to wounding or stimulation by growth factors. *See* callus; totipotency.

deficiency disease A disease caused by deficiency of a particular essential nutrient (such as a MICRONUTRIENT or trace element), usually with a characteristic set of symptoms. For example, iron deficiency results in CHLOROSIS between the veins, particularly in young leaves that are still synthesizing chlorophyll. Nitrogen deficiency causes stunting of growth and chlorosis of older leaves, as chlorophyll pigments are broken down so their components can be reused in the young growing parts of the plant.

definite growth *See* determinate growth.

definite inflorescence *See* cymose inflorescence.

definitive nucleus *See* polar nuclei.

deforestation (**disafforestation**) The permanent removal of forests, especially by means of logging for commercial timber and clearing for agriculture and human settlement. A major cause of loss of natural habitats and biodiversity worldwide, it can lead to soil erosion, flooding downstream, and increasing drought in neighboring areas.

degenerate code A part of the genetic code in which some amino acids are coded for by more than one triplet of nucleotide bases (codons). For some amino acids, only the first two bases appear to be important, and it matters little what the third base is. For example, leucine is coded for by CUU, CUC, CUA, and CUG, and by UUA and UUG.

dehiscent Describing a fruit or fruiting body that opens at maturity to release the seeds or spores. Dehiscence may be violent to aid seed dispersal. *Compare* indehiscent.

dehydration A process followed when tissues are prepared for permanent microscope slides. Water is removed by immersing the tissue in increasing strengths of ethyl alcohol. The alcohol concentration must be increased gradually as otherwise the cells would dehydrate too quickly and shrink. Dehydration is necessary because water does not mix with the chemicals used in clearing and mounting sections.

dehydrogenase An enzyme that catalyzes the removal of certain hydrogen atoms from specific substances in biological systems. Dehydrogenases are usually called after the name of their substrate, e.g. lactate dehydrogenase. Some dehydrogenases are highly specific, with respect to both their substrate and coenzyme, whilst others catalyze the oxidation of a wide range of substrates. Many require the presence of a coenzyme, which is often involved as a hydrogen acceptor. Dehydrogenases catalyze the transfer of two hydrogen atoms from substrates to NAD and NADP. *See also* alcohol dehydrogenase.

deletion *See* chromosome mutation.

deme A subpopulation of a species: a discrete interbreeding group of organisms, with recognizable cytological or genetic characters, that is spatially distinct from other such groups, although it may be adjacent to them. *Compare* subspecies.

denaturation A process that causes unfolding of the peptide chain of proteins or of the double helix of DNA. These changes may be brought about by a variety of physical factors: change in pH, temperature, violent shaking, and radiation. The primary structure remains intact. Denatured proteins and nucleic acids show changes in physical and biological properties; denatured proteins, for example, are often insoluble in solvents in which they were originally soluble.

dendrochronology A method of archaeological dating by the ANNUAL RINGS of trees, used when the lifespans of living and fossil trees in an area overlap. Exact dates for sites can be calculated and the method is more accurate than radiometric dating techniques. Bristlecone pines, which can live for up to 5000 years, have been used in such work.

denitrification The chemical reduction of nitrate by soil bacteria. Denitrifying bacteria such as *Pseudomonas*, *Micrococcus*, and *Clostridium* use nitrate as the terminal electron acceptor in anaerobic respiration. The process is important in terms of soil fertility since the products of denitrification (e.g. nitrites and ammonia) cannot be used by plants as a nitrogen source. *Compare* nitrification. *See* nitrogen cycle.

dentate Toothed: describing leaf margins that have a series of outward-pointing notches.

denticulate Describing a leaf margin that is finely toothed.

deoxyribonuclease *See* DNase.

deoxyribonucleic acid *See* DNA.

deoxyribose A pentose (five-carbon) sugar that is a component of the nucleotides that make up DNA.

deoxy sugar A SUGAR in which oxygen has been lost by replacement of a hydroxyl group (OH) with hydrogen (H). The most important example is DEOXYRIBOSE.

dermatogen (protoderm) *See* histogen theory.

desert An area in which the rate of evaporation exceeds the rate of precipitation for most of the time. A desert may form in any climate where average annual precipitation is less than 250 mm and intermittent. A 'true desert' has no plant life at all. The term is also used for semideserts in which plant and animal life is sparsely distributed, and adapted to long periods of drought. EPHEMERAL plants are common and there may also be succulent or xerophytic perennials. Cold deserts include parts of the TUNDRA and may also form in hot regions with annual rainfall of less than 400 mm.

desiccation The drying out of an organism due to evaporation.

desmids A group of green algae of the phylum GAMOPHYTA that are basically single-celled but in some species form filamentous or irregular colonies.

determinate growth (definite growth) A form of growth that has a finite limit, i.e. once a maximum size is reached, no further growth occurs. It occurs, for example, in annual and biennial plants and plant organs such as internodes. The term is sometimes applied to cymose inflorescences, in which growth ceases with the production of a terminal flower bud. *Compare* indefinite growth.

detritus Fragments of dead material, such as leaf litter and products of the breakdown of organic material by decomposers.

Deuteromycota *See* Fungi Anamorphici.

Devonian The geological period known as the 'Age of Fish', some 405–355 million years ago, between the Silurian and the Carboniferous periods of the Paleozoic era. It was also the time when the first major invasion of the land by plants occurred. The very first land plants date from the late Silurian, but during the Devonian many vascular land plants appeared, such as the rhyniophytes (all now extinct), clubmosses (Lycophyta), horsetails (Sphenophyta), and eusporangiate ferns (early members of the Filicinophyta. The gymnosperms probably arose in the Late Devonian. *See also* geological time scale.

De Vries, Hugo (1848–1935) Dutch plant physiologist and geneticist. An expert on Netherlands flora, De Vries showed that turgor pressure of the fluid in plant cells was partially responsible for growth and coined the term 'plasmolysis' for the effect on cells of water loss during wilting. His work in this area formed a basis for theories of osmosis.

In the 1880s he developed an interest in heredity, proposing that there were structures in the nucleus (which he called 'pangenes') that determine characteristics in the offspring. He carried out plant breeding experiments, and the results of his crosses showed segregation of characters in the progeny in a ratio of 3:1. He came across MENDEL's work in 1900 and immediately published his results.

In *The Mutation Theory* (1901–03), De Vries postulated that new species could arise from a single dramatic mutation. He had observed different types of evening primrose within a single colony and believed them to be the result of mutations. It has subsequently been discovered that the plants De Vries believed to be mutants were triploids or tetraploids, but his essential premise continues to be important, showing that variation is essential for evolution within a species.

dextrin Any of a class of intermediates produced by the hydrolysis of starch. Further hydrolysis eventually produces the monosaccharide glucose.

dextrose (**grape sugar**) An outdated name for GLUCOSE.

diakinesis The last stage of the PROPHASE in the first division of MEIOSIS. Chiasmata are seen during this stage, and by the end of diakinesis the nucleoli and nuclear membrane have disappeared. During diakinesis the sister chromatids of homologous pairs of chromosomes complete their separation, and the chromosomes coil tightly, shortening and thickening.

dialysis The separation of large molecules from smaller ones by DIFFUSION through a selectively permeable membrane that allows through only molecules up to a certain size. For example, proteins can be separated from amino acids or glucose in this way. *Compare* osmosis.

Diatoms A phylum of the Protoctista whose members are unicellular algae found in freshwater, the sea, and soil. Much of plankton is composed of diatoms and they are thus important in FOOD CHAINS. Diatoms are diploid unicells with silica cell walls (*frustules*) composed of two overlapping valves ornamented with perforations, which are arranged differently in each species. The frustules of dead diatoms are highly resistant to decay, and form deep sediments on the floors of oceans and lakes called *diatomaceous earth* or *kieselguhr*, which is used in insulating materials, fireproof cements, filters, and in the manufacture of explosives.

diatropism (**diageotropism**) *See* tropism; gravitropism.

dicaryon *See* dikaryon.

dichasial cyme (**dichasium**) *See* inflorescence.

dichogamy The condition in which the anthers and stigmas mature at different times thus helping prevent self-pollination. *Compare* homogamy. *See* protandry; protogyny.

dichotomy Forked branching produced by division of the growing point into two equal parts, seen for example in *Fucus*.

Dicotyledonae A class of flowering plants (phylum Angiophyta) characterized by having two cotyledons in the seed. They include herbs, shrubs, and trees and secondary growth is normal. Examples are the sunflower and the oak. The flower parts are arranged in fours or fives, or multiples thereof, and the leaf veins are branched. The vascular bundles of the stem are arranged in a ring within a single endodermis and pericycle, giving a eustele. *Compare* Monocotyledonae.

dictyosome A stack of membrane-bounded sacs (*cisternae*) that, together with associated vesicles (Golgi vesicles), forms the GOLGI APPARATUS. The term is usually only applied to plant cells, where many such stacks are found. In contrast, the Golgi apparatus of most animal cells is a continuous network of membranes. The term dictyosome is little used today.

dictyostele A modified solenostele that is broken up by large leaf gaps so crowded together that they overlap. The tube of a stelar tissue is thus broken up into a mesh, each small segment of remaining vascular tissue being called a *meristele*. It is found in certain fern stems (e.g. *Dryopteris*).

differentially permeable membrane *See* osmosis.

differentiation A process of change during which cells with generalized form become morphologically and functionally specialized to produce the different cell types that make up the various tissues and organs of the organism. *See also* totipotency.

diffuse-porous Describing wood in which vessels of approximately equal diameter tend to be evenly distributed so there is no obvious growth ring. Diffuse-porous wood is seen for example in yellow birch (*Betula lutea*). *Compare* ring-porous.

diffusion The movement of molecules along a concentration gradient from areas of high concentration to areas of lower concentration as a result of random movement. Gaseous exchange in plants take place by diffusion, and some of the movement of ions through plant fluids is also due to diffusion. *See* dialysis; transpiration. *Compare* active transport; osmosis.

diffusion pressure deficit (**DPD**) The net force that causes water to enter a plant cell. It is the difference between OSMOTIC PRESSURE (Π) and TURGOR PRESSURE:

$$DPD = \Pi - TP.$$

See water potential.

dihybrid (**dihybrid cross**) An organism that is heterozygous at two loci, formed by crossing homozygous parents with different alleles at two given loci: for example, Mendel's cross between yellow round (YYRR) and green wrinkled (yyrr) garden peas to give a yellow round dihybrid (YyRr). When a dihybrid is selfed, a characteristic *dihybrid ratio* of 9:3:3:1 is obtained in the offspring. Nine plants exhibit both dominant characters, six plants show one dominant and one recessive character, and one plant exhibits both recessives. *Compare* monohybrid.

dikaryon (**dicaryon**) A fungal hypha or mycelium whose cells each contain two different nuclei, arising from the fusion of two compatible cells, each with one nucleus. The nuclei do not fuse immediately, instead dividing independently but simultaneously. Dikaryosis is common in ascomycetes and basidiomycetes, and results from the fusion of two monokaryotic mycelia of different mating types.

dimorphism The existence of two distinct forms. Examples include male and female plants in dioecious species, aerial and submerged leaves, alternation of gametophyte and sporophyte stages of a life cycle, and mesophyll and bundle-sheath chloroplasts.

dinitrogenase The enzyme that catalyzes NITROGEN FIXATION.

dinoflagellate *See* Dinomastigota.

Dinomastigota (**dinoflagellates**) A phylum of the Protoctista containing mainly single-celled marine or freshwater algae that swim in a twirling manner by means of two UNDULIPODIA that lie at right angles to each other in two grooves within the organism's rigid body wall (*test*). The body wall is often stiffened by cellulose plates under the cell membrane. Many dinoflagellates possess stinging organelles (*trichocysts*) that they discharge to catch prey; and some produce potent toxins that are capable of killing fish. Planktonic dinoflagellates can multiply rapidly to form *red tides*. The phosphorescent seas of the tropics are usually due to bioluminescent dinoflagellates. Roughly half of all known dinoflagellates are capable of photosynthesis. Many form symbioses with marine invertebrates such as corals and clams.

dinucleotide A compound of two nucleotides linked by their phosphate groups. Important examples are the coenzymes NAD and FAD.

dioecious Denoting a plant species in which male and female reproductive organs are borne on separate individuals. *Compare* hermaphrodite; monoecious.

diplobiontic Describing life cycles showing a typical alternation of generations with haploid and diploid somatic bodies.

Ferns and mosses are diplobiontic organisms. *Compare* haplobiontic.

diploid A cell or organism containing twice the haploid number of chromosomes (i.e. 2n). In plants exhibiting an alternation of generations the sporophyte is diploid, and the gametophyte haploid. In lower plants such as mosses and liverworts the persistent vegetative plant is haploid, the sporophyte being a relatively short-lived phase. Higher plants are normally always diploid. Exceptions are those species in which polyploidy occurs.

diplont A diploid organism that represents the vegetative stage in life cycles in which haploidy is restricted to the gametes. Such *diplontic* life cycles are found in some protoctists, such as the diatoms and certain Phaeophyta (brown algae), including the wracks. *Compare* haplont; alternation of generations.

diplotene In MEIOSIS, the stage in late PROPHASE I when the pairs of chromatids begin to separate from the tetrad formed by the association of homologous chromosomes. Chiasmata can often be seen at this stage.

disaccharide A SUGAR with molecules composed of two monosaccharide units. Sucrose and maltose are examples. These are linked by a –O– linkage (*glycosidic link*). *See also* glycosidic bond.

discontinuous variation (**qualitative variation; qualitative inheritance**) A form of variation in which a character has two or more distinct forms in a population. Examples are Mendel's pea characters. It generally occurs when there are two or more allelic forms of a major gene in a population. *See* polymorphism.

disjunct Denoting a species or genus whose distribution is not continuous within its range. Examples are species that are found in the Alps, arctic regions, and north European mountains and magnolias that occur in eastern North America, Central America, and southeast Asia, but not in the intervening regions. Such disjunctions may arise where climate change reduces the range of a species, restricting them to certain isolated areas (*refugia*) where the local climate is more suitable for them, i.e. since the last Ice Age for the cold-loving arctic–alpine species now confined to isolated mountain tops. The other major cause of disjunction is CONTINENTAL DRIFT.

disk floret *See* inflorescence.

distal Denoting the part of an organ, limb, etc., that is furthest from the origin or point of attachment. *Compare* proximal.

distely Having two steles, e.g. the stem of *Selaginella kraussiana*. The steles are joined only at branches. *See also* polystely.

disulfide bridge In PROTEINS, a covalent bond formed between the sulfhydryl (–SH) groups of different cysteine molecules. Such bridges are important in stabilizing the tertiary structure of peptides and proteins.

diurnal 1. Describing an event or activity occurring during the day or a species or process that is active only by day and not by night.
2. Occurring at daily intervals. *See* circadian rhythm.

divergent evolution The evolution of different forms from a single basic structure in response to different selection pressures. For example, the basic angiosperm stem is vertical and elongated, but some cacti have swollen water-storing stems, while other xerophytic plants that have lost their leaves during the course of evolution have evolved flattened leaflike stems for photosynthesis, e.g. butcher's broom (*Ruscus aculeatus*).

division The equivalent of a phylum in the Five Kingdoms classification. The term division was retained for the plant kingdom until recently. Division names of plants (and hence phylum names too) end in *-phyta* (e.g. Bryophyta, Filicinophyta), while those of fungi end in *–mycota* (e.g.

Ascomycota, Basidiomycota). Divisions may be divided into subdivisions.

DNA (deoxyribonucleic acid) A nucleic acid, mainly found in the CHROMOSOMES, that contains the hereditary information of organisms. The molecule is made up of two helical polynucleotide chains coiled around

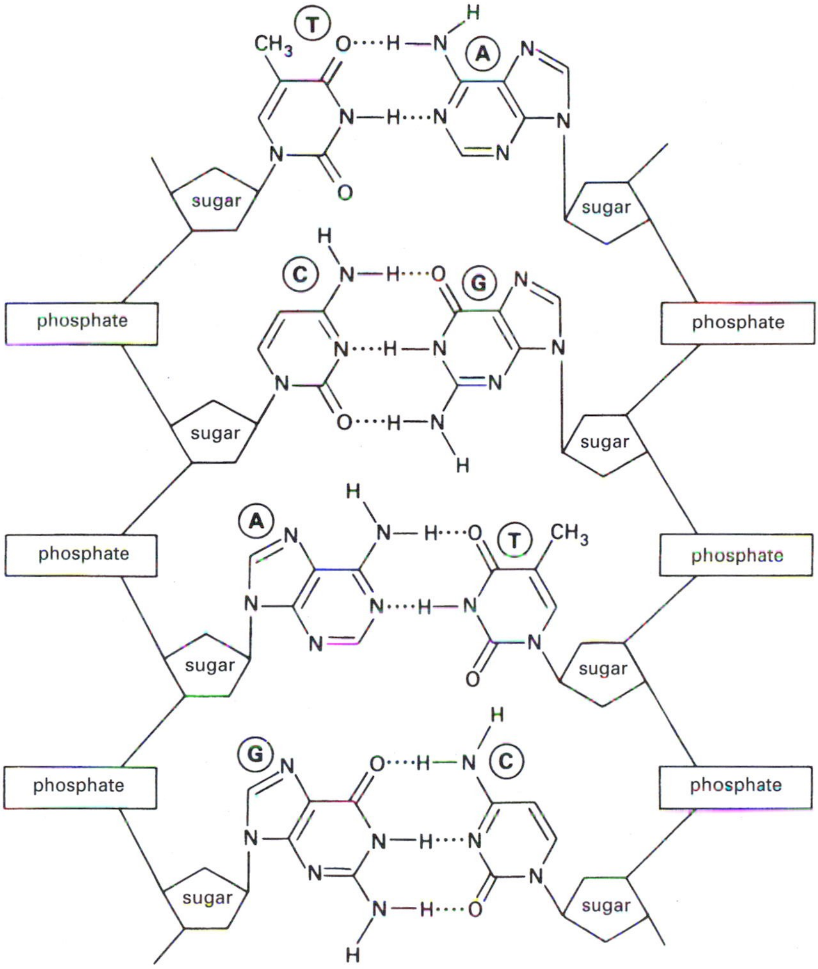

KEY

(A) adenine

(C) cytosine

(T) thymine

(G) guanine

Part of the structure of DNA showing hydrogen bonding (dotted lines) between complementary bases

DNA

each other to give a *double helix*. Phosphate molecules alternate with deoxyribose sugar molecules along both chains, linked by phosphodiester bonds. Each sugar molecule is also joined to one of four nitrogenous bases – adenine, guanine, cytosine, or thymine. The two chains are joined to each other by hydrogen bonding between bases. The sequence of bases along the chain makes up a code – the genetic code – that determines the precise sequence of amino acids in proteins (*see* messenger RNA; protein synthesis; transcription).

In DNA, the two purine bases (adenine and guanine) always bond with the pyrimidine bases (thymine and cytosine), and the pairing is quite specific: adenine with thymine and guanine with cytosine. DNA is the hereditary material of all organisms with the exception of RNA viruses. Together with RNA and histones it makes up the chromosomes of eukaryotic cells. Some DNA viruses contain single-stranded DNA rather than the double helix. *See also* junk DNA; replication; selfish DNA; RNA.

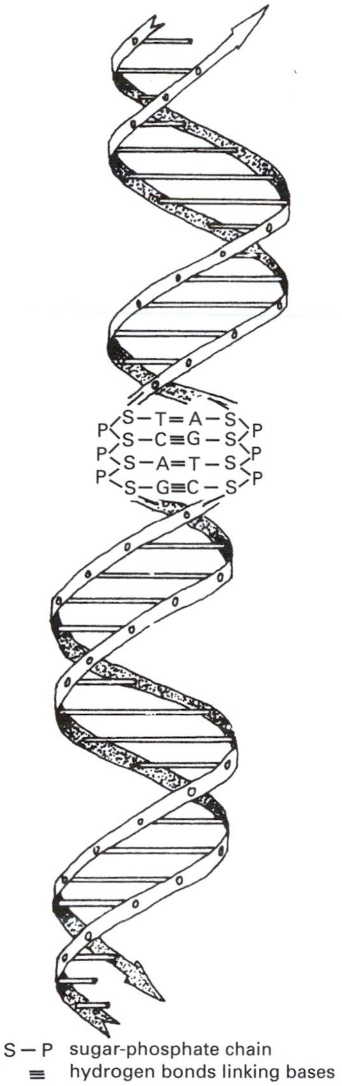

S — P sugar-phosphate chain
≡ hydrogen bonds linking bases

DNA: the double helix

DNA hybridization *See* nucleic acid hybridization.

DNA ligase An enzyme, used in DNA REPLICATION and the *in vitro* synthesis of DNA, that joins together segments of newly synthesized polynucleotide chains using energy from ATP. *See* replication.

DNA polymerase *See* polymerase.

DNA probe (**gene probe**) A nucleic acid consisting of a single strand of nucleotides whose base sequence is complementary to that of a particular DNA fragment being sought, for example a gene on a chromosome or a restriction fragment in a DNA digest. The probe is labeled (e.g. with a radioisotope or a fluorescent compound) so that when it binds to the target sequence, both it and the target can be identified (by autoradiography or FLUORESCENCE MICROSCOPY). DNA probes are used to analyze fragments of DNA or RNA generated by restriction enzymes and determine their sequence in the original nucleic acid, a process called restriction mapping (*see* restriction map). They are also used to detect gene and chromosome mutations. *See also* FISH.

DNase (**deoxyribonuclease**) Any enzyme that hydrolyzes the phosphodiester bonds of DNA. DNases are classified into two groups, according to their site of action in the DNA molecule (*see* endonuclease; exonuclease).

domain 1. A discrete part of the tertiary structure of a PROTEIN that has a particular function, e.g. as the binding site for the substrate of an enzyme.
2. A taxonomic grouping of organisms that ranks higher than kingdom. There are generally considered to be three domains: ARCHAEA, BACTERIA, and EUKARYA. The FIVE KINGDOMS CLASSIFICATION recognizes only two 'superkingdoms': Prokarya (which includes the single kingdom Bacteria comprising both archaea and the bacteria) and Eukarya.

domatium (*pl.* **domatia**) *See* ant plant.

dominant 1. An allele that, in a heterozygote, prevents the expression of another (recessive) allele at the same locus. Organisms with one dominant and one recessive allele thus appear identical to those with two dominant alleles, the difference in their genotypes becoming apparent only on examination of their progenies. The dominant allele usually controls the normal form of the gene, while mutations are generally RECESSIVE.
2. A plant species that is the most important member of a particular community and affects the structure of that community.

dormancy A period of minimal metabolic activity of an organism or reproductive body. It is a means of surviving a period of adverse environmental conditions, e.g. cold or drought. Seeds, spores, cysts, and perennating organs of plants are potentially dormant structures. *See* estivation.

dormin (**abscisin II**) A former name for ABSCISIC ACID.

dorsal 1. Designating the lower or abaxial surface of a lateral organ such as a leaf.
2. Designating the upper surface away from the substrate in thallose plants such as liverworts. *Compare* ventral.

dorsifixed A stamen in which the filament is fused to the back of the anther. *Compare* basifixed; versatile.

dorsiventral Describing a structure that has distinct upper and lower surfaces. A dorsiventrally flattened structure has flat dorsal and ventral surfaces and very narrow lateral surfaces.

double fertilization The fusion of one pollen nucleus with the egg to form the zygote and of the other pollen nucleus with a polar nucleus to form the triploid endosperm nucleus. The process is restricted to certain angiosperms but a primitive form is known from some gymnosperms.

double helix *See* DNA.

double recessive An organism containing both recessive alleles of a particular gene and thus expressing the recessive form of the gene in its phenotype. Double recessives, being of known genotype, are often used in test crosses to establish whether the organism to which it is crossed is heterozygous or homozygous for the same gene. *See* backcross.

DPD *See* diffusion pressure deficit.

drupe (**pyrenocarp**) A fleshy indehiscent fruit containing one or more seeds each surrounded by a hard stony wall, the endocarp. Drupes with one seed include plums and cherries while many-seeded drupes include holly and elder fruits. Blackberries and raspberries are collections of small drupes or drupelets.

duplex Double, or having two distinct parts. The term is particularly used to describe the double helix of the Watson–Crick DNA model.

duramen *See* heartwood.

dwarfism Stunted growth. This may be due to genetic mutation, causing GIBBERELLIN deficiency.

E

ecocline A CLINE that is due to a specific environmental factor, such as a gradient of heavy metal concentration in the soil.

ecological niche The functional role of an organism in a community. It includes the habitat in which the organism lives, the resources it uses, and the periods of time in which it is present and/or active there and its interactions with other organisms. It can normally be defined only by the presence of the organism and is used more extensively for animals than for plants. If two species occupy the same niche, then competition may occur until either one has replaced the other or the two divide the niche between them to some extent. A similar niche may be occupied by different species in different areas. For example different species of grasses are dominant in grasslands in different parts of the world.

ecology The study of the relationships of organisms to one another and to their living (*biotic*) and nonliving (*abiotic*) environment.

ecosystem (**ecological system**) A unit made up of all the living and nonliving components of a particular area that interact and exchange materials with each other. The concept of the ecosystem differs from that of the community in that more emphasis is placed on abiotic factors. Various studies have been made to attempt to itemize the energy flow of an entire ecosystem, taking into account factors such as incoming radiation, photosynthetic efficiency, etc. The term can be applied on various scales, from small ponds to the whole planet.

ecotype A population of individuals within a species adapted genetically to the combination of environmental factors in their local habitat, but still able to reproduce with other ecotypes belonging to the same species and produce fertile offspring. Differences between ecotypes may be physiological or morphological. *Compare* biotype. *See also* adaptive radiation; speciation.

ectomycorrhiza (**ectotrophic mycorrhiza**) (*pl.* **ectomycorrhizae**) *See* mycorrhiza.

ectoparasite *See* parasitism.

ectoplasm The outer gel-like layer of the cytoplasm in the cells of plants and some protoctists, which lies immediately beneath the cell membrane and contains a dense array of microtubules.

ectotrophic mycorrhiza *See* mycorrhiza.

edaphic factors The physical, chemical, and biological characteristics of the soil that together form an important component of the habitat because of their influence on plant distribution. The main edaphic factors are water content, pH, organic matter, and soil texture.

EDTA Ethylenediamine tetra-acetic acid, a compound that acts as a chelating agent, reversibly binding iron, magnesium, and other positive ions. It is used as a competitive inhibitor of certain enzymes that have metal ion cofactors, and as a source of iron, which it releases slowly into the solution.

effector A molecule that can combine with a repressor in an OPERON, making the

REPRESSOR inactive and thus enabling MESSENGER RNA and protein to be produced.

egg apparatus The three haploid nuclei that are situated at the micropylar end of the embryo sac in most flowering plants. The central nucleus is the female gamete and those to either side of it are called the synergids. *See* embryo sac. *See also* antipodal cells.

egg cell *See* ovum.

elaioplast (lipidoplast; oleoplast) A colorless plastid (leucoplast) storing lipids (fats or oils).

elaiosome *See* oil.

elater One of many elongated spirally thickened cells that are formed within the capsule of certain liverworts (e.g. *Marchantia* and *Pellia*). When the elaters are exposed to the air they dry out unevenly because of the differential thickening. The resulting twisting movements of the cells help in dispersing the spores.

electron carrier A molecule that can function as both an acceptor and a donor of electrons and/or protons in an electron-transport system. *See* coenzyme Q; cytochromes; electron-transport chain; ferredoxins.

electron micrograph *See* micrograph.

electron microscope *See* microscope.

electron-transport chain (respiratory chain) A chain of membrane-linked oxidation–reduction reactions involving proteins and enzymes, during which electrons are transferred from an initial electron donor through a series of intermediates to a final electron acceptor, resulting in the formation of ATP. The main electron-transport systems in plants are involved in respiration and the light reaction of PHOTOSYNTHESIS. Heterotrophic bacteria have a similar respiratory electron-transport chain.

In eukaryotic cells, the enzymes and other components of the respiratory electron-transport chain are located in the inner membrane of the MITOCHONDRIA. During aerobic respiration the reduced coenzyme NADH, produced by the KREBS CYCLE in the mitochondrial matrix gives up two electrons to the first component in the chain, NADH dehydrogenase, and two hydrogen ions (H^+) are discharged from the matrix of the mitochondrion into the intermembrane space. The electrons are transferred along the chain to a carrier molecule (COENZYME Q), and then in sequence to a series of cytochromes, finally acting with the enzyme cytochrome oxidase to reduce an oxygen atom, which combines with two H^+ ions to form water. During this electron transfer, a further two pairs of H^+ ions are pumped into the intermembrane space, making a total of six per molecule of NADH. If $FADH_2$ is the electron donor, only four H^+ ions are pumped across.

The function of electron transport in the mitochondrion is to phosphorylate ADP to ATP.

electrophoresis The migration of electrically charged particles toward oppositely charged electrodes in solution under an electric field – the positive particles to the cathode and negative particles to the anode. The rate of migration varies with molecular size and shape. The technique can be used to separate or analyze mixtures (e.g. of proteins or nucleic acids).

elicitors Molecules that initiate cell signaling pathways leading to the activation of plant defense genes and the production of PHYTOALEXINS and other defensive chemicals in response to infection by a pathogen.

elongation The enlargement of cells after mitotic division that results from the uptake of water by OSMOSIS until the cells become turgid. Elongation also involves the enzyme-mediated stretching of the cell walls. Hormones, especially AUXINS, are involved in the promotion of elongation. TROPISMS are due to differential elongation of cells on different sides of root or shoot, resulting in curvature.

elongation factor *See* translation.

elution The removal of an adsorbed substance in a CHROMATOGRAPHY column or ion-exchange column using a solvent (*eluent*), giving a solution called the *eluate*. The chromatography column can selectively adsorb one or more components from the mixture to ensure efficient recovery of these components graded elution is used. The eluent is changed in a regular manner starting with a nonpolar solvent and gradually replacing it by a more polar one. This will wash the strongly polar components from the column.

emasculation The removal of the stamens of a plant before the anthers burst to prevent self-pollination or unwanted crosses to nearby plants.

Embden–Meyerhof–Parnas pathway *See* glycolysis.

embedding The sealing of tissue prepared for permanent microscope slides in a solid block of paraffin wax prior to sectioning. After CLEARING, tissues are placed in two or three baths of molten paraffin wax. When the tissue is completely infiltrated by the wax it is allowed to harden. As the wax is opaque the block must be marked to insure correct orientation when sectioning. Tissue prepared for electron microscopy may be embedded in tougher epoxy resins such as Araldite and cut with a diamond knife. *See also* fixation.

embryo The organism that develops from the zygote of a vascular plant after the proembryo has differentiated into embryo and SUSPENSOR but before germination, i.e. while the young plant is still nutritionally dependent on parental tissues. In the Filicinophyta, Sphenophyta, and Lycophyta the embryo sporophyte produces a footlike structure that remains embedded in the tissues of the parent gametophyte plant, absorbing nutrients like the suspensor of higher plants. *See* embryo sac.

embryogeny (**embryony**) **1.** The development of an EMBRYO from the moment the ZYGOTE divides.
2. The development of embryos from diploid somatic cells in *embryo culture*. Such embryos are often termed *embryoids*, to distinguish them from embryos formed from zygotes. They are capable of giving rise to normal plants. Embryoids that develop from pollen grains grow into haploid plants.

embryology The study of the development of embryos, usually from fertilization.

embryony *See* embryogeny.

embryo sac A large oval thin-walled sac in the nucellus of flowering plants in which egg fertilization and subsequent embryo development occurs. It corresponds to the female GAMETOPHYTE of lower plants and contains a number of nuclei derived by division of the MEGASPORE nucleus.

There are commonly eight nuclei in the embryo sac: the egg apparatus at the micropylar end, made up of an egg nucleus and two synergid nuclei; three antipodal cells at the opposite chalazal end that probably aid embryo nourishment; and two polar nuclei in the center that fuse to form the *primary endosperm nucleus*. Other plants have four nuclei or sixteen nuclei in their embryo sacs. At fertilization one male

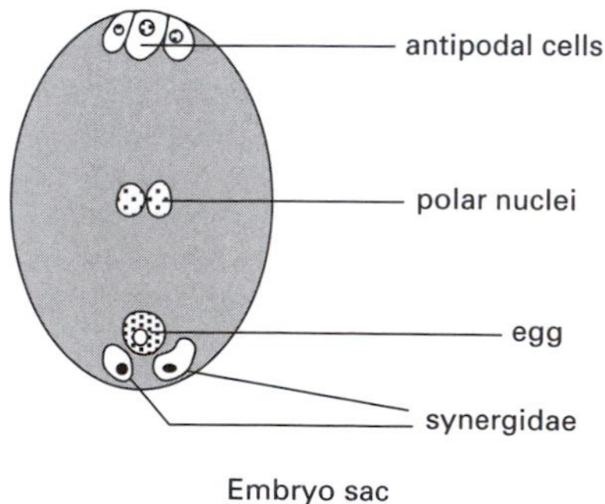

Embryo sac

nucleus fuses with the egg nucleus to form the zygote, while the second male nucleus fuses with the primary endosperm nucleus to form a triploid cell, or more rarely a second diploid or a pentaploid cell that later gives rise to the ENDOSPERM. In the gymnosperms the megaspore gives rise to a cell that is termed the embryo sac because of its similarity to the angiosperm structure. *See* nucellus; micropyle.

Emerson effect The observation by Robert Emerson in 1957 that PHOTOSYNTHESIS proceeds much faster when chloroplasts are illuminated with light at a wavelength of 650 nm as well as at 700 nm (the absorption wavelength of chlorophyll a molecules). It contributed to the discovery of PHOTOSYSTEMS I AND II.

endangered species *See* CITES; Red Data Book.

endarch Denoting a stele in which metaxylem develops to the outside of protoxylem. *Compare* centrarch; exarch; mesarch.

endemic Describing a population or species that is restricted geographically. Some endemics have evolved in geographical isolation on islands or mountain tops; others may be relics of once widespread species that have become restricted owing to climate change, geological change, or human activity.

endocarp *See* pericarp.

endocytosis The bulk transport of materials into cells across the plasma membrane by processes not involving diffusion or active transport. It is described as *pinocytosis* (cell drinking) or *phagocytosis* (cell eating) depending on whether the material is fluid, containing molecules in solution, or solid respectively. The process involves extension and invagination of the plasma membrane to form small vesicles in pinocytosis (*pinocytotic vesicles*) or vacuoles in phagocytosis (*food vacuoles*). The contents are often digested by enzymes from lysosomes. *Compare* exocytosis. *See* lysosome.

endodermis The innermost part of the cortex, consisting of a single layer of cells that controls the passage of water and solutes between the cortex and the stele. A clearly defined endodermis is seen in all roots and in the stems of the non seed-bearing vascular plants and some dicotyledons. *See* Casparian strip; passage cells.

endogenous Produced or originating within an organism. *Compare* exogenous.

endomitosis The duplication of chromosomes without division of the nucleus, causing polyploidy and leading to an increase in nuclear and cytoplasmic volume. Endomitosis occurs regularly in certain tissues, such as the phloem cells of some leguminous plants. *Compare* amitosis; mitosis.

endomycorrhiza (**endotrophic mycorrhiza**) (*pl.* **endomycorrhizae**) *See* mycorrhiza.

endonuclease An enzyme that catalyzes the hydrolysis of internal bonds of polynucleotides such as DNA and RNA, producing short segments of linked nucleotides (oligonucleotides). *See also* DNase; restriction endonuclease.

endophyte A bacterium, protoctist, or fungus that lives inside a plant or other photosynthetic organism, but is not a parasite. *Compare* epiphyte.

endoplasm The inner layer of cytoplasm in plant cells, which contains the main organelles. *Compare* ectoplasm.

endoplasmic reticulum (**ER**) A system of membranes forming tubular channels and flattened sacs (*cisternae*), running through the cytoplasm of all eukaryotic cells and continuous with the nuclear envelope. Although often extensive, it was only discovered with the advent of electron microscopy. Its surface is often covered with ribosomes, forming *rough ER*. The proteins they make can enter the cisternae for transport to other parts of the cell or for secretion via the GOLGI APPARATUS. ER lacking ribosomes is called *smooth ER* and is involved with lipid synthesis, including phospholipids and steroids.

endopolyploidy *See* endomitosis.

endoscopic Describing the type of development of a plant embryo in which the inner cell formed by the first division of the zygote develops into the embryo, while the outer cell develops into the suspensor. It is seen in many ferns and in all seed plants. *Compare* exoscopic.

endosperm The nutritive tissue that surrounds the embryo in angiosperms. In nonendospermic seeds most of the endosperm is absorbed by the developing embryo and the food stored in the cotyledons. In endospermic seeds the endosperm replaces the nucellus and is often a rich source of growth-regulating substances. Many endospermic seeds (e.g. cereals and oil seeds) are cultivated for their food reserves. The endosperm is usually triploid and develops from the *primary endosperm nucleus*, the triploid nucleus resulting from the fusion of one of the male gametes from the pollen tube with the polar nuclei in the embryo sac.

endospore A resting stage produced by certain bacteria under unfavorable conditions. Endospores are formed within the cell, one in each parent cell, and are surrounded by a thick coat containing dipicolinic acid. On germination the wall is lysed and one vegetative cell is produced. Endospores can remain viable for several centuries and are resistant to heat, desiccation, and x-rays.

endosporium (**intine**) *See* pollen.

endosymbiont theory The theory that eukaryotic organisms evolved from symbiotic associations between bacteria. It proposes that integration of photosynthetic bacteria, for example purple bacteria and cyanobacteria, into larger bacterial cells led to their permanent incorporation as forerunners of the plastids (e.g. chloroplasts) seen in modern eukaryotes. Similarly, other symbiotic aerobic bacteria gave rise to the mitochondria.

endotrophic mycorrhiza (*pl.* **endotrophic mycorrhizae**) *See* mycorrhiza.

entire Describing the margin of a leaf, sepal, or petal that is smooth and undivided.

entomophily Pollination by insects. Various structures and mechanisms have evolved to attract insects (e.g. showy petals and nectar) and to insure that they carry pollen away on their bodies.

environment The complete range of external conditions under which an organism lives, including physical, chemical, and biological factors, such as temperature, light, the availability of food and water, and the effects of other organisms.

enzyme A compound that catalyzes biochemical reactions. Enzymes are proteins, which act with a given compound (the substrate) to produce a complex, which then forms the products of the reaction. The enzyme itself is unchanged in the reaction; its presence allows the reaction to take place by lowering the activation energy. The names of most enzymes end in -ase, added to the substrate (e.g. lactase) or the reaction (e.g. hydrogenase).

Enzymes are extremely efficient catalysts for chemical reactions, and specific to particular reactions. They may have a nonprotein part (*cofactor*), which may be an inorganic ion or an organic constituent (*coenzyme*). The mechanism of action of most enzymes appears to be by *active sites* on the enzyme molecule. These have very specific shapes and charge distributions. The substrate acting with the enzyme changes shape to fit the active site, and the reaction proceeds. Enzymes are very sensitive to their environment – e.g. temperature, pH, and the presence of other substances. *See* activation energy. *See also* ribozyme.

Eocene The second oldest epoch of the Tertiary period, 55–38 million years ago. It was a period of widespread temperate and subtropical forests. The extensive grasslands of the late Tertiary had not yet developed, but large-hoofed grazing mammals, such as horses and elephants, were evolving rapidly. *See also* geological time scale.

eosin An acid stain that colors cellulose red and cytoplasm pink. *See* staining.

ephemeral A plant that has a very short life cycle and may complete more than one life cycle within a year. Such plants are often referred to as annuals. Examples are shepherd's purse (*Capsella bursa-pastoris*) and certain desert plants that grow, flower, and set seed in brief periods of rain. *Compare* annual; biennial; perennial.

epicarp *See* pericarp.

epicotyl The part of the plumule above the cotyledons. *Compare* hypocotyl.

epidermis The outer protective layer of cells in plants. In aerial parts of the plant the outer wall of the epidermis is usually covered by a waxy cuticle that prevents desiccation, protects the underlying cells from mechanical damage, and increases protection against fungi, bacteria, etc. The cells are typically platelike and closely packed together except where they are modified for a particular function, as are guard cells. The epidermis arises from the tunica meristem. In damaged stems and roots, and in those undergoing secondary growth, it is replaced by a secondary layer, the *periderm*. The specialized epidermal area of the roots from which the root hairs arise is termed the *piliferous layer*. The term *epiblem* is sometimes used instead to denote the outermost layer of cells in the root.

epigeal germination Seed germination in which the cotyledons form the first photosynthetic organs above the ground, e.g. sunflower (*Helianthus*). *Compare* hypogeal germination.

epigyny The type of flower structure in which the perianth and androecium are inserted above the gynoecium, giving an inferior ovary, fused with the receptacle. It is seen in the Asteraceae and Rosaceae. *Compare* hypogyny; perigyny.

epilimnion In a stratified lake (a lake in which there are distinct layers of different temperature densities), the upper layer of warm water. *See* stratification.

epinasty (epinastic movements) A nastic movement that involves the curving of a plant organ away from the axis (i.e. downward) as a result of greater growth on the upper surface, as in the opening of flowers. *See* nastic movements; leaf mosaic.

epiphylly **1.** The growth of a plant on the leaf of another plant. Such a plant is termed an *epiphyll*, or is described as *epiphyllous*. Examples include many mosses.
2. The production of adventitious buds on a leaf, e.g. *Bryophyllum*, where the buds form along the leaf margins, developing into small plantlets with adventitious roots that eventually become detached from the leaf.

epiphyte Any plant growing upon or attached to another plant or object merely for physical support. Examples of tropical epiphytes are ferns, orchids, and bromeliads. In temperate regions vascular epiphytes are rare: here most epiphytes are lichens, mosses, liverworts, and algae.

epithelium (*pl.* **epithelia**) A tissue consisting of a sheet (or sheets) of cells that covers a surface. Epithelia may line resin canals in gymnosperms, or, sometimes, gum ducts in dicotyledons, and usually have a secretory role.

equatorial plate An arrangement of the chromosomes in which the centromeres become aligned in a single plane at the center of the spindle during metaphase of mitosis and meiosis.

Equisetales (horsetails) The single order

of the phylum SPHENOPHYTA, which contains the horsetails.

ER *See* endoplasmic reticulum.

era A unit of the geological time scale that is made up of several periods, e.g. the Mesozoic Era comprises the Triassic, Jurassic, and Cretaceous periods. When used formally, the initial letter is capitalized.

ergot A fungal disease of many cereals. The hard black sclerotia of *Claviceps purpurea*, an ascomycete, replace the grains (*see* sclerotium). They contain alkaloids that can cause severe poisoning and even death if taken in with food by humans and other animals.

Escherichia coli A rod-shaped Gram-negative bacterium, some strains of which are normally present in animal intestines, soil, or water. It is widely used in genetic research and genetic engineering as a repository for fragments of DNA incorporated into plasmids in its cytoplasm.

essential amino acid *See* amino acids.

essential element An element that is indispensable for the normal growth, development, and maintenance of a living organism. Some, the *major elements*, are required in relatively large quantities and may be involved in several different metabolic reactions (*see* carbon; hydrogen; oxygen; nitrogen; sulfur; phosphorus; potassium; magnesium; calcium). Others, such as iron, manganese, molybdenum, boron, zinc, copper, cobalt, iodine, and selenium, are required in only small or minute amounts. *See* micronutrient.

essential oil Any of a number of volatile OILS secreted by aromatic plants that are the source of characteristic odors or tastes.

ester A compound formed by the condenstion reaction of a carboxylic acid with an alcohol:

$$RCOOH + HOR_1 \rightarrow RCOOR_1 + H_2O.$$

estivation The way in which sepals and petals are folded in the flower bud before expansion. *Compare* ptyxis; vernation.

ethanedioic acid *See* oxalic acid.

ethanoic acid *See* acetic acid.

ethene *See* ethylene.

ethylene (**ethene**) A gaseous hydrocarbon (C_2H_4), produced in varying amounts by many plants, that functions as a plant HORMONE. Its production may be stimulated by other plant hormones. The amino acid methionine is a precursor. Ethylene is involved in the control of germination, cell growth, fruit ripening, abscission, and senescence, and it inhibits longitudinal growth and promotes radial expansion. In ripening fruit a rapid rise in ethylene production precedes respiration to reach the climacteric ripeness: production of ethylene stimulates further production. This is why ripe fruit stimulate other fruits to ripen quickly and this process can be controlled to some extent for fruit storage and transport.

etiolation The type of growth exhibited by plants grown in darkness, usually from seed. They lack chlorophyll and therefore appear white or yellow. They show less differentiation and contain reduced amounts of supporting material, such as lignin, concentrating resources on elongation of internodes. The plant can then reach the light faster and photosynthesize and synthesize more chlorophyll.

etioplast A modified chloroplast formed from proplastids in leaves grown in total darkness.

Eubacteria A major subkingdom of the Bacteria, containing a large and diverse group of bacteria, principally distinguishable from the other major subkingdom, the Archaea, by differences in the base sequences of RNA subunits of the ribosomes, which are thought to reflect the very early evolutionary divergence between the two groups. Most are unicells that divide by binary fission. The cells can be spherical, rod-shaped, or helical, and some form assemblages of cells, such as branching filaments. Most are immotile, but some possess flagella. They are a ubiquitous group, some being found in extreme conditions. Cell-wall structure, morphology, and metabolic features are used to distinguish the different phyla. The Eubacteria contain both Gram-negative and Gram-positive bacteria, and bacteria with no cell walls at all.

eucaryote *See* eukaryote.

euchromatin *See* chromatin.

Euglenophyta Formerly, a phylum of aquatic single-celled protoctists that swim using one or more UNDULIPODIA (flagella), now classified as a class, the Euglenida, of the phylum Discomitochondria. It contains both photosynthesizing and nonphotosynthesizing members. For example, members of the genus *Euglena* possess a flexible pellicle surrounding the cell, and a single undulipodium. Many species contain chloroplasts, with pigments similar to those found in plants, although these organisms may also consume dissolved or particulate food from their surroundings.

Eukarya In taxonomy, a DOMAIN of living organisms characterized by having eukaryotic cells (*See* eukaryote). In the Five Kingdoms classification, the Eukarya constitutes a 'superkingdom' containing the kingdoms Animalia, Fungi, Plantae, and Protoctista.

eukaryote (**Eukaryotae**) All the living kingdoms except the Bacteria (Archaea and Eubacteria) are the eukaryotes. They are defined by the presence of a much more elaborate cell than the prokaryotes. Eukaryotes are characterized by: the genetic material packaged in chromosomes within a membrane-bound nucleus; possession of mitochondria, and, in photosynthetic eukaryotes, chloroplasts; a quite different and much elaborated membrane structure, including internal membranes such as the endoplasmic reticulum and Golgi apparatus; different-sized ribosomes; and complex UNDULIPODIA composed of arrays of MICROTUBULES. Some of these features probably arose through endosymbiosis of prokaryotes. *See* endosymbiont theory.

Euphorbiaceae A large family of dicotyledonous plants that include the spurges. There are over 8000 species. They are mostly trees and shrubs, with but a few herbaceous species, and are found worldwide, but especially in the tropics. Commercially important species include *Hevea brasiliensis*, the main source of natural rubber; cassava (*Manihot esculenta*); and castor oil (*Ricinus communis*).

euphotic zone *See* photic zone.

euploidy The normal state in which an organism's chromosome number is an exact multiple of the haploid number characteristic of the species. For example, if the haploid number is 7, the euploid number would be 7, 14, 21, 28, etc., and there would be equal numbers of each different chromosome. *Compare* aneuploidy.

eusporangiate Describing the condition, found in clubmosses, horsetails, and certain ferns in which the sporangia develop from a group of initial cells and do not have an elaborate spore dispersal mechanism. *Compare* leptosporangiate. *See also* Filicinophyta.

eustele The STELE arrangement found in most gymnosperms and dicotyledons in which the vascular tissue is arranged into a ring of discrete bundles separated by medullary rays, all contained within a single ring of endodermis and pericycle. It is a form of SIPHONOSTELE.

eutrophic Describing lakes or ponds that are rich in nutrients and consequently are able to support a dense population of plankton and littoral vegetation. *Eutrophication* is the process that results when an excess of nutrients enters a lake, for example as sewage or from water draining off land treated with fertilizers. The nutrients stimulate the growth of the algal population giving a great concentration or 'bloom' of such plants. When these die they are decomposed by bacteria, which use up the oxygen dissolved in the water, so that aquatic animals such as fish are deprived of oxygen and die from suffocation. *Compare* oligotrophic.

evergreen Describing plants that retain their leaves throughout the year or through several years. Many tropical species of broadleaved flowering plants are evergreen and their leaves are thicker and more leathery than those of deciduous trees. Water conservation is not a problem in tropical evergreen forests, so there is no need to shed leaves in a particular season. In polar and cold temperate regions many of the evergreens are shrubs or trees with needle-like or scalelike leaves, adaptations that prevent water loss by transpiration. *Compare* deciduous.

evolution The process of genetic change that occurs in populations of organisms over a period of time. It manifests itself as new characteristics in a species, and eventually the formation of new species. *See* Darwinism; natural selection.

exarch Denoting a stele in which the metaxylem develops to the inside of the protoxylem. *Compare* centrarch; endarch; mesarch.

excretion The process by which excess, waste, or harmful materials, resulting from the chemical reactions that occur within the cells of living organisms, are eliminated from the body. In plants most waste products accumulate in the vacuoles, whose selectively permaeable membranes prevent them interfering in cell reactions. Gaseous excretion takes place by diffusion through the stomata and lenticels. In fungi and small protoctists waste products are excreted by diffusion through the cell or body surface into the external medium.

exine (**exosporium**) *See* pollen.

exocarp (**epicarp**) *See* pericarp.

exocytosis The bulk transport of materials out of the cell across the plasma membrane. It involves fusion of vesicles or vacuoles with the plasma membrane in a reversal of endocytosis. *See also* lysosome. *Compare* endocytosis.

exodermis An outermost layer of thickened or suberized cortical cells that sometimes replaces the epidermal layer in the older parts of roots if the epidermal cells have died.

exogamy The fusion of gametes produced by organisms that are not closely related. *See* outbreeding.

exogenous Produced or originating outside an organism. *Compare* endogenous.

exon A segment of a gene that is both transcribed and translated and hence carries part of the code for the gene product. Most eukaryotic genes consist of exons interrupted by noncoding sequences (*see* intron). Both exons and introns are transcribed to heterogeneous nuclear RNA (hnRNA), an intermediary form of messenger RNA (mRNA); the introns are then removed leaving mRNA, which has only the essential sequences and is translated into the protein. Bacteria do not have introns.

exonuclease An enzyme that catalyzes the hydrolysis of the terminal linkages of polynucleotides such as DNA and RNA, thereby removing terminal nucleotides. *See also* DNase; restriction endonuclease.

exoscopic Describing the type of development of a plant embryo in which the apex of the sporophyte develops from the outer cell formed by the first division of the zygote, the inner cell giving rise to the foot. *Compare* endoscopic.

exosporic The formation of spores outside the spore-producing organ. In the basidiomycete fungi, for example, the basidiospores are borne on the tips of the sterigmata, which are outgrowths of the basidium.

exosporium (exine) *See* pollen.

explantation The culture of isolated tissues of adults or embryos in an artificial medium for the study of maintenance, growth, and/or differentiation. Such a piece of transplanted tissue is called an *explant*, and its placement on the culture medium is called *inoculation*. Explants are also a means of plant propagation, producing clones and offering a more rapid means of multiplication than seed production.

exponential growth A type of growth in which the rate of increase in numbers at a given time is proportional to the number of individuals present. Thus, when the population is small multiplication is slow, but as the population gets larger, the rate of multiplication also increases. An exponential growth curve starts off slowly and increases faster and faster as time goes by (called the *log phase*). However, at some point factors such as lack of nutrients, accumulated wastes, etc., limit further increase, when the curve of number against time begins to level off and the curve becomes sigmoid (S-shaped). When the population remains stable, it has reached the *stationary phase*. If organisms are introduced into a new medium, there is often a period during which the population remains constant before it begins to increase (called the *lag phase*). During this phase the cells are increasing in size and synthesizing materials necessary for growth under the new conditions.

extensin *See* glycoprotein.

extracellular Occurring or situated outside a cell.

extrachromosomal DNA In eukaryotes, DNA found outside the nucleus of the cell and replicating independently of the chromosomal DNA. It is contained within organelles in the cytoplasm, e.g. mitochondria, chloroplasts, and plastids, and is responsible for cytoplasmic inheritance. *See* cytoplasmic inheritance.

extrorse Denoting anthers in which dehiscence lines are to the outside of the flower, promoting cross-pollination. *Compare* introrse.

eyespot (stigma) A light-sensitive structure of certain protoctists and invertebrate animals. The eyespot of unicellular and colonial algae and their gametes and zoospores contains globules of orange or red carotenoid pigments. It controls locomotion, ensuring optimum light conditions for photosynthesis. Its location varies. In *Chlamydomonas* it is just inside the chloroplast; in *Euglena* it is near the base of the flagellum.

F

F_1 The first filial generation; i.e. the first generation that results from a particular cross.

F_2 The second filial generation, obtained by crossing or selfing within the F_1 generation. If the characteristic is governed by simple Mendelian genes, the typical monohybrid and dihybrid ratios become apparent in the F_2 generation.

Fabaceae A large family of dicotyledonous flowering plants, commonly known as legumes. There are about 18,000 species, which include herbs, trees, shrubs, climbers, and some aquatics. Most species have root nodules containing nitrogen-fixing bacteria. The leaves are usually compound and pinnate, with stipules. The Fabaceae have a worldwide distribution and many are of economic importance, including peas (*Pisum sativum*), beans (*Phaseolus* and *Vicia*), lentils (*Lens culinaris*), and peanuts (*Arachis hypogea*). Others are used for forage, e.g. clovers (*Trifolium*) and lucerne (*Medicago sativa*). There are also ornamental species such as lupins (*Lupinus*).

facilitated diffusion A passive transport of molecules across a cell membrane along a concentration gradient, mediated by carrier molecules or complexes, usually proteins. No energy is expended in this process, but it enables the passage through the membrane of molecules that otherwise could not pass through.

facultative Describing an organism that can utilize certain conditions but is not dependent on them, i.e. an organism that can adopt an alternative mode of living. For example, a facultative anaerobe is an organism that can grow under anaerobic conditions but is also able to survive in aerobic conditions. *Compare* obligate.

FAD (**flavin adenine dinucleotide**) A derivative of riboflavin that is a coenzyme in electron-transfer reactions. It acts as a prosthetic group to various dehydrogenase enzymes, the substrate being oxidized as FAD accepts electrons and is reduced to $FADH_2$. *See also* flavoprotein.

Fagaceae A family of deciduous or evergreen trees and shrubs that includes the oaks (*Quercus*), beeches (*Fagus*), southern beeches (*Nothofagus*), and chestnuts (*Castanea*). Members of this family dominate broad-leaved forests, especially of temperate regions.

false fruit *See* pseudocarp.

family A taxonomic category involving a collection of similar genera. Families may be subdivided into subfamilies, tribes, and subtribes. Plant family names generally end in *aceae*. Similar families are grouped into orders.

far-red light Electromagnetic radiation of wavelength approximately 740 nm. *See* phytochrome; photoperiodism.

fascicular cambium (*pl.* **cambia**) *See* intrafascicular cambium.

fast green *See* staining.

fat A triglyceride of a long-chain carboxylic acid (fatty acid) that is solid below 20°C. Fats commonly serve as energy storage material in some plants, especially in seeds. *See also* lipid.

fatty acid *See* carboxylic acid.

feedback inhibition The inhibition of the activity of an enzyme (often the first) in a multienzyme reaction sequence by the product of that sequence. When the product accumulates beyond an optimal amount it binds to a site (ALLOSTERIC SITE) on the enzyme, changing the shape so that it can no longer react with its substrate. However, once the product is utilized and its concentration drops again, the enzyme is no longer inhibited and further formation of product results. The mechanism is used to regulate the concentration of certain substances within a cell.

feedback loop (**feedback mechanism**) A regulatory mechanism in a system. This may involve negative feedback, for example, an increase in the number of herbivores in an ecosystem leads to overgrazing and reduction of the herbivores population by starvation or increased susceptibility to disease, which allows the plants to grow back, and so on – the population of both species fluctuates about a mean. Positive feedback has the opposite effect, reinforcing a change in the system. In certain metabolic pathways, for example, the substrate of the first enzyme in the pathway stimulates subsequent reactions in the sequence.

fermentation The breakdown of organic substances, particularly carbohydrates, under anaerobic conditions. It is a form of anaerobic respiration and is seen in certain bacteria and yeasts, and also in plants under waterlogged conditions. *See* alcoholic fermentation. *See also* glycolysis; lactic acid bacteria.

ferns *See* Filicinophyta.

ferredoxins A group of red-brown proteins found in green plants, many bacteria, and certain animal tissues. They contain nonheme iron in association with sulfur at the active site. They are strong reducing agents (very negative redox potentials) and function as electron carriers, for example in photosynthesis and nitrogen fixation. They have also been isolated from mitochondria. In photosynthesis ferredoxin is involved in electron transfer between photosystem I and the final electron acceptor, NADP.

fertilization (**syngamy**) The fusion of a male gamete with a female gamete to form a zygote, the essential process of sexual reproduction. *External fertilization* occurs when gametes are expelled into a watery substrate from the parental bodies before fusion; it is typical of sporebearing plants. *Internal fertilization* takes place in the seed plants and some clubmosses and ferns, often by means of a pollen tube. Internal fertilization is an adaptation to life in a terrestrial environment.

Feulgen's stain A test used to show the presence of DNA, especially in dividing nuclei. A sample of tissue is placed in dilute hydrochloric acid at 60°C to hydrolyze the DNA. This exposes the aldehyde groups of deoxyribose. The tissue is then soaked in Schiff's reagent: the development of a magenta color indicates the location of DNA. *See* staining.

fiber A form of SCLERENCHYMA cell that is often found associated with vascular tissue. Fibers are long narrow cells, with thickened walls and finely tapered ends. Their function is more as supporting tissue than as conducting tissue. Where they occur interspersed with the xylem they may be distinguished from tracheids by their narrower lumen.

fiber-tracheid An elongated cell with bordered pits found in wood, intermediate in form between a fiber and a tracheid.

fibrous root system A branching root system in which there is no main root. In fibrous root systems the radicle or primary root either branches or dies and is replaced by adventitious roots, e.g. grasses (Poaceae). Fibrous root systems are characteristic of monocotyledons. *Compare* taproot.

field capacity The point at which the soil contains all the water it can hold by capillary and chemical attraction. Any more water added to soil at field capacity would drain away by gravity. Soil at field capacity has a high water potential. Water content at field capacity is usually expressed as a percentage of the weight or volume of oven-dry soil. This is affected by the soil texture, clay having a high field capacity and sand a low field capacity.

field layer The herb layer of a plant community. *Compare* canopy; ground layer.

filament 1. The stalk of the stamen bearing the anther in angiosperms.
2. The vegetative body of the filamentous algae (e.g. *Spirogyra*), composed of a line of similar cells joined by their end walls.
3. A strand of protein of diameter 4–15 nm found inside a cell. Such filaments are usually composed of the contractile proteins actin and/or myosin, and are thought to play a role in cell motility, and perhaps also change in shape.

Filicinophyta (**ferns**) The largest phylum of spore-bearing vascular plants comprising the ferns and including about 12,000 species. It is divided into two main groups, the LEPTOSPORANGIATE ferns and the EUSPORANGIATE ferns. In leptosporangiate ferns, the sporangia are thin-walled and develop from one cell. In the Filicales the leptosporangium has a specialized spore dispersal mechanism involving an annulus of cells with thick walls that ejects s pores explosively as it dries out. Leptosporangiate ferns include the typical ferns (Filicales) and the water ferns (Marsileales and Salviniales). The Filicales is much the largest living group of ferns, and they normally have large spirally arranged leaflike fronds (MEGAPHYLLS) derived from lateral branches of the main axis, some or all of which bear sporangia on their margins or undersurfaces. In eusporangiate ferns, the sporangia are thick-walled and develop from a number of cells. Eusporangiate ferns include the adderstongues (Ophioglossales), Marattiales, and royal ferns (Osmundales), which are to some extent intermediate between leptosporangiate and eusporangiate ferns. The whisk ferns (Psilotales) may also be included here, but in the Five Kingdoms classification they have the status of a phylum (Psilophyta). There are also a number of fossil genera, leptosporangiate ferns dating back to the Cretaceous era, eusporangiate ferns to the Devonian. Ferns do not show secondary thickening, but in large ferns such as tree ferns bands of sclerenchyma and the overlapping bases of the fronds help to support the plant.

fimbriae (*sing.* **fimbria**) *See* pili.

fine structure *See* ultrastructure.

FISH (**fluorescence in situ hybridization**) A technique for locating specific sequences of DNA using fluorescent probes that are viewed under the MICROSCOPE. The probes bind to (hybridize with) specific sequences of DNA; different colors of probe may be used to distinguish different DNA sequences. *See* DNA probe.

fission A type of asexual reproduction in which a parent cell divides into two daughter cells (*binary fission*) genetically identical to the parent. Binary fission occurs in many unicellular organisms, such as protoctists, bacteria, and fission yeasts. Fission begins with division of the nucleus by mitosis, followed by cytoplasmic division, and sometimes sporulation. *Compare* budding; fragmentation.

fitness In an evolutionary context, the ability of an organism to produce a large number of offspring that survive to a reproductive age.

Five Kingdoms classification A classification system that recognizes five kingdoms containing at least 96 phyla. It comprises two superkingdoms, the Prokarya (PROKARYOTES) and Eukarya (EUKARYOTES). The Prokarya contains a single kingdom, the BACTERIA, which is divided into two subkingdoms, the ARCHAEA and the EUBACTERIA. The Eukarya contains four kingdoms, the Animalia, Plantae, Fungi, and Protoctista.

fixation 1. The situation in a usually small population when an allele reaches a 100% frequency due to the complete loss of all other alleles of that gene. All members of the population are homozygous for this particular allele.
2. In the preparation of microscope slides, the process by which tissues are rapidly killed and preserved by chemicals to prevent decay or autolysis with minimal distortion of structure. Fixing agents (fixatives) should also render cell organelles and inclusions more visible and harden the tissue to prevent shrinkage and distortion during dehydration, embedding, and sectioning. Examples of fixatives are formaldehyde, osmium tetroxide, ethanol, and acetic acid. *See also* dehydration; embedding.
3. The incorporation of inorganic materials such as carbon and nitrogen into organic compounds by living organisms. An example is the incorporation (fixation) of carbon from carbon dioxide into new organic compounds during photosynthesis. *See also* nitrogen fixation.

flaccid Lacking turgor. When a plant loses water to the extent that its cells become flaccid, wilting occurs. *See* plasmolysis.

flagellum (*pl.* **flagella**) A whiplike extension of prokaryote cells, with a basal body at its base, whose beat causes locomotion of the cell. Strictly the term is now reserved for the bacterial flagellum. The flagella and cilia of eukaryote cells have a quite different structure and are called UNDULIPODIA. Bacterial flagella are much simpler than undulipodia, being hollow cylinders about 15 nm in diameter, consisting of subunits of a protein (*flagellin*) arranged in helical spirals.

flavin A derivative of riboflavin occurring in the flavoproteins; i.e. FAD or FMN. Flavins are light-sensitive yellow pigments with an absorption peak around 370 nm. They are involved in phototropism, and in PHOTOPERIODISM, where they are thought to affect phytochrome by transferring light energy from blue to red wavelengths. *See* phytochrome.

flavin adenine dinucleotide *See* FAD.

flavin mononucleotide *See* FMN.

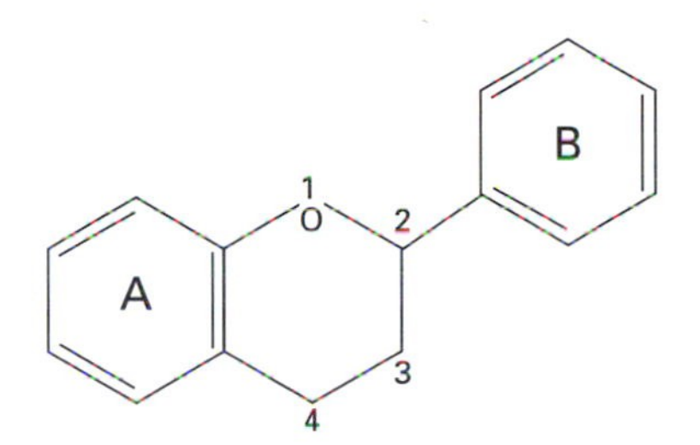

Flavonoid structure

flavonoid One of a common group of plant compounds having the C_6–C_3–C_6 chemical skeleton in which C_6 is a benzene ring. They are an important source of nonphotosynthetic pigments in plants.

flavoprotein A conjugated protein in which a FLAVIN (FAD or FMN) is the prosthetic group joined to a protein component. Flavoproteins are enzymes and many are dehydrogenases. They act as electron-transport agents and play an important role in the electron-transport chains of respiration, photosynthesis, and the microsomal system responsible for the desaturation of fatty acids in heterotrophic bacteria.

flocculation The aggregation of soil particles into crumbs. Compacted structureless clay soils can be flocculated by the addition of neutral salts, particularly of calcium. The addition of lime to saturated clay soils (*liming*) is a common agricultural practice, improving soil structure by encouraging crumb formation and making heavy soils more workable.

floral formula A series of symbols used to describe the structure of a flower. The various whorls of structures are abbreviated as follows: K (calyx), C (corolla), P (perianth), A (androecium), and G (gynoecium). These letters are each followed by a number indicating the number of parts in the whorl. If the number exceeds 12 then the symbol ∞ is used to denote an indefinite number. Fusion of parts of the whorl is indicated by placing the number of parts in brackets. The position of the gynoecium is shown by a line above or below the letter G denoting an inferior or superior ovary respectively. The formula is preceded by ⊕ to indicate actinomorphic flowers and ·|· for zygomorphic flowers. The floral formula for the buttercup would be written:

$$\oplus K5\ C5\ A\infty\ \underline{G\infty}$$

floral province (**floral kingdom; floral region**) One of the six main regions and a few minor ones into which the world is divided on the basis of plant distribution. Each province has one or more centers of diversity and each contains unique families and genera of plants. The main provinces are the Boreal, which covers all of the north temperate; the Neotropical, covering tropical Central and South America; the Paleotropical, covering tropical Africa and Asia; the Australian; the Cape Province, covering a small area around the Cape of Good Hope; and the Antarctic, including New Zealand and temperate South America. Oceanic islands, particularly in the Pacific, form minor provinces. There are floral affinities and connections between the provinces, particularly near their boundaries.

floret **1.** An individual small flower of a clustered inflorescence, e.g. daisy (*Bellis*). **2.** In grasses, the lemma, palea, and the flower they enclose.

florigen (**flowering hormone**) A hypothetical plant hormone that has been postulated to account for the transfer of the photoperiod stimulus from the leaves to the apex where flowering is induced. The stimulus can be transferred from one plant to another by grafting, but attempts to isolate florigen have so far been unsuccessful. *See* photoperiodism.

flower The characteristic reproductive structure of an angiosperm. It usually consists of an axis or receptacle bearing the sepals, petals, and stamens. The gynoecium is borne either above the receptacle or enclosed within it. One or more of these whorls may be absent. Flower structures are extremely variable and show numerous adaptations to promote pollination and seed dispersal. *See also* floral formula; inflorescence.

flowering hormone *See* florigen.

flowering plants *See* angiosperms.

fluid-mosaic model *See* plasma membrane.

fluorescence in situ hybridization *See* FISH.

fluorescence microscopy A form of microscopy in which fluorescent probes are added to the material being investigated. These probes bind to specific parts of cells or to particular molecules, such as particular sequences of DNA or protein receptors.

FMN (**flavin mononucleotide**) A derivative of riboflavin that is a coenzyme in electron-transfer reactions. It acts as a prosthetic group to various dehydrogenases, such as NADH dehydrogenase, which catalyzes the transfer of electrons from NADH to coenzyme Q in the respiratory electron-transport chain. *See also* flavoprotein.

folic acid (**pteroylglutamic acid**) One of the water-soluble B-group of VITAMINS. Folic acid is important in metabolism in various coenzyme forms, all of which are specifically concerned with the transfer and utilization of the single carbon (C_1) group. It is important in the growth and reproduction of cells, participating in the synthesis of purines and thymine.

follicle A dry dehiscent fruit formed from one carpel that splits along one edge to release its seed, for example *Delphinium* fruit.

food chain The chain of organisms existing in any natural community, through which food energy is transferred. Each link in the food chain obtains energy by eating the one preceding it and is in turn eaten by the organisms in the link following it. At each transfer a large proportion (80–90%) of the potential energy is lost as heat, therefore the number of links in a sequence is limited, usually to 4 or 5. The shorter the food chain, the greater the available energy, so total energy can be increased by cutting out a step in the food chain, for example if people consume cereal grain instead of consuming animals that eat cereal grains.

Food chains are of two basic types: the grazing food chain, which goes from green plants, to grazing herbivores, and finally to carnivores; and the detritus food chain, which goes from dead organic matter, to microorganisms, and then to detritus-feeding organisms. The food chains in a community are interconnected with one another, because most organisms consume more than one type of food, and the interlocking pattern is referred to as a *food web* or *food cycle*. *See* trophic level.

food web *See* food chain.

foot The basal portion of an embryo, sporophyte, or spore-producing body, which is embedded in the tissue of the parent.

forest 1. A plant formation dominated by trees whose crowns touch, forming a continuous canopy.
2. A major plant community in which the dominant plants are trees.

form The lowest taxonomic group, ranking below the variety level, normally referring to the recognizably different morphology found in several members within a plant species. Subforms may also be recognized.

formalin A mixture of about 40% formaldehyde, 8% methyl alcohol, and 52% water (the methyl alcohol is present to prevent polymerization of the formaldehyde). It is a powerful reducing agent and is used as a disinfectant, germicide, and fungicide and also as a general preserving solution.

fossil The remains of, or impressions left in rocks by, long-dead animals and plants. Most fossils consist of hard skeletal material because soft tissues and organs rot away very quickly. Plants seldom contain hard mineral material, so they are preserved well only under certain conditions. With rapid burial in a moist anoxic environment, such as the coal swamps of the Carboniferous period, carbonaceous films of plant material like fern fronds may survive. Alternatively, mineral salts from surrounding rocks gradually replace the hard organic material, to give a *cast* in a process termed *petrification*, as in the various petrified forests dotted around the globe. The hard coats of spores and pollen grains are preserved better, forming *microfossils*. Alternatively the organic material dissolves away leaving an impression or mold in the surrounding rocks. *See also* chemical fossils; palynology.

fragmentation 1. A form of ASEXUAL REPRODUCTION in which the plant or alga breaks into fragments, each fragment growing into a new, independent plant. Examples include filamentous algae such as *Spirogyra*, and waterweeds like the Canadian pondweed (*Elodea*). *Compare* budding; fission.
2. The separation through human activity of once continuous habitats into fragments that may or may not contain viable populations of the organisms present.

frameshift mutation *See* gene mutation.

freeze drying A technique for dehydrating a specimen by the sublimation of ice under a vacuum. It avoids damage that would result from heating.

freeze fracturing A method of preparation of material for electron microscopy, particularly useful for studying membranes. Material is frozen rapidly (e.g. by immersion in liquid nitrogen) thus preserving it in lifelike form. It is then fractured, usually with a sharp knife. The fracture plane tends to follow lines of weakness, such as between the two lipid layers of membranes, revealing their internal surfaces. Carbon or platinum replicas made of the surfaces are shadowed for examination in the electron microscope. In *freeze etching* the fractured surface is etched, i.e. some ice is allowed to sublime away before SHADOWING. This exposes more detailed structure, such as the outer surface of the membrane.

Fries, Elias Magnus (1794–1878) Swedish mycologist. Fries, working at Lund University, published *Systema Mycologicum*, his three-volume work on the classification of fungi, in 1921–22. He also published flora covering all the Scandinavian countries and an important work on European lichens.

frond A term usually applied to large well-divided leaves as found in ferns, palms, and cycads. The leaflike thalli of certain algae and lichens may also be termed fronds. *See also* megaphyll.

fructification A seed- or spore-bearing structure. The term is used especially for the aerial fruiting bodies of fungi, e.g. mushrooms.

fructose A sugar ($C_6H_{12}O_6$) found in fruit juices, honey, and cane sugar. It is a ketohexose, existing in a pyranose form when free. In combination (e.g. with glucose in the disaccharide sucrose) it exists in the furanose ring form. Fructose forms polymers called *fructosans*, which include INULIN, an important storage polysaccharide in the Asteraceae.

fruit The ripened OVARY of a flower that is usually formed following fertilization of the ovule. It may consist of the ripened ovary only or include other parts of the flower. Fruits vary according to the method of seed dispersal, succulent fruits normally being distributed by animals while dry fruits may be dispersed by wind or water, and dehiscent fruits split to release the seeds with the fruit remaining on the plant. Fruits are classified according to how the ovary wall (pericarp) develops, depending on whether it becomes fleshy or hard. Fruits are further classified according to whether or not the fruit wall opens to release the seeds. *See also* dehiscent; indehiscent; multiple fruit; pseudocarp.

Fucales An order of Phaeophyta (brown algae) in which the haploid generation consists only of the gametes. The gametes are produced in special cavities called conceptacles, often in the inflated tips of branches of the thallus, e.g. wracks (*Fucus*). *See* alternation of generations.

Fuchs, Leonhard (1501–66) German botanist and physician. His beautifully illustrated herbal manual *De historia stirpum* was published in 1542. It contained about four hundred German plants and about a hundred foreign plants, arranged in alphabetical order. The illustrations were accompanied by a description of each plant and its habitat and even information about the best time to collect them. The book also included a glossary of terms. Fuchs is best known for the genus of ornamental shrubs *Fuchsia*, which was named for him.

fucoxanthin A brownish xanthophyll pigment found in Diatoms, Chrysomonada, Haptomonada, Phaeophyta, and some Dinomastigota. The light absorbed is used with high efficiency in photosynthesis, the energy first being transferred to chlorophyll *a*. It has three absorption peaks covering the blue and green parts of the spectrum, which penetrate to greater depths under water than the red end of the spectrum. *See* carotenoids.

fumaric acid (*trans*-butenedioic acid) An unsaturated dicarboxylic acid, which occurs in many plants. The fumarate ion participates in several important metabolic pathways, e.g. the Krebs cycle, purine pathways, and the urea cycle.

dry fruit

indehiscent

monocarpellary

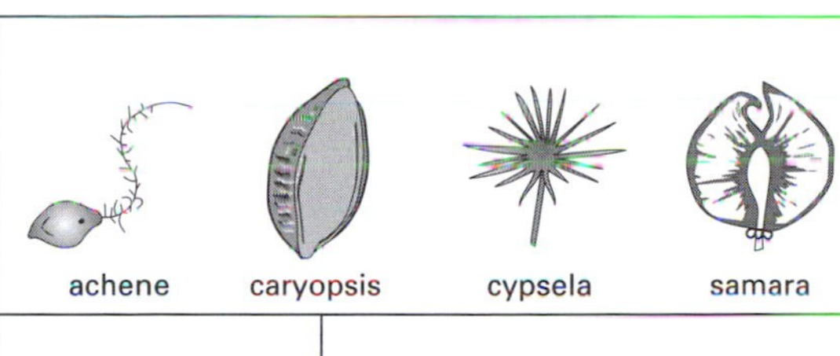

polycarpellary

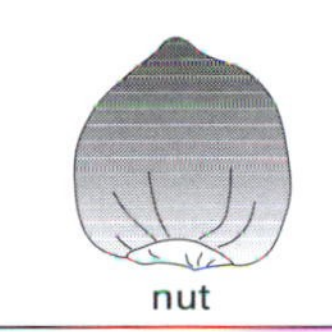

dehiscent

monocarpellary

polycarpellary

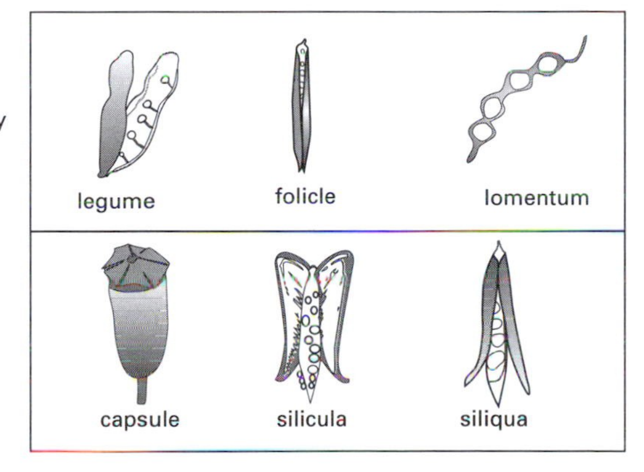

schizocarpic

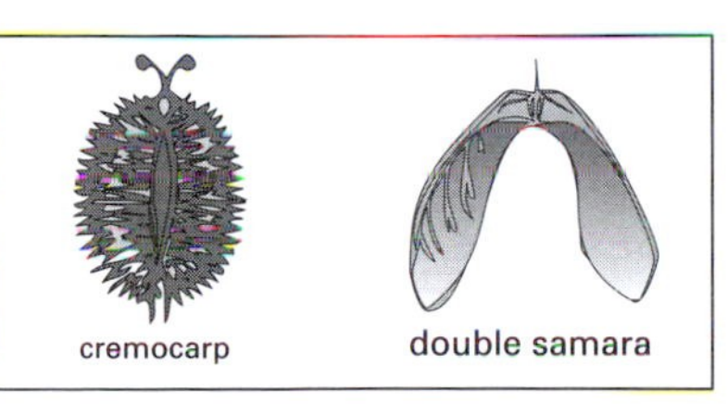

succulent fruit

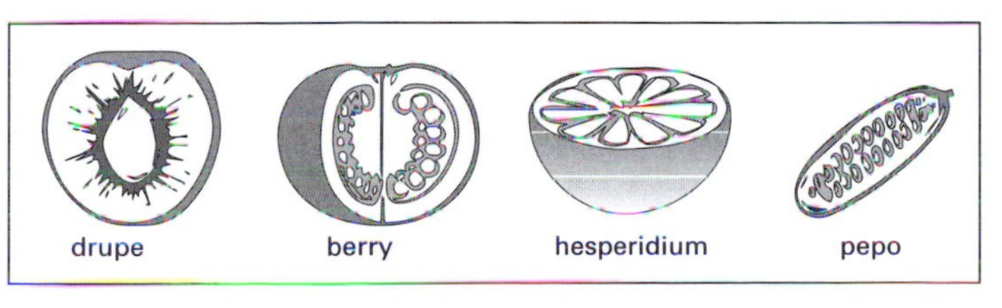

Fruit: a classification of types of fruit

Fungi A kingdom of nonphotosynthetic mainly terrestrial organisms that are quite distinct from plants and animals, and placed in their own kingdom, Fungi. They are characterized by having cell walls made chiefly of chitin, not the cellulose of plant cell walls, and they all develop directly from spores without an embryo stage. Moreover, undulipodia are never found in any stage of their life cycles. Fungi are generally saprophytic or parasitic, and may be unicellular or composed of filaments (termed *hyphae*) that together comprise the fungal body or *mycelium*. Hyphae may grow loosely or form a compacted mass of *pseudoparenchyma* giving well-defined structures, as in toadstools. Fungi are major decomposers in soil, and fresh water. Some form symbiotic associations with algae or cyanobacteria to form lichens, while others form extremely important mycorrhizae with the roots of many plants, including most forest trees. Some parasitic fungi cause diseases in plants and animals, while others are sources of antibiotics. Yeasts and other fungi are used in commercial fermentation processes such as brewing. There are three main phyla of Fungi, based on the nature of their fruiting structures: the Zygomycota, Ascomycota, and Basidiomycota. In the modern Five Kingdoms classification, certain funguslike organisms that were previously considered to be fungi, are now assigned to phyla in the Kingdom Protoctista.

Fungi Anamorphici (**Deuteromycota; Fungi Imperfecti; mitosporic fungi**) A phylum of so-called 'imperfect' fungi in which sexual reproduction is unknown or has been lost during evolution. Its members, for example the *Penicillium* molds, are assigned as a taxonomic convenience, rather than by any true criteria, and some authorities prefer to allocate the deuteromycotes to either the Ascomycota (the majority) or the Basidiomycota, on the basis of available evidence. Indeed *Penicillium* is now known to have a sexual stage, formerly regarded as a quite distinct ascomycete fungus (*Talaromyces*). Most of these fungi reproduce by conidia produced by mitosis (*see* conidium). There are about 15 000 species.

fungicide A chemical that inhibits fungal growth. They are used to prevent fungal infection as well as to deal with existing infections.

Fungi Imperfecti *See* Fungi Anamorphici.

funicle (**funiculus**) The stalk attaching the ovule (later the seed) to the placenta in angiosperm ovaries. Nutrients pass through the funicle to the ovule and seed.

furanose A sugar that has a five-membered ring (four carbon atoms and one oxygen atom). *See also* sugar.

fusiform initial *See* initial.

G

galactose A sugar found in lactose and many polysaccharides. It is an aldohexose, isomeric with glucose. *See also* sugar.

gall An abnormal outgrowth or swelling on a root, stem, or leaf in response to attack by a parasite or to invasion by a symbiont. ROOT NODULES induced by the nitrogen-fixing bacterium *Rhizobium* are part of a symbiotic relationship (*see* symbiosis).

gametangium (*pl.* gametangia) A cell or organ in which sexual cells (gametes) are produced. When the gametes are readily distinguished as male or female, other terms may be used such as ANTHERIDIUM, ARCHEGONIUM, or OOGONIUM.

gamete A cell capable of fusing with another cell to produce a zygote, from which a new individual organism can develop. Gametes may have similar structure and behavior (*isogametes*), as in many simple organisms, but are usually dissimilar in appearance and behavior (*anisogametes*). The typical male gamete is small, motile (by means of UNDULIPODIA), and produced in large numbers. The typical female gamete is large because of the food reserves it contains, and is produced in smaller numbers than the male gametes. In all except the most primitive organisms, the female gamete is nonmotile (OOGAMY). Gametes are usually haploid, so fusion of gametes results in the nucleus of the zygote having the diploid number of chromosomes.

gametogenesis The formation of sex cells or gametes.

gametophyte The gamete-producing generation in the life cycle of a plant or multicellular alga. The gametophyte is HAPLOID and produces sex organs. It derives from the germination of haploid spores produced by the SPOROPHYTE generation. It is the main generation in the life cycle of the Bryata, in which the sporophyte is completely or partially dependent upon it. The PROTHALLUS of Lycophyta (clubmosses), Sphenophyta (horsetails), and Filicinophyta (ferns), which may be a green structure on the soil surface or a subterranean mycorrhizal structure, is the gametophyte. In these groups and in the bryophytes, some species produce unisexual gametophytes and some produce hermaphrodite gametophytes. In the Coniferophyta the female gametophyte is reduced to a nonphotosynthetic tissue with up to 2000 nuclei in the ovule, which produces archegonia (*see* archegonium). While in the Angiophyta the female gametophyte is simply the EMBRYO SAC. The male gametophytes of conifers and angiosperms are the POLLEN grains. *See also* alternation of generations.

Gamophyta A phylum of green algae in which sexual reproduction is by conjugation, e.g. *Spirogyra* and DESMIDS. Amoeboid gametes are produced, which cross to adjacent filaments through conjugation bridges and fuse to form a zygote, which usually forms a resistant zygospore. On germination, the zygospore undergoes meiosis to produce haploid cells that give rise to a new thallus. Asexual reproduction is by FRAGMENTATION.

garrigue Scrub woodland, a secondary formation derived from the original mixed forest, characteristic of limestone areas of the Mediterranean region, with low rainfall and thin, dry soils. It is dominated by low-growing prickly dwarf shrubs and aromatic herbs (most commonly of the Lamiaceae), such as lavender (*Lavandula vera*), sage (*Salvia officinalis*), and thymes (*Thymus*), many with drought-resistant foliage. There are also bulbous plants such as species of *Iris*. Compare chaparral; maquis.

gas chromatography A technique widely used for the separation and analysis of mixtures. Gas chromatography employs a column packed with either a solid stationary phase (*gas–solid chromatography* or *GSC*) or a solid coated with a nonvolatile liquid (*gas–liquid chromatography* or *GLC*). The whole column is placed in a thermostatically controlled heating jacket. A volatile sample is introduced into the column using a syringe, and an unreactive carrier gas, such as nitrogen, passed through it. The components of the sample will be carried along in this mobile phase. However, some of the components will cling more readily to the stationary phase than others, either because they become attached to the solid surface or because they dissolve in the liquid. The time taken for different components to pass through the column is characteristic and can be used to identify them. The emergent sample is passed through a detector, which registers the presence of the different components in the carrier gas.

gas–liquid chromatography *See* gas chromatography.

gas–solid chromatography *See* gas chromatography.

gel filtration (**gel-permeation chromatography**) A chromatographic method using a column packed with porous carbohydrate-gel beads of standard size and porosity. It is a standard technique used for separating and identifying macromolecules of various sizes, e.g. proteins or nucleic acids. A solution of the mixture of macromolecules is added to the top of the column and allowed to flow through by gravity. The smaller molecules are hindered in their passage down the column because they are better able to penetrate the hydrated pores within the particles of the gel. Molecules too large to penetrate the pores are excluded, and thus flow more rapidly through the column. By analyzing the liquid that drips from the bottom of the column (the *eluate*) at set intervals and comparing it with a standard (obtained by running a known macromolecule through the column) information about the sizes and molecular weights of the components of the mixture is gathered. The most frequently used commercial gel is Sephadex.

gemma (*pl.* gemmae) **1.** A vegetative propagule produced by mosses, liverworts, and certain Lycophyta (clubmosses). Gemmae are small dispersal units consisting of one to several cells, often forming disks, plates, or filaments. They often form in groups in receptacles called *gemma-cups* and eventually become detached from the parent to form new plants.
2. *See* chlamydospore.

gemmation A type of asexual reproduction, seen in mosses and liverworts, involving the production of a group of cells (a *gemma*) that develops into a new individual, before or after separation from the parent.

gene In classical genetics, a unit of hereditary material located on a chromosome that, by itself or with other genes, determines a characteristic in an organism. It corresponds to a segment of the genetic material, usually DNA (although the genes of some viruses consist of RNA). Genes may exist in a number of forms, termed *alleles*. For example, a gene controlling the characteristic 'height' in peas may have two alleles, one for 'tall' and another for 'short'. In a normal diploid cell only two alleles can be present together, one on each of a pair of homologous chromosomes: the alleles may both be of the same type, or they may be different. The segregation of alleles at meiosis and their dominance relationships are responsible for the particulate nature of inheritance. Genes can occasionally undergo changes, called MUTATIONS, to new allelic forms.

Although the DNA molecules of the chromosomes account for the great majority of genes, genes are also found as PLASMAGENES in certain DNA-containing cytoplasmic bodies (e.g. mitochondria, plastids). A gene can be defined as the smallest hereditary unit capable either of recombination or of mutation. alternatively it may be described as a unit of function, a CISTRON, which may be much larger than a unit of recombination or mutation, and controls the synthesis of a single polypeptide or a messenger RNA molecule whose genetic code is contained within the gene. Research with bacteria has shown that the smallest unit of recombination or mutation is one base pair, while the size of a unit of function can be determined by the CIS–TRANS TEST.

gene bank 1. A place where plant material is stored in a viable condition in order to preserve plants that are in danger of becoming extinct in the wild or cultivars that are being lost from cultivation. The material also provides a source of genes for breeding new varieties.
2. *See* gene library.

gene cloning (**DNA cloning**) A technique of GENETIC ENGINEERING whereby a gene sequence is replicated, giving many identical copies. The gene sequence is isolated by using restriction endonucleases, or by making a complementary DNA from a messenger RNA template using a reverse transcriptase. It is then inserted into the circular chromosome of a cloning vector, i.e. a plasmid or a bacteriophage. The hybrid is used to infect a bacterium, usually *Escherichia coli*, and is replicated within the bacterial cell. A culture of such cells produces many copies of the gene, which can subsequently be isolated and purified.

genecology The study of ecological genetics, or population genetics in relation to environment.

gene flow The movement of alleles within and between populations of the same species through interbreeding.

gene frequency The proportion of an allele in a population in relation to other alleles of the same gene.

gene library A collection of cloned DNA fragments derived from the entire GENOME of an organism. The genetic material of the organism is first broken up randomly into fragments using restriction enzymes, for example. Then each fragment is cloned using a VECTOR (e.g. plasmid or bacteriophage) inside a suitable host, such as the bacterium *E. coli*. Particular genes or DNA sequences are identified by a suitable DNA PROBE. *See also* gene bank; gene cloning.

gene mutation An alteration in a single gene resulting from a change in the number, type, or sequence of nucleotide bases. Unlike CHROMOSOME MUTATIONS, gene mutations are not visible under the microscope. A *point mutation* is a change at a single locus. It may be an addition, deletion, inversion, or substitution of one or more bases. A substitution of one or two bases may still result in the same amino acid, or at worst to just one amino acid change, and may have little effect on the resulting protein. However, if the mutation results in the loss or gain of one or two bases, it causes the remainder of the gene to be misread during translation, as it changes the 'reading frame'; for example, CUC/CAG/GCA becomes UCC/AGG/CA. Such mutations are called *frameshift mutations*, and are more likely to be deleterious. *See* mutation; chromosome mutation.

gene pool The total number and variety of genes existing within a breeding population or species at a given point in time.

gene probe *See* DNA probe.

generative cell One of the haploid cells in the POLLEN TUBE of seed plants. In angiosperms it gives rise directly to the two male gametes or generative nuclei. In gymnosperms it gives rise to the body cell and stalk cell.

generative nuclei The two gametic nuclei formed by the division of the generative cell in the pollen tube of angiosperms. One fuses with the egg cell to form a zygote while the other either degenerates or, in certain angiosperms, fuses with the polar nuclei to give the primary endosperm nucleus.

gene sequencing Determination of the order of bases making up a gene on a DNA molecule. Sequencing requires multiple-cloned copies of the gene; long DNA sequences are cut into more manageable lengths using restriction enzymes. Since these cleave DNA at specific points, it is possible to reconstitute the overall sequence once the constituent fragments have been analyzed individually.

gene splicing **1.** The joining of EXONS after the INTRON sequences have been removed, to produce functional messenger RNA that leaves the nucleus to undergo TRANSCRIPTION. In the nucleus this is performed by a special assemblage of RNA and proteins called a *spliceosome*.
2. In GENETIC ENGINEERING, the joining of DNA fragments by the action of the enzyme DNA ligase.

genetic code The sequence of bases in either DNA or messenger RNA that conveys genetic instructions to the cell. The basic unit of the code consists of a group of three consecutive bases, the base triplet or CODON, which specifies instructions for a particular amino acid in a polypeptide, or acts as a start or stop signal for translation of the message into polypeptide assembly. For example, the DNA triplet CAA (which is transcribed as GUU in mRNA) codes for the amino acid valine. There are 64 different triplet combinations but only 20 amino acids; thus many amino acids can be coded for by two or more triplets. The code is said to be *degenerate*, since some amino acids are coded for by more than one triplet: in certain cases only the first one or two bases, are necessary to insure the coding of a specific amino acid. Three triplets, termed '*nonsense triplets*', do not code for any amino acid and have other functions, e.g. marking the beginning and end of a polypeptide chain.

genetic drift (**Sewall Wright effect**) The fluctuation of allele frequencies in a small population due entirely to chance. If the number of matings is small, then the actual numbers of different types of pairing may depart significantly from the number expected on a purely random basis. Genetic drift is one of the factors that can disturb the HARDY–WEINBERG EQUILIBRIUM.

genetic engineering The direct introduction of foreign genes (from other individuals or species or artificially synthesized) into an organism's genetic material by micromanipulation at the cell level. Genetic engineering techniques bypass crossbreeding barriers between species to enable gene transfer between widely differing organisms. The commonest method of genetic engineering is RECOMBINANT DNA TECHNOLOGY. Genes may be take from living tissues by isolating the RNA produced when the gene is active, and recreating the DNA sequence of the gene using the enzyme reverse transcriptase. Gene transfer can be achieved by various methods, many of which employ a replicating infective agent, such as a virus, plasmid, or phage for DNA segments up to 24 000 base pairs in length, a *cosmid* for segments up to 45 000 base pairs, or a YAC (*yeast artificial chromosome*) for pieces up to 50 000 base pairs, as a VECTOR (*see* gene cloning). Other methods include microinjection of DNA into cell nuclei and direct uptake of DNA through the cell membrane. Recognizing whether or not transfer has occurred may be difficult unless the new gene confers an obvious visual or physiological characteristic. Consequently the desirable gene may be linked to a *marker gene*, e.g. a gene conferring resistance to an antibiotic in the growth medium. The transferred gene must also be linked to appropriate regulatory DNA sequences to insure that it works in its new environment and is regulated correctly and predictably.

Initial successes in DNA transfer were achieved with bacteria and yeast. Human genes coding for medically useful proteins have been transferred to bacteria. Human insulin, growth hormone, and interferon are now among a wide range of therapeutic substances produced commercially from genetically engineered bacteria. Genetically engineered vaccines have also been produced by transfer of antigen-coding genes to bacteria. Genes have been introduced to crop plants for various reasons, for instance to reduce damage during harvest or to make them resistant to the herbicides used in controlling weeds. Genetically modified tomatoes and soya beans are now widely available. There is also hope that in future many genetic diseases will be treatable by manipulating the faulty genes responsible. However, genetic engineering raises many legal and ethical issues, and the introduction of genetically modified organisms into the environment requires strict controls and monitoring.

first base	second base: U	second base: C	second base: A	second base: G	third base
U	UUU Phe	UCU Ser	UAU Tyr	UGU Cys	U
U	UUC Phe	UCC Ser	UAC Tyr	UGC Cys	C
U	UUA Leu	UCA Ser	UAA stop codon	UGA stop codon	A
U	UUG Leu	UCG Ser	UAG stop codon	UGG Trp	G
C	CUU Leu	CCU Pro	CAU His	CGU Arg	U
C	CUC Leu	CCC Pro	CAC His	CGC Arg	C
C	CUA Leu	CCA Pro	CAA Gln	CGA Arg	A
C	CUG Leu	CCG Pro	CAG Gln	CGG Arg	G
A	AUU Ile	ACU Thr	AAU Asn	AGU Ser	U
A	AUC Ile	ACC Thr	AAC Asn	AGC Ser	C
A	AUA Ile	ACA Thr	AAA Lys	AGA Arg	A
A	AUG Met (start codon)	ACG Thr	AAG Lys	CUG Arg	G
G	GUU Val	GCU Ala	GAU Asp	GGU Gly	U
G	GUC Val	GCC Ala	GAC Asp	GGC Gly	C
G	GUA Val	GCA Ala	GAA Glu	GGA Gly	A
G	GUG Val	GCG Ala	GAG Glu	GGG Gly	G

KEY

Ala	Alanine	Leu	Leucine
Arg	Arginine	Lys	Lysine
Asn	Asparagine	Met	Methionine
Asp	Aspartic acid	Phe	Phenylalanine
Cys	Cysteine	Pro	Proline
Gln	Glutamine	Ser	Serine
Glu	Glutamic acid	Thr	Threonine
Gly	Glycine	Trp	Tryptophan
His	Histidine	Tyr	Tyrosine
Ileu	Isoleucine	Val	Valine

Genetic code

genetic fingerprinting A technique for identifying individuals by means of their DNA. The DNA being tested is extracted from cells (from blood, semen, tissue fragments, etc.) and broken into fragments of 600–700 bases each, using restriction enzymes. The human genome contains many loci where short base sequences are repeated in tandem, with great variation between individuals in the number of such repeats. These so-called *variable number tandem repeats* (VNTR) can be identified using special DNA probes, thus providing a virtually unique set of markers for any given individual. This technique is used in veterinary and human medicine to establish the parentage of individuals, and in forensic science to identify individuals from traces of body tissue or fluids. Even minute amounts of DNA can now be amplified, using the POLYMERASE CHAIN REACTION, to provide sufficient material for genetic fingerprinting.

genetic map A map showing the sequence of particular genes or segments of DNA on a chromosome. Restriction enzymes and DNA probes are used to determine the exact sequence, but it is possible to get a rough estimate by analyzing the frequency of recombination between the alleles of linked genes. *See* chromosome map; DNA probe; restriction endonuclease.

genetics The term coined by the British biologist Bateson (1861–1926) to describe the study of inheritance and variation and the factors controlling them. Today the subject has four main subdivisions – Mendelian genetics (classical genetics), population genetics, cytogenetics, and molecular or biochemical genetics.

genome A complete haploid chromosome set. Haploid organisms have one genome, diploid organisms have two homologous sets, polyploid organisms have many sets sometimes from the same ancestor (*autopolyploids*) and sometimes from different ancestors (*allopolyploids*). Interspecific hybrids also have two or more difference genomes.

genotype The genetic make-up of an organism. The actual appearance of an individual (the *phenotype*) depends on the dominance relationships between alleles in the genotype and the interaction between genotype and environment. *Compare* phenotype.

genus (*pl.* **genera**) A taxonomic category involving a collection of similar species. Genera may be subdivided into subgenera, and also, especially in plant taxonomy, into sections, subsections, series, and subseries. The scientific name of a species always includes the genus name (or its abbreviation) as the first word of the binomial. Similar genera are grouped into families. *See* binomial nomenclature.

geological time scale A system of measuring the history of the earth by studying the rocks of the earth's crust. Since new rocks are generally deposited on top of existing material, those lower down are oldest, although this is often disrupted by e.g. volcanic activity, tectonic movements and erosion. The strata of rock are classified according to their age, and a time scale corresponding to this can be constructed. The main divisions (eras) are the PALEOZOIC, MESOZOIC, and CENOZOIC. These are further subdivided into periods and epochs.

geophyte A plant whose perennating buds are situated below ground as, for example, in plants with bulbs, corms, tubers, or rhizomes. *See* Raunkiaer's plant classification.

geotropism *See* gravitropism. cell

germ cell Any cell in a cell lineage that eventually produces gametes.

germination The first outward sign of growth of a reproductive body, such as a spore or pollen grain. The term is most commonly applied to seeds, in which germination involves the emergence of the radicle, plumule, or coleoptile through the testa. Both external conditions (e.g. water availability, temperature, and light) and internal biochemical status must be appro-

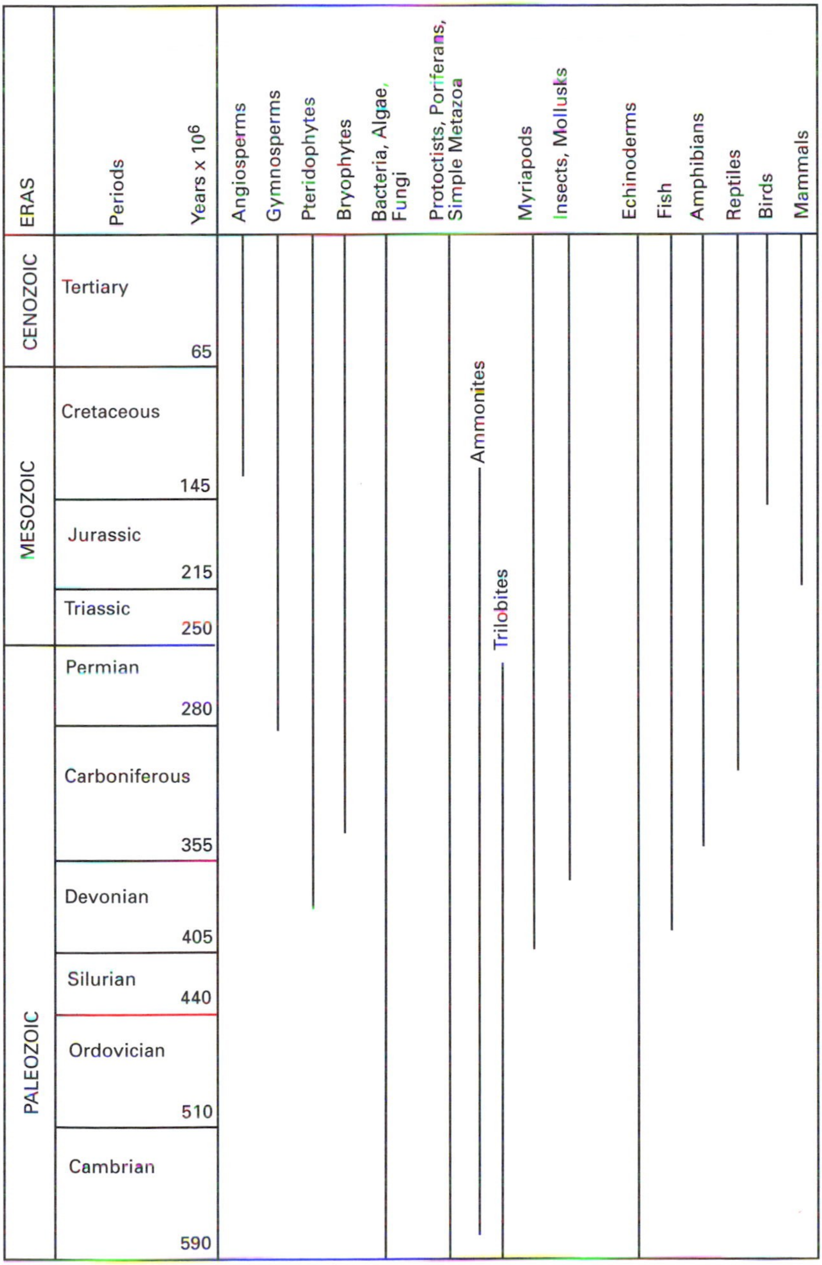

ERAS	Periods	Years x 10^6
CENOZOIC	Tertiary	65
MESOZOIC	Cretaceous	145
	Jurassic	215
	Triassic	250
PALEOZOIC	Permian	280
	Carboniferous	355
	Devonian	405
	Silurian	440
	Ordovician	510
	Cambrian	590

Angiosperms
Gymnosperms
Pteridophytes
Bryophytes
Bacteria, Algae, Fungi
Protoctists, Poriferans, Simple Metazoa
Ammonites
Trilobites
Myriapods
Insects, Mollusks
Echinoderms
Fish
Amphibians
Reptiles
Birds
Mammals

Geological time scale

priate before germination can occur. Seed germination may be either *epigeal*, in which the cotyledons appear above ground, or *hypogeal*, in which the cotyledons remain below ground.

germ line The lineage of cells from which gametes arise, continuous through generations.

germ plasm The part of an organism that, according to Weissmann at the beginning of the 20th century, passed its characters on to the next generation. It is now known that most of this information is carried by DNA in the chromosomes.

giant ferns *See* Marattiaceae.

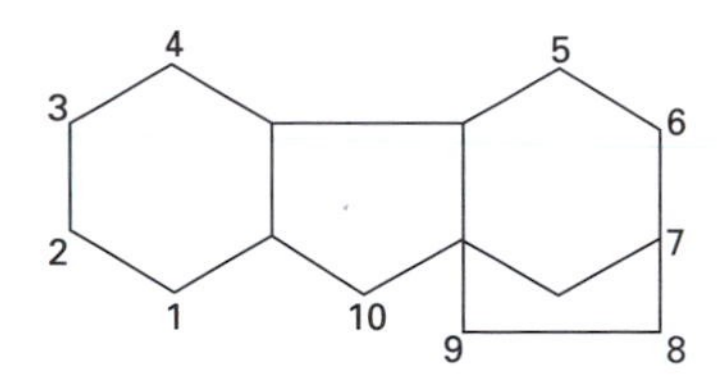

Gibberellic acid: the skeleton of the molecular structure

gibberellic acid (GA_3) A common gibberellin and one of the first to be discovered. Together with GA_1 and GA_2 it was isolated from *Gibberella fujikuroi*, a fungus that infects rice seedlings causing abnormally tall growth. Gibberellic acid is a terpenoid synthesized from mevalonic acid. Some growth retardants are thought to act by blocking this synthesis pathway. *See* gibberellin.

gibberellin A plant hormone involved chiefly in shoot extension. Gibberellins are diterpenoids chemically related to gibberellic acid; their molecules have the gibbane skeleton. More than thirty have been isolated, the first and one of the most common being GIBBERELLIC ACID, GA_3. Gibberellins stimulate elongation of shoots of various plants, especially the extension to normal size of the short internodes of genetically dwarf pea or maize plants. This is achieved by increasing cell-wall expansibility, but it requires the presence of auxin, too. Such dwarf varieties may be used in bioassays for gibberellin. Increased gibberellin levels can mimic or mediate the effect of long days. Thus they stimulate internode extension and flowering in LONG-DAY PLANTS such as lettuce and spinach. They are also effective in inhibiting tuber development or breaking tuber dormancy, e.g. in the potato, and breaking bud dormancy in woody species. Gibberellins can induce PARTHENOCARPY, and are used in the production of seedless fruits. They may be produced in both shoots and roots and travel in both xylem and phloem. They work antagonistically with ABSCISIC ACID, and in some situations work cooperatively with AUXINS as, for example, in the control of sex expression in dioecious plants.

gill One of many thin platelike spore-producing structures radiating outward from the stalk on the undersurface of the cap (*see* pileus) of agaric fungi such as mushrooms and toadstools (*see* Basidiomycota). The spores (basidiospores) are produced on the outer layer (hymenium).

Ginkgophyta (**ginkgoes**) A phylum of deciduous gymnosperms with fan-shaped, stalked leaves that have dichotomously branching veins, a feature unique among living gymnosperms. There is a single class, order, genus and species, the maidenhair tree or ginkgo (*Ginkgo biloba*) from China, known only in cultivation. Fossil remains of other species have been found in many parts of the world, dating back to the Middle Jurassic period. The ginkgo is dioecious.

glabrous Smooth, with no hairs or other projections.

gland A specialized cell or group of cells concerned with the secretion of various substances produced as byproducts of plant metabolism. The secretions may pass to the exterior or be contained in cavities or canals in the plant body.

GLC *See* gas–liquid chromatography.

gley (**glei**) A waterlogged soil lacking in oxygen, in which raw humus accumulates as a result of lack of decomposition by bacteria. The defining feature is the gley horizon below the humus – blue-grey clay whose color is due to ferrous iron compounds that have been reduced by microorganisms, flecked with localized areas of rust-colored oxidized ferric compounds. The formation of a gley is known as *gleying*. Gley soils are typical of tundra, meadows, and boggy areas.

global warming *See* greenhouse effect.

globulin A kind of globular protein that dissolves in dilute salt solutions but is relatively insoluble in pure water. Globulins are found in plant seeds as storage proteins, particularly in the seeds of legumes (Fabaceae).

glucan A polysaccharide made up entirely of glucose units, e.g. cellulose, starch. *See* polysaccharide.

gluconeogenesis The synthesis of glucose from noncarbohydrate precursors, such as pyruvate, amino acids, and fatty acids (which are converted to glucose by the KREBS CYCLE and GLYOXYLATE CYCLE), glycerol, and intermediates of the Krebs cycle. In plants, glucose is also formed by photosynthesis.

glucose A monosaccharide ($C_6H_{12}O_6$) occurring widely in nature as a storage compound: it occurs as glucose units in sucrose (which is hydrolyzed to glucose and fructose) and GLYCOGEN (which is hydrolyzed to glucose), as well as in STARCH and CELLULOSE. It is the main energy source of most organisms, and the starting point for synthesis of carbohydrates and nucleotide sugars. Glucose is synthesized directly from carbon dioxide in the Calvin cycle, and from fatty acids via β-oxidation to acetyl CoA, which then enters the GLYOXYLATE CYCLE (a synthetic pathway in germinating oil-rich seeds). Glucose is oxidized in respiration through glycolysis and the KREBS CYCLE. *See* glycolysis; photosynthesis.

glume *See* bract.

glutamic acid *See* amino acids.

glutamine An amino acid formed by the addition of ammonia to glutamic acid, using energy from ATP. Glutamine is important in the storage and neutralization of free ammonia. It also donates an amino group in synthesis reactions, e.g. the synthesis of purines and pyrimidines. *See* amino acids.

gluten A mixture of proteins found in wheat flour. It is composed mainly of two proteins (gliaden and glutelin), the proteins being present in almost equal quantities. Certain people are sensitive to gluten (*celiac disease*) and must have a gluten-free diet.

glycan **1.** A POLYSACCHARIDE made up of a single type of sugar unit (i.e. >95%). As a class the glycans serve both as structural units (e.g. cellulose in plants and chitin in fungi) and energy stores (e.g. starch). The most common homoglycans are made up of D-glucose units and called GLUCANS.
2. Any polysaccharide.

glyceraldehyde (CHOCH(OH)-CH_2OH) A simple monosaccharide bearing an aldehyde group. The phosphory- lated form, glyceraldehyde 3-phosphate is derived from fructose 1,6-bisphosphate during GLYCOLYSIS, and from glycerate 3-phosphate in the Calvin cycle (*See* photosynthesis). Glyceraldehyde 3-phosphate is involved in many metabolic pathways.

glycerate 3-phosphate (**phosphoglyceric acid**) The phosphorylated form of glyceric acid, a 3-carbon atom molecule. Glycerate 3-phosphate is the first product of the dark reactions of PHOTOSYNTHESIS, formed by the carboxylation and cleavage of ribulose bisphosphate. It is also formed from glyceraldehyde 3-phosphate in GLYCOLYSIS, with the release of ATP, after which it may be converted to phosphoenol pyruvate. Glycerate 3-phosphate is a precursor of the amino acid serine.

glyceride *See* acylglycerol.

glycerol (**glycerin; 1,2,3-propanetriol**) An alcohol with three OH groups. Glycerol is biologically important as the alcohol involved in LIPID formation (these particular lipids being called *acylglycerols*). It is also used to protect living tissues such as corneas and red blood cells for freezing, and as a mounting medium in microscopy. *See* acylglycerol; phosphoglyceride.

glycerolipid Lipids derived from GLYCEROL. Glycerolipids are synthesized from glycerol 3-phosphate by the addition of fatty acid chains and a head group. Plants synthesize glycerolipids in mitochondria, plastids and the endoplasmic reticulum. Glycerolipids include triacylglycerols (glycerol linked to three fatty acid chains), important storage compounds in seeds, being broken down with the aid of lipases into fatty acids, which can then undergo β-oxidation. *Compare* glycolipid.

glycine (**aminoethanoic acid; aminoacetic acid**) The simplest AMINO ACID, with the formula NH_2CH_2COOH. It may be synthesized directly from ammonia, carbon dioxide, and a donated methyl group; derived from serine; or derived from glycerate 3-phosphate by TRANSAMINATION of glyoxylate. It is broken down by a reversal of the first two pathways, or by deamination to glyoxylate, followed by conversion to malic acid and further breakdown in the Krebs cycle. Glycine is necessary for the synthesis of purines, and is a basic component of PORPHYRINS.

glycogen A polysaccharide composed of many GLUCOSE units, with a branching structure similar to that of amylopectin. It is used as a storage compound in fungi and certain bacteria as well as in animals, being hydrolyzed to glucose with the aid of amylase.

glycolipid (**glycosyldiacylglycerol**) An acylglycerol containing a carbohydrate group or an amino sugar. Glycolipids are the major lipids in chloroplasts. *See* glycerolipid.

glycolysis (**Embden–Meyerhof–Parnas pathway**) The conversion of glucose into pyruvate, with the release of some energy in the form of ATP. Glycolysis occurs in cell cytoplasm. In ANAEROBIC RESPIRATION, breakdown proceeds no further and pyruvate is converted into ethanol (or lactic acid in animals and some bacteria) for storage or elimination. In AEROBIC RESPIRATION, glycolysis is followed by the KREBS CYCLE. Glycolysis alone yields only two molecules of ATP per molecule of glucose in anaerobic respiration. In aerobic respiration there is a net yield of six (the conversion of NADH back to NAD yields a further four ATP molecules, and can occur only when oxygen is present). The main regulatory enzyme in glycolysis is phosphofructokinase, which catalyzes the phosphorylation

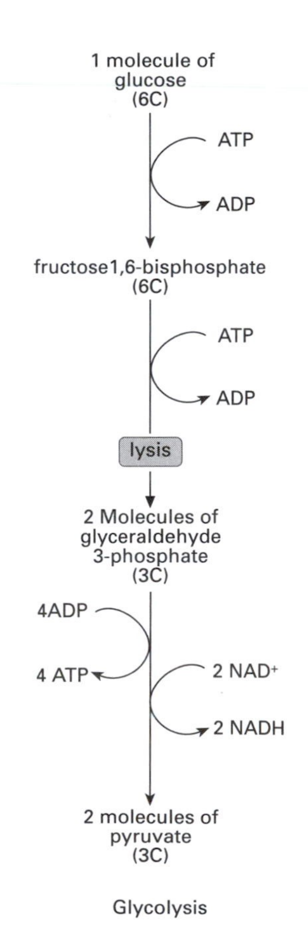

Glycolysis

of fructose 6-phosphate to fructose 1,6-bisphosphate, and is inhibited by high concentrations of ATP. *See also* respiration; alcoholic fermentation.

glycoprotein A conjugated protein formed by the combination of a protein with carbohydrate side chains. Certain antigens, enzymes, and hormones are glycoproteins. One of the most important plant glycoproteins is *extensin*, a structural component of cell walls, in which the side chains are tetrasaccharides of arabinose.

glycoside A derivative of a pyranose sugar (e.g. glucose) in which there is a group (R) attached to the carbon atom that is joined to the –CHO group. In a glycoside the C–OH is replaced by C–OR. The linkage –O– is a GLYCOSIDIC BOND, joining monosaccharides in polysaccharides. Glycosides are formed by the reaction of a pyranose sugar with an alcohol or phenol.

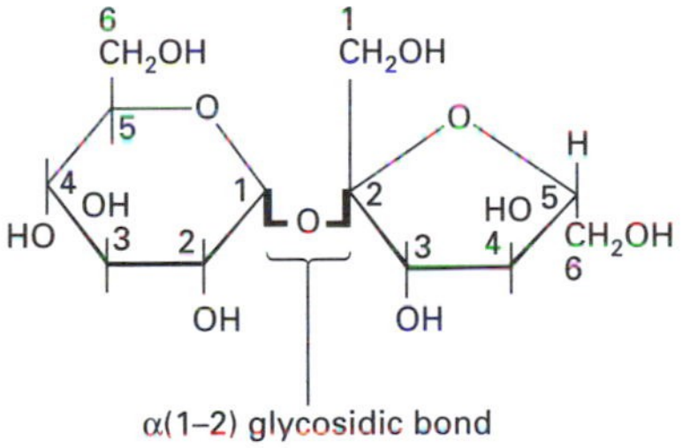

6 CH_2OH 1 CH_2OH O 5 O H 4 OH 1 O 2 HO 5 HO 3 2 3 4 CH_2OH 6 OH OH

α(1–2) glycosidic bond

Glycosidic bond: the glycosidic bond in sucrose

glycosidic bond (**glycosidic link**) A bond formed between monosaccharides by removal of water (condensation reaction). The most common glycosidic bond occurs between the carbon-1 of one sugar and the carbon-4 of the other (a 1-4 glycosidic bond); sucrose has a 1-2 glycosidic bond between the carbon-1 of glucose and carbon-2 of fructose. If the bond lies below the plane of the glucose ring, the bond is termed an α-glycosidic bond (e.g. starch), if above this plane, a β-glycosidic bond (e.g. cellulose).

glycosyldiacylglycerol *See* glycolipid.

glyoxylate cycle A modification of the KREBS CYCLE occurring in plants, protoctists, and some microorganisms, in regions where fats are being rapidly metabolized, e.g. in germinating fat-rich seeds. Acetyl groups formed from the fatty acids are passed into the glyoxylate cycle, with the eventual formation of mainly carbohydrates. The glyoxylate cycle bypasses the carbon dioxide-releasing steps of the Krebs cycle, allowing carbohydrates to be synthesized from fatty acids via succinate. See illustration overleaf.

glyoxysome *See* microbody.

Gnetophyta A phylum of gymnosperms with compound cones, cells in the secondary xylem almost identical to angiosperm vessels, and the tendency for the tip of the nucellus to be drawn out into a micropylar tube; there are no resin canals. The embryo has two cotyledons. Often more than one embryo is produced initially, but usually only one reaches maturity. There are three very different orders, each with only one genus: Gnetales (*Gnetum*), Ephedrales (*Ephedra* or joint-firs), and Welwitschiales (*Welwitschia*).

Golgi apparatus (**Golgi body; Golgi complex**) An organelle of eukaryotic cells discovered by Camillo Golgi in 1898. It is associated with the endoplasmic reticulum but lacks ribosomes. It consists of stacks of flattened membrane-bounded sacs (*cisternae*) that bud off vesicles (*Golgi vesicles*) containing the products of the secretory activity in the Golgi apparatus. The Golgi apparatus is rich in secretory cells. The cisternae are spread randomly (as *dictyosomes*) in plant cells but form a single network in most animal cells. In the cisternae, materials (e.g. enzymes and polysaccharides) are processed and leave in Golgi vesicles for transport, often to the plasma membrane for secretion. If the vesicles release their contents on the cell surface, their membranes become part of the plasma membrane; thus the Golgi apparatus contributes to cell membrane synthesis.

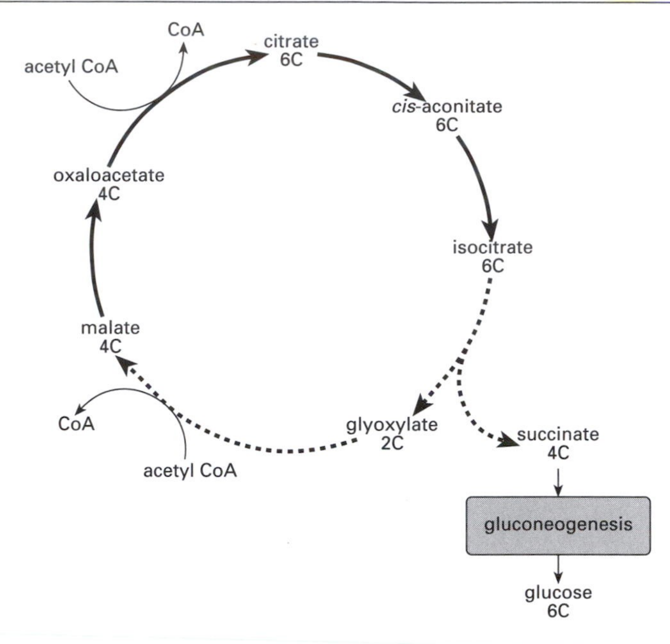

Glyoxylate cycle

graft The transplantation of an organ or tissue in plants and animals. In plants, grafting is an important horticultural technique in which an above-ground part (the *scion*) of one individual is united with the root of another (the *stock*) of the same or a different species. (Where more than two parts are grafted together, the middle piece is called the *interstock*.) Usually the shoot or bud of the scion is grafted onto the lower part of the stock. The cambia of scion and stock merge, allowing the vascular tissues of the two plants to combine. Production of callus around the graft area helps to secure the union. Grafting is usually successful only between closely related species. It is often used to propagate varieties derived from mutations or hybridization that cannot be grown from seed, especially woody species such as apples, pears, and roses.

graft hybrid *See* chimera.

grain *See* caryopsis.

Gram, Hans Christian Joachim (1853–1938) Danish bacteriologist. He devised the important method of staining bacteria that bears his name (*see* Gram's stain) in Berlin in 1884, building on the method of Paul Erlich. The discovery that some bacteria retain the stain (Gram positive), while others do not (Gram negative), has been important in identifying and classifying bacteria and also in deciding upon the treatment of bacterial diseases.

Gramineae *See* Poaceae.

Gram's stain A stain containing the basic dye crystal violet or gentian violet that forms the basis for the division into Gram positive and Gram negative bacteria. The bacteria are first killed and stained with the violet dye, then treated with an organic solvent such as acetone or ethanol. Gram positive bacteria (e.g. Actinobacteria) retain the deep purple color of crystal violet; Gram negative bacteria (e.g. Cyanobacteria) are decolorized, and may be counterstained with other dyes, such as safranin, carbol fuchsin, or neutral red. A number of physiological differences are associated with the difference in staining, including nutrition, and susceptibility to antibiotics.

granum (*pl.* **grana**) A stack of membranes (resembling a pile of coins) in a CHLOROPLAST. With the light microscope these stacks are just visible as grains (grana). The photosynthetic pigments and accessory pigments are bound to these membranes, which also contain the components of the electron-transport chain and enzymes needed for the light-dependent reactions of photosynthesis.

grape sugar *See* glucose.

grassland A major world plant community (biome) dominated by grasses (Poaceae). Grasslands occur in many temperate regions where annual rainfall is 250–500 mm and in tropical regions with 750–1500 mm. They are widespread in the interiors of continents, where there is a distinct rainy season in spring and summer. Many grasslands, especially in the wetter areas, are naturally maintained by grazing mammals and, to some extent, fire.

graticule *See* micrometer.

gravireceptor A receptor that can sense gravity. This is usually an object that moves in response to a gravitational field (a STATOLITH). Starch-filled plastids (AMYLOPLASTS) are thought to act as gravireceptors, moving in response to gravity and pressing on the endoplasmic reticulum near the outer edge of specialized cells called STATOCYTES located in the root cap or young root, and in a layer (termed the STARCH SHEATH) adjacent to the vascular tissues. Other possible gravireceptors are stretch-activated ion channels in membranes, which may sense movement of the protoplasm in response to gravity. Such channels may be involved in the response of the endoplasmic reticulum to moving amyloplasts. *See* gravitropism. *See also* photoreceptor.

gravitropism (**geotropism**) A directional growth movement of part of a plant in response to gravity. Primary roots (tap roots) grow vertically toward gravity (*positive gravitropism*) whereas primary shoots grow vertically away from gravity (*negative gravitropism*), though the direction of shoot growth may also be modified by light. Dicotyledon leaves and some stem structures (e.g. rhizomes and stolons) grow horizontally (*diatropism*). Secondary (lateral) roots and stem branches may grow at an intermediate angle with respect to gravity (*plagiotropism*). Gravitropic responses are believed to involve hormones, maybe auxins or gibberellins. *See also* gravireceptor; klinostat; tropism.

green algae *See* Chlorophyta; Gamophyta.

greenhouse effect The rise in temperature of the atmosphere, analogous to that in a greenhouse. Solar (short-wave) radiation passes easily through the atmosphere (or glass in a greenhouse) and is absorbed by the earth's surface. It is re-emitted in the form of infrared (long-wave) radiation, which is absorbed by water vapor, carbon dioxide, methane, chlorofluorocarbons (CFCs), and other so-called *greenhouse gases* in the atmosphere with a consequent increase in the atmospheric temperature. Many scientists believe that increasing atmospheric pollution by carbon dioxide, mainly due to the burning of fossil fuels, and many other gases is leading to a rise in global temperatures (*global warming*), which will eventually affect other aspects of climate and have profoundly damaging effects on natural ecosystems and agriculture, as well as raising sea level.

Grew, Nehemiah (1641–1712) British physician and botanist. Grew carried out important work in plant anatomy and was a contemporary of MALPIGHI. His work culminated in his publication of *The Anatomy of Plants*, in which he was among the first to observe chloroplasts and stomata, describing the latter as 'breathing pores'.

ground layer The lowermost layer of a plant community, usually comprising mosses, lichens, fungi, and low-growing herbs, especially rosette plants and creeping, prostrate forms. *Compare* field layer.

ground meristem The central region of the apical meristem from which the ground tissues of pith, cortex, medullary rays, and mesophyll differentiate. In root meristem it is also called *periblem.*

ground tissues The unspecialized tissues that are found in any region of the plant not occupied by the specialized tissue of vascular bundles, cambium, epidermis, etc. The pith and cortex of the root and stem are ground tissue, as are the mesophyll layers of the leaf.

growth An irreversible increase in size and/or dry weight. It excludes certain developmental processes that involve no size change, such as cleavage and uptake of water by seeds (IMBIBITION). Growth involves cell division and cell expansion through synthesis of new materials, and is closely related to subsequent developmental processes. If some measure of growth of an organism, such as height or weight, is plotted throughout its life, a characteristic S-shaped (sigmoid) *growth curve* is obtained for most organisms. In some organisms, including many plants growth never stops entirely, though it may become extremely slow (*indeterminate* or *indefinite growth*). This allows plants to continue growing and many plants grow laterally showing clonal growth. The result is that some can be extremely long-lived, with clones, such as those of the creosote bush (*Larrea tridentata*) known to have lived for over 10 000 years.

growth factor Any substance that affects the growth of a plant. This includes GROWTH SUBSTANCES (HORMONES) and GROWTH INHIBITORS, as well as food reserves and minerals.

growth inhibitor A substance that slows the growth of a plant, for example the plant hormones ABSCISIC ACID and ETHYLENE (ethene), which act at very low concentrations.

growth ring *See* annual ring.

growth substance *See* hormone.

GSC Gas–solid chromatography. *See* gas chromatography.

guanine A nitrogenous base found in DNA and RNA. Guanine has a purine ring structure. The nucleoside guanosine triphosphate (GTP) is derived from guanine, and supplies energy for certain enzyme-catalyzed metabolic reactions.

guanosine (**guanine nucleoside**) A nucleoside present in DNA and RNA and consisting of guanine linked to D-ribose via a β-glycosidic bond.

guard cell A specialized kidney-shaped epidermal cell, located to the side of a STOMA. Two guard cells together encircle each stoma and control the opening and closing of the stomatal aperture. The control is effected through changes in turgidity. The wall of the guard cell bordering the pore is heavily thickened while the opposite wall is comparatively thin. Thus when the guard cell is turgid the thin wall becomes distended, bulging out away from the pore, and causes the thickened wall, which cannot distend, to be drawn outwards with it. This results in an aperture being formed between adjacent guard cells. When osmotic pressure of the guard cell drops the pore closes. Most plants exhibit a stomatal rhythm whereby the stomata open in light and close in darkness.

gum Any substance that swells in water to form a gel or a sticky solution. Similar compounds that produce slimy solutions are called *mucilages*. Gums and mucilages are not distinguishable chemically. Most are heterosaccharides, being large, complex, flexible, and often highly branched molecules.

guttation Loss of water as liquid from the surface of a plant. Water is normally lost as vapor during transpiration but, if the atmosphere is very humid, water may also be forced from the leaves through hydathodes as a result of root pressure. Drops of guttated water are often seen at the tips of grasses in the early morning, when the low temperature reduces transpiration. It is also common in tropical plants living in environments with high humidity that reduces transpiration rates.

gymnosperm A general term for any seed plant except the angiosperms, characterized by bearing naked seeds. There are four widely different living groups: the cycads, ginkgo, conifers, gnetophytes, and several fossil groups. Gymnosperms lack endosperm and the precise double fertilization characteristic of the angiosperms, although some show a primitive form of it. Cycads and ginkgo have motile sperm, unlike gnetophytes and conifers, which produce pollen. The female gametophyte is reduced, but larger than in angiosperms; it is not autotrophic. Most gymnosperms retain archegonia except for the most advanced genera *Gnetum* and *Welwitschia*. These genera and *Ephedra* also have advanced tracheids with a structure reminiscent of angiosperm vessels, whereas in other gymnosperms the xylem is composed solely of more primitive tracheids.

gynoecium (*pl.* **gynoecia**) The CARPEL or carpels of a flower. The term pistil or compound pistil is also used to mean one or many carpels respectively. A gynoecium with only one carpel is called *monocarpellary*; with two or more separate carpels, *apocarpous*; and with fused carpels *syncarpous*. The gynoecium is denoted by the letter G in the FLORAL FORMULA. *Compare* androecium. *See* ovary.

gynomonoecious Denoting a species that has female and hermaphrodite flowers on the same plant. *Compare* andromonoecious.

H

habitat The place where a particular organism lives, described in terms of its climatic, vegetative, topographic, and other relevant factors.

hair (**trichome**) An outgrowth from a plant epidermal cell. Plant hairs may take a wide variety of forms, often characteristic of particular species or genera. They may be unicellular or multicellular, simple, forked, stellate (starlike), downy, silky, long, or short. Some, such as the hairs on the leaves of geraniums (*Pelargonium*) are glandular. *See also* root hair.

halophyte A plant that grows in soils with a high concentration of salt, as found in salt marshes or other littoral habitats. Examples are species of *Spartina*. Halophytes are adapted to obtain water from soil water with a higher osmotic pressure than normal soil water, so they need to be able to accumulate a high concentration of salts in their root cells. Many have a succulent growth form with swollen stems or leaves.

halosere A plant community in a SUCCESSION that starts on land periodically inundated by the sea. Haloseres include salt- and flooding-tolerant pioneer communities on mudflats and saltmarshes.

haplobiontic Describing life cycles in which only one type of somatic body is formed, which may be either haploid or diploid, i.e. either the sporophyte or the gametophyte generation is missing. For example, in the unicellular *Chlamydomonas* the vegetative cell is haploid, and the only diploid cells are the zygospores, which germinate to produce haploid zoospores. In diatoms the vegetative cells are diploid, and are formed directly by the fusion of gametes. Haplobiontic is thus a collective term for haplontic and diplontic. *Compare* diplobiontic. *See* alternation of generations.

haploid (**monoploid**) Describing an organism, cell, or nucleus containing only one representative from each of the pairs of homologous chromosomes found in the normal DIPLOID cell. Haploid chromosomes are thus unpaired and the haploid chromosome number (*n*) is half the diploid number (2*n*). Meiosis, which usually precedes spore or, sometimes, gamete formation, halves the chromosome number to produce haploid gametes. The diploid condition is restored when the nuclei of two gametes fuse to give the zygote. Gametes may develop without fertilization, or meiosis may substantially precede gamete formation, leading to the formation of haploid organisms, or haploid phases in the life cycles of organisms.

haplont A haploid organism that represents the vegetative stage in life cycles in which diploidy is restricted to the zygote. *Haplontic* life cycles are typical of the filamentous green algae. *Compare* diplont.

haplostele A type of protostele having a solid strand of stele with the xylem on the inside encircled by phloem, pericycle, and endodermis. This type of stele was present in the earliest known vascular plants, such as *Rhynia*.

hapteron (*pl.* **haptera**) **1.** *See* holdfast. **2.** In species of *Equisetum* (horsetails), the outer wall of the spore, which forms an X-shaped structure, the arms of which are wrapped around the spore. On drying out, the arms, or *haptera*, act in similar manner to the elaters of liverworts, and are thus termed elaters. *See* elater.

Haptomonada (**Haptophyta**) A phylum of the PROTOCTISTA that contains mostly unicellular algae, most of which have a threadlike structure called a *haptonema* between their two undulipodia. The haptonema is thought to be used for temporary anchorage rather than for locomotion and may be coiled while the cell is motile. Haptomonads are common in marine plankton and freshwater.

haptonasty (**haptonastic movements**) A NASTIC MOVEMENT in response to contact. An example is the sensitive plant (*Mimosa pudica*), whose leaflets progressively collapse, sometimes throughout the whole plant, once one leaflet is touched. *See* pulvinus.

haptophyta *See* Haptomonada.

haptotropism *See* thigmotropism; tropism.

hardpan 1. A hardened soil horizon, usually in the middle or lower part of a soil profile, A hardpan is typical of PODSOLS, where iron compounds leached from the upper layers accumulate in the B horizon, staining it red-brown and forming an impermeable layer (*iron pan*) that can prevent drainage, resulting the in waterlogging of higher horizons.
2. Also called *calcrete*, a hardened layer in or on a soil formed on calcareous substrata, formed in arid and semiarid regions as a result of fluctuating climatic conditions by calcite that precipitates out of solution as water evaporates at the surface.

Hardy–Weinberg equilibrium The situation in a large, randomly mating, stable population under specific conditions, in which the proportion of dominant to recessive genes remains constant from one generation to the next. If a pair of alleles, A and a, have the frequencies p and q in a population, then the Hardy-Weinberg equilibrium is described by the equation $p^2 + 2pq + q^2 = 1$, where p^2 and q^2 are the frequencies of the double dominant and double recessive respectively, and $2pq$ is the frequency of the heterozygote. The *Hardy–Weinberg equation* $p^2 + 2pq + q^2 = 1$ is obtained from the expansion of $(p + q)^2$, the total of the frequencies making up the gene pool being unity. The law was formulated in 1908 and disproved the then current theory that dominant genes always tend to increase in a population at the expense of their equivalent recessive alleles. The equilibrium holds only when the population is sufficiently large to avoid chance fluctuations of allele frequencies in the gene pool (GENETIC DRIFT) and providing there is no mutation, natural selection, or migration into or out of the population. The fact that allele frequencies may be seen to change fairly rapidly in large populations that show minimal mutation and migration emphasizes the important role natural selection must play. Until the Hardy–Weinberg law was formulated, the extent of natural selection was not fully appreciated.

Hartig net *See* ectomycorrhiza.

Hatch–Slack pathway *See* C_4 plant.

haustorium (*pl.* **haustoria**) A specialized outgrowth of many parasitic fungi and certain parasitic plants such as dodder (*Cuscuta*) and broomrape (*Orobanche*), that penetrates into and withdraws food material from the cells of the host plant.

heartwood (**duramen**) The hard central region of a tree trunk made up of xylem vessels that are no longer involved in water transport. Such vessels are often blocked by substances (e.g. resins and tannins) that give the wood a darker color. *Compare* sapwood.

heath An area dominated by dwarf shrubs of the heath family (Ericaceae). Legumes (family Fabaceae) may also be common components of heath floras. Heaths normally develop on poor, sandy, well-drained soils in temperate climates, and are usually subclimax communities maintained by grazing, burning, and sometimes, cutting.

heavy metal tolerance Biochemical and physiological adaptation to concentrations of heavy metals, such as copper, lead, and zinc, that would prevent the growth of most plant species or genotypes. Such adaptation usually involves converting the metal to a harmless form; some species exclude the metal from the plant altogether, while others confine it to the roots, thus protecting the more sensitive shoots. Certain species or strains have evolved such tolerance, and some, for example certain strains of bent grass (*Agrostis capillaris*), benefit from the reduced competition in soil containing heavy metal ions.

helical thickening *See* spiral thickening.

helicase Any of a group of enzymes that catalyze the unwinding of double-stranded nucleic acids such as DNA, using energy from ATP.

heliophyte A plant typical of sunny habitats.

heliotropism A form of PHOTOTROPISM in which flowers move to point toward the sun, tracking it as it moves across the sky. It is a daily movement, mainly in spring flowers that are pale and cup-shaped like wood anemones and buttercups.

helophyte A perennial marsh plant that has its overwintering buds under water. An example is bulrush (*Typha*).

hematoxylin A blue dye derived from the logwood tree (*Haemotoxylon campechianum*), which when oxidized to hematin stains nuclei and cellulose cell walls blue. *See* staining.

heme (**haem**) An iron-containing porphyrin that is the prosthetic group in cytochromes. *See* porphyrins.

hemicellulose One of a group of substances that make up the amorphous matrix of plant cell walls together with pectic substances (and occasionally, in mature cells, with lignin, gums, and mucilages). They are *heteropolysaccharides*, i.e. POLYSACCHARIDES built from more than one type of sugar, mainly the hexoses (mannose and galactose) and the pentoses (xylose and arabinose). Galacturonic and glucuronic acids are also constituents. They vary greatly in composition between species. The hemicelluloses and pectic substances interact with the cellulose microfibrils of the cell wall to give extra strength and at the same time retain some plasticity. In some seeds (e.g. the endosperm of dates) hemicelluloses are a food reserve. *See* cell wall.

hemicryptophyte A perennial plant, usually nonwoody, with its over-wintering buds at or just below soil level. The buds are protected by the soil or surface litter. *See also* Raunkiaer's plant classification.

hemiparasite A parasitic plant that has chlorophyll and can photosynthesize, but depends on another plant for its water and mineral salts, e.g. mistletoe.

hemizygous Describing genetic material that has no homologous counterpart and is thus unpaired in the diploid state. Both single genes and chromosome segments may be hemizygous; for example, the X chromosome in the heterogametic sex, and whole chromosomes in aneuploids.

Hepatophyta (**Marchantiopsida; liverworts**) A phylum containing leafy and prostrate thallose forms, commonly known as liverworts. Thallose liverworts are prostrate and dichotomously branched with thalli several cells thick. There are five orders of liverworts, including the Metzgeriales, e.g. *Pellia*, and the Marchantiales, e.g. *Marchantia*.

herbaceous perennial *See* perennial.

heritability The proportion of phenotypic variation due to genetic factors. It can be estimated from measurements of individuals from different generations, and is used in plant and animal breeding to predict how successful genetic selection will be in improving a particular trait.

hermaphrodite (**bisexual**; **monoclinous**) A plant bearing stamens and carpels in the same flower. In many such plants, self-fertilization and inbreeding are prevented by specific self-incompatibility systems. *Compare* monoecious; dioecious.

hesperidium (*pl.* **hesperidia**) A type of berry with a leathery epicarp, as in oranges and lemons. Each carpel forms a segment of the fruit, filled with fluid-filled trichomes. *See* berry.

heteroecious Denoting rust fungi that require two host species to complete their life cycle. An example is *Puccinia graminis*, the stem rust of cereals and grasses, which overwinters on the barberry. *Compare* autoecious.

heterogametic sex The sex with dissimilar SEX CHROMOSOMES, one (in mammals the Y chromosome) being shorter than the other (the X chromosome).

heterogamy *See* anisogamy.

heterokaryosis The presence of two or more nuclei with differing genotypes within a single cell (termed a *heterokaryon*). Fusion of fungal hyphae of differing genetic complements leads to heterokaryosis. Heterokaryons often grow better than homokaryons, since any genetic deficiencies, for example in the ability to synthesize certain compounds, in one type of nucleus may be overcome by alleles present in the other nuclei.

heteromorphism The existence of more than one form, used especially with reference to life cycles in which the alternating generations are markedly different morphologically, as in ferns, bryophytes, and algae. *Compare* isomorphism; polymorphism. *See* alternation of generations.

heterophylly Denoting plant species that have more than one form of foliage leaf on the same individual. Examples include certain species of *Lycopodium* and *Selaginella*, which have two rows of lateral leaves and one or two rows of smaller abaxial or adaxial leaves.

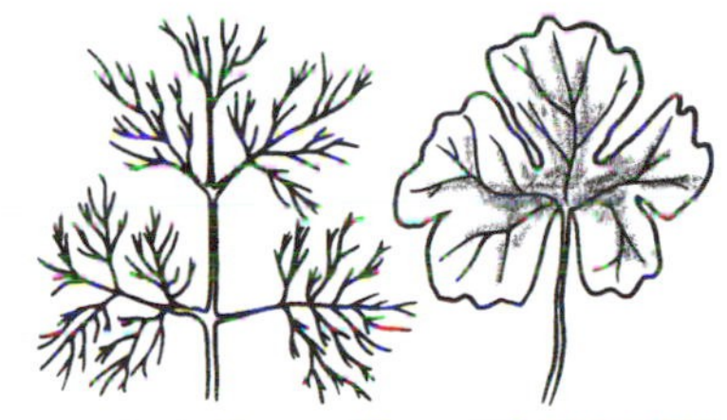

water buttercup (*Ranunculus aquatilis*)

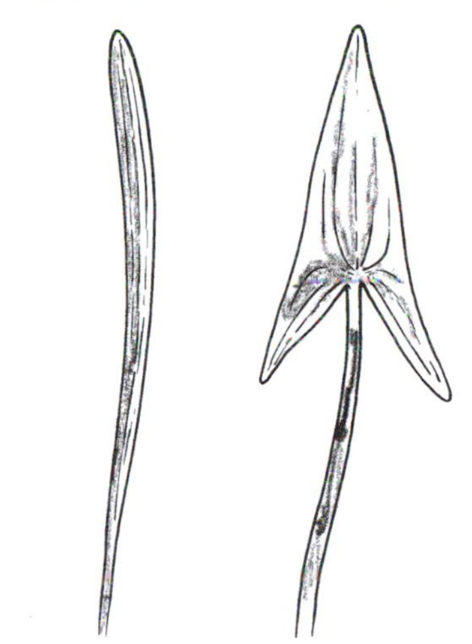

arrowhead (*Sagittaria sagittifolia*)

Heterophylly: examples of heterophylly in two different aquatic plants

heterosis (**hybrid vigor**) The condition in which the expression of a characteristic is greater in the heterozygous offspring than in either of the homozygous parents. The effect arises from an accumulation of dominant genes in the F_1. Thus, if height is controlled by two genes, A and B, and tall and short forms are determined by dominant and recessive alleles respectively, then the cross AAbb × aaBB would give an F_1 AaBb, containing both dominant genes for tallness. Usually the more unlike the parents are the more hybrid vigor is released, but the effect diminishes in subsequent generations as more recessive homozygotes reappear. For crop plants, hybrid vigor is particularly important for characteristics such as growth rate, yield, and disease resistance.

heterospory The production of two different sizes of spore: MICROSPORES and MEGASPORES. The larger megaspores usually contain more food reserves. *Compare* homospory.

heterostyly A DIMORPHISM in which the styles of flowers of the same species are of different lengths, thus dividing the species into groups. An example is the primrose, which is divided into pin-eyed and thrum-eyed types. Pin-eyed plants have long styles and short stamens and thrum-eyed plants conversely have short styles and long stamens. Such differences promote pollination between groups.

heterothallism A condition found in algae and fungi in which sexual reproduction occurs only between genetically different self-incompatible mating types (strains) of the same species. The strains may vary morphologically or in the size of the gametes they produce. When heterothallism is purely physiological, morphologically identical strains exist (often designated plus and minus strains). *Compare* homothallism.

heterotrophism A type of nutrition in which the principal source of carbon is organic, i.e. the organism cannot make all its own organic requirements from inorganic starting materials. Most heterotrophic organisms are chemotrophic (i.e. show *chemoheterotrophism*); these comprise all animals and fungi, most bacteria, and parasitic plants. A few heterotrophic organisms are phototrophic (i.e. show *photoheterotrophism*). The nonsulfur purple bacteria, for instance, require organic molecules such as ethanol and acetate. *Compare* autotrophism. *See* chemotrophism; phototrophism.

heterozygous Having two different alleles at a given locus. Usually only one of these, the dominant allele, is expressed in the phenotype. On selfing or crossing heterozygotes some double recessives may appear, giving viable offspring. Selfing heterozygotes halves the heterozygosity, and thus outbreeding maintains heterozygosity and produces a more adaptable population. *Compare* homozygous.

hexose A SUGAR that has six carbon atoms in its molecules.

hexose monophosphate shunt *See* pentose phosphate pathway.

high-energy phosphate bond A phos-

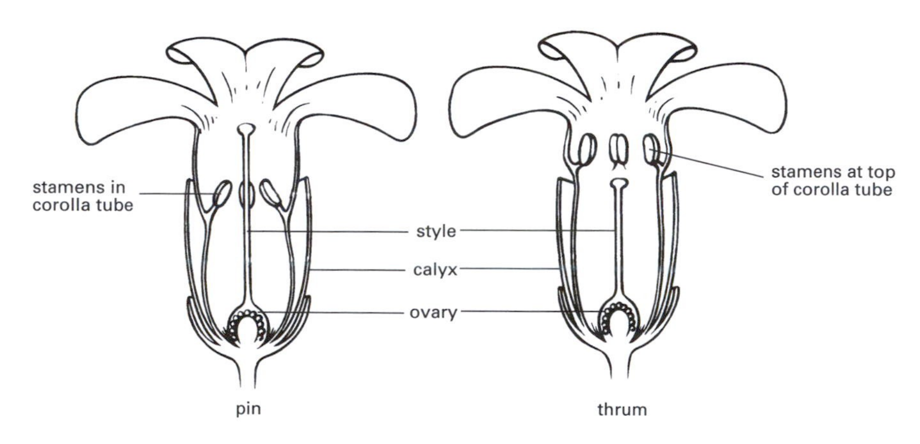

Heterostyly: example in *Primula*

phate linkage that hydrolyzes spontaneously, releasing energy, for example the last two phosphate bonds of ATP. ATP is the main form in which chemical energy is transported from one part of the cell to another.

Hill reaction The reaction, first demonstrated by Robert Hill in 1937, by which isolated illuminated chloroplasts bring about the reduction of certain nonphysiological electron acceptors with accompanying evolution of oxygen. For example, the blue dye dichlorophenol indophenol (DCPIP) may be reduced to a colorless substance. The reaction involves part of the normal light reaction of photosynthesis. Electrons from water involved in noncyclic photophosphorylation are used to reduce the added substance. It provided support for the idea that a light reaction preceded reduction of carbon dioxide in photosynthesis. *See* cyclic photophosphorylation.

hilum (*pl.* **hila**) **1.** A scar on the testa of a seed marking the point at which it was attached to the ovary wall by the funicle. It is a feature that distinguishes seeds from fruits.
2. The center of a starch grain around which the layers of starch are deposited.
3. A small projection at the base of a basidiospore near its attachment to the sterigma.

hip A type of PSEUDOCARP (false fruit) consisting of a cup-shaped receptacle containing achenes, each of which bears small hooklike hairs. Hips are typical of the genus *Rosa*.

histidine A basic AMINO ACID found only rarely in proteins. It is formed from ATP and 5-phosphoribosyl pyrophosphate in a complex biosynthetic pathway. Glutamate is an intermediate in its breakdown.

histochemistry The location of particular chemical compounds within tissues by the use of specific staining techniques, for example phloroglucinol to stain lignin.

histogen theory A theory, proposed by Hanstein, in which the apical meristem is considered to consist of three main zones, the *dermatogen*, *periblem*, and *plerome*, which differentiate into the epidermis, cortex, and stele respectively. This concept has now been replaced, for stem apices, by the tunica–corpus theory. In roots, however, the concept is still applied, and in some angiosperm roots a fourth histogen zone is recognized, the *calpytrogen*, which gives rise to the root cap. *Compare* tunica–corpus theory.

histology The study of tissues and cells at microscopic level.

histone One of a group of relatively small proteins found in chromosomes of eukaryotic cells, where they organize and package the DNA. When hydrolyzed, they yield a large proportion of basic amino acids, as they are rich in the basic amino acids arginine and lysine. They dissolve readily in water, dilute acids, and alkalis but do not coagulate readily on heating.

hnRNA (**heterogeneous nuclear RNA**) *See* messenger RNA.

Hofmeister, Wilhelm Friedrich Benedict (1824–77) German botanist who first demonstrated the ALTERNATION OF GENERATIONS in lower plants. Hofmeister also showed that the gametophytes and sporophytes of seed-bearing plants were homologous with those of ferns and mosses and that the gymnosperms formed a plant group between the latter and the angiosperms.

holdfast (**hapteron**) In algae, the cell or organ that attaches the plant to the substrate. It is often disklike, sometimes divided into many fingerlike processes, and is particularly prominent in the brown algae (e.g. *Fucus*).

Holocene (**Recent**) The present epoch in the geological time scale, being the second epoch of the Quaternary period, dating from the end of the last glaciation, about 10 000 years ago, to the present day.

holoenzyme A catalytically active complex made up of an APOENZYME and a COFACTOR (nonprotein prosthetic group). The former is responsible for the specificity of the holoenzyme whilst the latter determines the nature of the reaction. Some holoenzymes, such as pyruvate dehydrogenase, are highly complex, with several cofactors. *See* coenzyme.

holophytic The type of nutrition in which complex organic molecules are synthesized from inorganic molecules using light energy. It is another term for *phototrophic.*

holozoic (**heterotrophic**) Designating organisms that feed on other organisms or solid organic matter, i.e. insectivorous and parasitic plants, fungi, many bacteria and most animals. *Compare* holophytic.

homeobox A segment of DNA found in many so-called homeotic genes concerned with controlling the development of organisms. It consists of 180 base pairs, and the sequence of bases is remarkably similar across a wide range of eukaryotic species, from yeasts to human beings. This suggests it arose early in evolutionary time and has been little changed since. The sequence encodes the amino acids of a peptide sequence (the *homeodomain*) that enables the parent protein to bind to DNA. This is consistent with the suggested role of homeotic proteins as genetic switches, binding to genes to control their expression. *See* differentiation.

homeotic gene (**homeogene**) Any of a class of genes that are crucial in determining the DIFFERENTIATION of tissues in different parts of the body during development. They encode proteins that regulate the expression of other genes by binding to DNA. This binding capability can be pinpointed to a characteristic base sequence known as a HOMEOBOX. Such genes often occur in complexes in which the physical order of the genes corresponds to the order in which they are expressed from anterior to posterior in the developing embryo.

homogametic sex The sex with homologous SEX CHROMOSOMES, in mammals, designated XX.

homogamy The condition in flowers in which the anthers and stigmas ripen at the same time, so encouraging self-pollination. Homogamy occurs in the closed cleistogamous flowers that appear late in the season in certain plants. *Compare* dichogamy.

homologous **1.** Describing structures that, though in different species, are believed to have the same origin in a common ancestor. Thus the stamens of angiosperms and the MICROSPOROPHYLLS of lower vascular plants are said to be homologous, being derived from a leaflike structure bearing the MICROSPORANGIA (anthers). *Compare* analogous.
2. *See* isomorphism.

homologous chromosomes Chromosomes that pair at meiosis. Each carries the same genes as the other member of the pair but not necessarily the same alleles for a given gene. One member of each pair of maternal origin, the other of paternal origin. With the exception of the sex chromosomes (X and Y) they are morphologically similar, being of similar size and shape with centromeres at the same location. During the formation of the germ cells only one member of each pair of homologs is passed on to the gametes. At fertilization each parent contributes one homolog of each pair, thus restoring the diploid chromosome number in the zygote.

homospory The production of only one kind of asexual spore, which then develops into a hermaphrodite gametophyte. In the vascular plants the condition is seen in most ferns, psilophytes, and *Lycopodium* species. *Compare* heterospory.

homostyly The usual condition found in plants in which the styles of all flowers of the same species are about the same length. *Compare* heterostyly.

homothallism A condition found in some algae and fungi in which each thallus is self-compatible. Homothallic species may produce distinctly different sizes of gametes that can fuse with each other and are thus effectively hermaphrodite. *Compare* heterothallism.

homozygous Having identical alleles for any specified gene or genes. A homozygote breeds true for the character in question if it is selfed or crossed with a similar homozygote. An organism homozygous at every locus produces offspring identical to itself on selfing or when crossed with a genetically identical organism. Homozygosity is obtained by inbreeding, and homozygous populations may be well adapted to a certain environment, but slow to adapt to changing environments. *Compare* heterozygous. *See* pure line.

honey guides Lines or spots on petals that direct a pollinating insect to the nectaries.

Hooke, Robert (1635–1703) English physicist. In 1665 he published his *Micrographia*, a beautifully illustrated book that described both his theories of the nature of light and the investigations he had made with his improved microscope. While observing cork, he described it as resembling a honeycomb and was the first to refer to the cavities as 'cells'.

Hooker, Sir Joseph Dalton (1817–1911) Plant taxonomist and explorer. Son of the British botanist William Jackson Hooker (1785–1865), the first director of Kew Gardens, Hooker studied medicine at Glasgow University. His famous voyage to the Antarctic (1839–43) on HMS *Erebus* resulted in a six-volume flora of Antarctica, New Zealand, and Tasmania that attracted the attention of Charles DARWIN. Hooker was one of those in whom Darwin confided his theory and was instrumental in encouraging Darwin to present his joint paper with Wallace on the theory of evolution by natural selection.

In 1865 he succeeded his father as director of Kew Gardens, which became an international center for botanical research. With George BENTHAM, he produced the world flora *Genera Plantarum*.

hormone **(growth substance; plant hormone; phytohormone)** One of a group of essential organic substances produced in plants. They are effective in very low concentrations and control growth, development, and responses to external stimuli by their interactions. They affect physiological processes, but are not actually part of major metabolic pathways. The same hormone may have many different effects, depending upon the balance between the concentrations of different hormones, the particular tissues concerned, and the age of the plant. *See* auxin; gibberellin; cytokinin; abscisic acid; ethylene.

horsetails *See* Sphenophyta.

host An organism used as a source of nourishment by another organism, the parasite, which lives in or on the body of the host. *See* parasitism.

humidity The moisture content of the atmosphere. The *absolute humidity* is the amount of water vapor present in a unit volume of air (usually 1 m^3 of air). The *relative humidity* is the ratio of the amount of water vapor actually present in the air to the greatest amount that the air could hold at the same temperature, and is usually expressed as a percentage.

humus The nonliving finely divided organic matter in soil derived from the decomposition of animal and plant substances by soil bacteria. Humus consists of 60% carbon, 6% nitrogen, and small amounts of phosphorus and sulfur, and is valued by horticulturalists and farmers as it improves the fertility, water-holding capacity, and workability of the soil. Humus has colloidal properties that enable it to retain water, so it can improve the moisture content of sandy soils. It aids the formation of crumbs in the soil, and is often added to clay soil to increase the particle size, promoting drainage and aeration.

hyaline cells Any leaf cell with no chlorophyll that is normally used for storage of water or solutes.

hyaloplasm (**cell matrix; ground substance**) *See* cytoplasm.

hybrid An organism derived from crossing genetically dissimilar parents. Thus most individuals in an outbreeding population could be called hybrids, and geneticists apply the term to the F_1 generation resulting from a monohybrid cross, where the offspring differ in the alleles of a single gene. However, the term is usually reserved for the product of a cross between individuals that are markedly different. If two different species are crossed, the offspring is often sterile, owing to the nonpairing of the chromosomes necessary for gamete formation. In plants this is sometimes overcome by the doubling of the chromosome number, giving an allopolyploid. By contrast, hybrids derived from different varieties of the same species are often more vigorous than their parents (*hybrid vigor*), and are selected and propagated by vegetative means by agriculturists and horticulturists. *See* chimera; heterosis.

hybrid sterility The reduced ability of some HYBRIDS to produce viable gametes. It is due to the lack of homologous pairs of chromosomes, so chromosomes cannot pair in meiosis, and the resulting gametes may have extra or missing chromosomes. Hybrid sterility is sometimes overcome by polyploidy. *Compare* heterosis.

hybrid vigor *See* heterosis.

hydathode A specialized secretory structure involved in the removal of excess water from plants. It may be a modified stoma with the guard cells permanently open so that water is lost passively by hydrostatic pressure, as in sea lavenders (*Limonium*), or a glandular hair from which water is actively secreted, as in runner bean (*Phaseolus*). Hydathodes are found at the leaf tips or along the leaf margins.

hydrocarbon An organic compound that contains carbon and hydrogen only, e.g. acetylene, butane. Many of the components of natural gas, petroleum, and coal are hydrocarbons.

hydrogen An essential element in living tissues. It enters plants, with oxygen, as water and is used in building up complex reduced compounds such as carbohydrates and fats. Water itself is an important medium, making up 70–80% of the weight of organisms, in which chemical reactions of the cell can take place. Hydrogenated compounds, particularly fats, are rich in energy and on breakdown release energy for driving living processes.

hydrogen bond A type of bond occurring between molecules. Hydrogen bonding takes place between oxygen, nitrogen, or fluorine atoms on one molecule, and hydrogen atoms joined to oxygen, nitrogen, or fluorine on the other molecule. The attraction is due to electrostatic forces. Hydrogen bonding is responsible for the properties of water. It is important in many biological systems for holding together the structure of large molecules, such as proteins and DNA.

hydrolase An enzyme that catalyzes a HYDROLYSIS reaction. Glycosidases, peptidases, and phosphatases are examples. Hydrolases play an important part in rendering insoluble food material into a soluble form, which can then be transported in solution. They are important in the mobilization of stored food, and in the uptake of nutrients by fungi and other heterotrophic organisms.

hydrolysis The breakdown of a compound due to a reaction with water (H^+ or OH^- ions).

hydrophilic Describing a molecule or surface that has an affinity for water. Such molecules are usually polar, for example proteins. *Compare* hydrophobic.

hydrophily Pollination in which water carries the pollen from anther to stigma.

The pollen may be transported on the water surface or through the water. Pollen that floats is light and water repellent, and may have a coating of oil that alters the surface tension of the water.

hydrophobic Describing a molecule or surface that has no affinity for water. Such molecules are nonpolar, e.g. benzene. *Compare* hydrophilic.

hydrophyte A plant found growing in water or in extremely wet areas. Hydrophytes show certain adaptations to such habitats, notably development of aerenchyma, reduction of cuticle, root system, and mechanical and vascular tissues, and divided leaves. Large intercellular air spaces in leaves, stems, and roots allow oxygen to diffuse through the plant. Hydrophytes with floating leaves can exchange gases with the atmopshere; many submerged hydrophytes, such as spiked water milfoil (*Myriophyllum spicata*) have no stomata, absorbing water and gases over their entire surface. *Compare* mesophyte; xerophyte.

hydroponics (**water culture**) The growth of plants in liquid culture solutions rather than soil. The solutions must contain the correct balance of all the essential mineral requirements. The method is used commercially, especially for glasshouse crops, and also in experimental work in determining the effects of mineral deficiencies.

hydrosere Any plant community in a SUCCESSION that starts in fresh water.

hydrotropism (**hydrotropic movement**) A TROPISM in which the stimulus is water. It is a special kind of chemotropism. Roots are positively hydrotropic and hypocotyls negatively hydrotropic. If water is in short supply, the stimulus of water is stronger than the stimulus of gravity in determining response.

hygroscopic Able to absorb water from the surroundings, including moisture from the atmosphere.

hymenium (*pl.* **hymenia**) A layer of the fruiting body of certain ascomycete and basidiomycete fungi in which the asci or basidia are borne. The hymenium may be directly exposed to the air, as in the gills of the mushroom (a basidiomycete) or may open into a flask-shaped cavity, as in the perithecia of certain ascomycetes.

hyperplasia Enlargement of a tissue due to an increase in the number of its cells. In plants this usually occurs in response to a disease-causing organism.

hypertonic Designating a solution with an osmotic pressure greater than that of a specified other solution, the latter being hypotonic. When separated by a selectively permeable membrane (e.g. a cell membrane) water moves by osmosis into the hypertonic solution from the hypotonic solution. *Compare* hypotonic; isotonic.

hypertrophy Enlargement of a tissue or organ due to an increase in the size of its cells. An example is the swelling of roots of plants from the Brassicaceae family infected with club root (*Plasmodiophora brassicae*). *Compare* hyperplasia.

hypha (*pl.* **hyphae**) In fungi, a fine nonphotosynthetic tubular filament that spreads to form a loose network termed a *mycelium* or aggregates into fruiting bodies (e.g. toadstools). Hyphae may be branched or unbranched and may or may not have cross walls (*septa*) dividing them into cells. Many fungi have incomplete septa. Hyphae are parasitic or saprophytic and the tips secrete enzymes to digest and penetrate the food supply. The hyphal walls of most species of fungi differ from those of plants in being composed of microfibrils of a nitrogenous compound called chitin or a form of fungal cellulose. Hyphae also differ in lacking plastids, any pigment being contained in the walls, cytoplasm, or oil globules.

hypocotyl The stem below the cotyledons, occupying the region between the cotyledon stalks and the point where lateral roots arise. Rapid elongation of the hypocotyl after germination pushes the cotyledons above ground in plants showing epigeal germination. *Compare* epicotyl.

hypodermis One or more layers of cells that may be found immediately below the epidermis of plants. It may be composed of thin-walled colorless cells and functions as water-storing tissue as in certain succulent leaves and the aerial roots of epiphytes.

hypogeal germination Germination in seed plants during which the cotyledons remain underground, as in broad bean (*Vicia faba*). The cotyledons thus act only as food storage organs and not as photosynthetic organs and the hypocotyl does not elongate. *Compare* epigeal germination.

hypogyny The simple arrangement of flower parts in which the receptacle is expanded at the top of the pedicel in such a way that the androecium and the perianth arise from beneath the gynoecium giving a superior ovary. *Compare* epigyny; perigyny.

hypolimnion The lower noncirculating water in a thermally stratified lake.

hyponasty (**hyponastic movements**) The curving of a plant organ upward and toward the axis, caused by greater growth on the lower side. *See* nastic movements.

hypostasis The situation in which the expression of one gene (the *hypostatic gene*), is prevented in the presence of another, nonallelic, gene (the *epistatic gene*).

hypotonic Designating a solution with an osmotic pressure less than that of a specified other solution, the latter being hypertonic. When separated by a selectively permeable membrane (e.g. a cell membrane) water is lost by osmosis from the hypotonic to the hypertonic solution. *Compare* hypertonic; isotonic.

I

IAA (indole acetic acid) A naturally occurring auxin. *See* auxin.

ICBN *See* International Code of Botanical Nomenclature.

Ice Age A period in the latter part of the Pleistocene characterized by successive coolings and warmings of the earth. In at least four major glaciations (cold periods), ice caps spread south from the Arctic and north from the Antarctic.

idioblast Any specialized plant cell that is dispersed among cells of a very different kind. Idioblasts may contain a variety of materials, e.g. tannins, oils, crystals, and waste products. Isolated sclereids can be called idioblasts.

idiogram *See* karyogram.

imbibition The phenomenon in which a substance absorbs a liquid and swells, but does not necessarily dissolve in the liquid. The process is reversible, the substance contracting on drying. Water is imbibed by many biological substances: cellulose, hemicelluloses, pectic substances, lignin (all plant cell wall constituents); starch; certain proteins, especially in seeds; etc. Dry seeds absorb water by imbibition, initially via the testa, and as seed volume increases great imbibitional pressures develop.

immunity The ability of plants and animals to withstand harmful infective agents and toxins. In most plants this is achieved by physical barriers preventing entry of pathogens and by physiological reactions to isolate the pathogen and its effects.

immunoelectrophoresis *See* electrophoresis.

inbreeding Breeding between closely related individuals. The most extreme form of inbreeding is self-fertilization, which occurs in some plants and algae (e.g. *Spirogyra*). Inbreeding increases homozygosity so that deleterious recessive genes are expressed more often in the phenotype, and decreases heterozygosity and hence the potential genetic variability of the population. There is also a general lowering of vigor in inbred stock (*inbreeding depression*), which is especially pronounced amongst normally outbreeding populations. *Compare* outbreeding. *See* cleistogamy.

incipient plasmolysis The condition of a cell that results when it is surrounded by a solution having the same osmotic pressure as the cell contents. *See* plasmolysis.

incompatibility **1.** The rejection of grafts, transfusions, or transplants between plants or animals of different genetic composition.
2. A mechanism in flowering plants that prevents fertilization and development of an embryo following pollination by the same or a genetically identical individual. It results in self-sterility, thus preventing inbreeding. It is due to interaction between genes in the pollen grain and those in the stigma, in such a way that the pollen is either unable to grow or grows more slowly on the stigma.
3. A genetically determined mechanism in some fungi that prevents sexual fusion between individuals of the same race or strain (e.g. *Mucor*). *See also* heterothallism.

incomplete dominance (**co-dominance**) The situation in which two different alleles are equally dominant. If they occur together the resulting phenotype is intermediate between the two respective homozygotes. For example, if white antirrhinums (AA) are crossed with red antirrhinums (A′A′) the progeny (AA′) will be pink. Sometimes one allele may be slightly more dominant than the other (*partial* or *incomplete dominance*) in which case the offspring, though still intermediate, will resemble one parent more than the other.

indefinite growth (**indeterminate growth**) A form of GROWTH that has no finite limit: the plant or organ continues to grow indefinitely. All plants are somewhat indeterminate, as they have meristems that can differentiate into almost any part given the right conditions.

indehiscent Describing a fruit or fruiting body that does not open at maturity to release the seeds or spores. The fruit wall either decays releasing the seeds *in situ*, or the fruit is eaten by an animal and the seeds pass out intact in its droppings. *Compare* dehiscent.

independent assortment The law, formulated by MENDEL (Mendel's Second Law), that genes segregate independently at meiosis so that any one combination of alleles is as likely to appear in the offspring as any other combination. The work of T. H. Morgan later showed that genes are linked together on chromosomes and so tend to be inherited in groups. The law of independent assortment therefore applies only to genes on different chromosomes. The term can also be applied to whole chromosomes. *See* linkage; Mendel's laws.

indeterminate growth *See* indefinite growth.

indicator A substance used to test for acidity or alkalinity of a solution by a color change. Examples are litmus and phenolphthalein. A *universal indicator* shows a range of color changes over a wide range from acid to alkaline, and can be used to estimate the pH.

indicator species An organism that can be used to measure the environmental conditions that exist in a locality. For example, lichen species are indicators of levels of pollution, as different species are sensitive to different levels and types of pollutants.

indigenous Describing an organism that is native to an area, rather than introduced.

indole acetic acid (**IAA**) A naturally occurring auxin. *See* auxin.

inducer A substance that activates a structural gene or block of genes by combining with a REPRESSOR. *See* operon.

inducible enzyme (**adaptive enzyme**) An enzyme that is produced by a cell only in the presence of high concentrations of its substrate or a structurally similar substance.

indusium The flap of tissue, derived from the placenta, that encloses the developing sporangia in the sorus of a fern.

inferior Below. In botany, the term is generally applied to the position of the OVARY of a flower in which the sepals, petals, and stamens arise above it. The ovary appears to have sunk into and fused with the cup-shaped receptacle. This condition occurs only in epigynous flowers. In a floral formula, an inferior ovary is denoted by a line above the gynoecium symbol and number. *Compare* superior. *See also* epigyny.

inflorescence A collection of flowers sharing a common stalk, the *peduncle* and usually subtended by a bract. Each flower usually arises in the axil of a small bract, or bracteole, and may or may not be borne on its own individual stalk, called a *pedicel*. The development of an inflorescence may represent the end of vegetative growth of that apex (definite growth), or allow it to continue (indefinite growth). There are many types of inflorescence, determined mainly by the method of branching. If the stem ends in a flower and subsequent growth is then from lateral buds below the apex, which themselves form flowers and more lateral shoots, the inflorescence is termed a *cymose inflorescence* or *cyme*, e.g. lime (*Tilia*). If one shoot develops behind each axis a *monochasial cyme*, or *monochasium*, is formed, as in avens (*Geum*). Variations in the inflorescence arrangement occur, according to the direction of the lateral branches, for example in forget-me-not (*Myosotis*), all the branches arise on the same side of the parent stem while in buttercup (*Ranunculus*), the branches arise on alternate sides of the parent stem. If two shoots develop below each axis this gives a *dichasial cyme*, or *dichasium*, as in the campions and catchflies (*Silene*). Cymose inflorescences typically open from the apex downwards; in flat-topped forms the oldest flowers are in the center. In a *raceme* (*racemose inflorescence*) the apex continues growing, and subsequent flowers develop in sequence up the stem, e.g. foxglove (*Digitalis*). In a *compound raceme* each branch of the inflorescence bears a smaller raceme of flowers, e.g. fescues (*Festuca*). If the lateral branches of a raceme are themselves branched, as in many grasses, the inflorescence is called a *panicle*. This term is often applied to any sort of branched racemose inflorescence, for example the horse chestnut, in which each branch is actually a cyme (*Aesculus*). A *spike* is a type of racemose inflorescence having sessile flowers (they have no pedicels) borne on an elongated axis, as in wheat (*Triticum*). The *catkin* and *spadix* are modifications of the spike. A catkin is a short, densely packed raceme bearing unisexual flowers with highly reduced or absent perianth, e.g. oak (*Quercus*). A spadix is a type of inflorescence found in the family Araceae, e.g. cuckoopint (*Arum maculatum*). It is a modified spike with a large fleshy axis on which are borne small hermaphrodite or, more usually, unisexual flowers. The inflorescence is enclosed by a large bract, the *spathe*, which may be foliose or petalloid and has been shown to attract insects in certain species. In a *capitulum* (*pl.* capitula), as in many Asteraceae, e.g. daisy (*Bellis*), the inflorescence

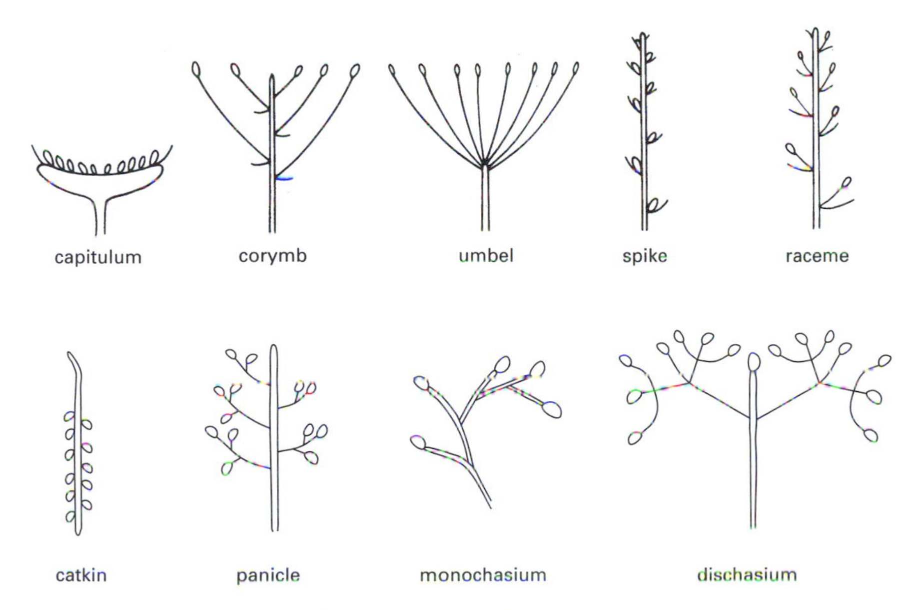

Inflorescence: types of inflorescence

comprises many unstalked florets inserted on the flattened disklike end of the peduncle and surrounded by a ring of sterile bracts, the involucre. The internodes are not recognizable. Each floret may or may not be borne in the axil of a bract on the disk. Many species of composite have two distinct types of floret in the capitulum: *disk florets* in the center are tubular florets ending in five short teeth, while *ray florets* have a straplike extension to the tube and occur around the edge of the capitulum, rather like petals. Some species, such as thistles, have only ray florets, others have only disk florets, e.g. chicory (*Cichorium*), and many have both disk and ray florets, e.g. sunflower (*Helianthus*). In racemose inflorescences the flowers typically open from below upwards. An *umbel* is a type of inflorescence in which the stem axis is not elongated and individually stalked flowers appear to arise from same point on the stem. These flowers are massed on one plane, giving the appearance of an umbrella, with the oldest flowers on the outside and the youngest in the middle. The umbel is typical of the carrot family (Apiaceae). Umbels may be grouped into a *compound umbel*, composed of umbels of smaller umbels, as in onion (*Allium*). A *corymb* is an inflorescence with flower stalks of different lengths, the lowest being the longest. This gives a flat-topped cluster of flowers at the same level that is characteristic of many brassicas, e.g. candytuft (*Iberis*). Umbels, corymbs, and capitula can in turn be grouped into cymose inflorescences, e.g. onion, *Viburnum*, and scabious (*Scabiosa*) respectively.

initial A cell permanently in a MERISTEM, actively dividing and adding new cells to the plant body. An initial never becomes differentiated. One of its daughters remains in the meristem as an initial, while the other differentiates to form a nonmeristematic cell. Two basic groups exist: apical initials at root and shoot apices, and lateral meristem initials whose position depends on the location of the meristem, e.g. intrafascicular CAMBIUM between xylem and phloem. Apical initials are rarely single cells except in certain lower plants, and initiate more than one structure, e.g. lateral shoots, flowers, and leaves, as well as new stem tissues. Vascular cambium possesses two distinct types of initials, *ray initials* forming medullary rays, and *fusiform initials* producing xylem and phloem elements.

initiation factor *See* translation.

inositol An optically active cyclic sugar alcohol, synthesized from glucose 6-phosphate. A derivative of iniositol, phytic acid, is an important storage compound in seeds, which is believed to be converted during germination to inositol, and thence to glucuronic and uronic acids for the synthesis of cell-wall components.

insectivorous plant *See* carnivorous plant.

integument A layer surrounding the nucellus in the ovules of gymnosperms and angiosperms. Most angiosperm ovules possess two integuments, while gymnosperms usually have only one. Enclosure of the nucellus by the integuments is incomplete, the micropyle remaining to allow access to the embryo sac or, in gymnosperms, to the archegonium.

intercalary meristem A region of actively growing primary tissue clearly separate from the apical meristem. Intercalary meristems occur at the internode and leaf-sheath bases (joints) of many monocotyledons including grasses, and at the internodes of horsetails, where they serve to increase longitudinal growth.

intercellular Describing materials found and processes occurring between cells. *Compare* intracellular.

interfascicular cambium A single layer of actively dividing cells between the vascular bundles in stems. It is formed when parenchyma cells resume meristematic activity. The interfascicular cambium and INTRAFASCICULAR CAMBIUM (within the vascular bundles) link into a complete cambium cylinder that cuts off secondary xylem tissue to the inside, secondary phloem tissue to the outside, and parenchyma cells both sides (forming *medullary rays*). *See* vascular cambium.

International Code of Botanical Nomenclature (**ICBN**) A set of rules for the scientific naming of wild plants, algae, fungi, and slime molds, which allocates a single unambiguous name to each taxon. Naming of families and lower taxa is based on the first valid name published, the starting-point being taken as the publication of *Species Plantarum*, by LINNAEUS, in 1753. But if a plant is later assigned to different taxa or rank, a different name may be given. Thus a plant may legitimately have two names, according to which genus different authors place it in. If several taxa are combined into one, that TAXON takes on the name of the former taxon with the oldest name. Different starting dates apply to fossil plants and bacteria. There is a separate code for cultivated plants, the International Code of Nomenclature of Cultivated Plants (ICNCP). *See* classification.

International Union for the Conservation of Nature and Natural Resources (**IUCN**) *See* World Conservation Union.

internode The region of the stem between two NODES. *See* intercalary meristem.

interphase The stage in the CELL CYCLE when the nucleus is not in a state of division. Interphase is divisible into various stages each characterized by a differing physiological activity.

intine (**endosporium**) *See* pollen.

intracellular Describing the material enclosed and processes occurring interior to the cell membrane. *Compare* intercellular.

intrafascicular cambium (**fascicular cambium**) The part of the VASCULAR CAMBIUM within the vascular between the xylem and phloem.

intraspecific selection Natural selection acting on individuals of the same species.

introgression (**introgressive hybridization**) The introduction of genetic material from one gene pool to another by hybridization and subsequent back-crossing to one or other of the parents. It forms the basis of the evolution of most crop plants, e.g. sunflower (*Helianthus annuus*). Where the ranges of two species overlap, and fertile hybrids can be produced, these hybrids will tend to backcross with the most abundant species, so the offspring down many generations come to resemble the most abundant parents.

introgressive hybridization *See* introgression.

intron A noncoding DNA sequence that occurs between coding sequences (exons) in many eukaryote genes. Messenger RNA (mRNA) does not contain introns, these being removed during the transcription process. Intron removal is now thought to be an autocatalytic process in which the RNA acts as its own enzyme (*see* ribozyme). However, in the case of mRNA in the nucleus the process is regulated by a complex of proteins called a *spliceosome*. *Compare* exon.

introrse Denoting anthers in which dehiscence lines are toward the center of the flower. Pollen is consequently shed toward the carpels, which favors self-pollination. *Compare* extrorse.

intussusception The incorporation of cellulose molecules into the existing cell wall, giving an increase in wall area. The process is usually preceded by water uptake, increasing the pressure on the cell wall, which stretches. The new wall is then thickened by APPOSITION.

inulin A polysaccharide food reserve of some higher plants, particularly the Asteraceae, e.g. *Dahlia* root tubers. It is a polymer of fructose.

inversion *See* chromosome mutation.

invertase *See* sucrase.

invert sugar *See* sucrose.

in vitro Literally 'in glass'; describing experiments or techniques performed in laboratory apparatus rather than in the living organism. Cell tissue culture is an example. *Compare* in vivo.

in vivo Literally 'in life'; describing processes that occur within the living organism. *Compare* in vitro.

involucre **1.** A protective structure consisting of a ring of bracts arising below the INFLORESCENCE in angiosperms with condensed inflorescences (e.g. the capitulum and umbel).
2. A sheathlike outgrowth of tissue in bryophytes protecting the archegonia or antheridia in certain liverworts.
3. A tubular extension of the thallus of hornworts that rises up and surrounds the base of the sporophyte.

iodine *See* staining.

ion-exchange chromatography A type of CHROMATOGRAPHY in which the compounds are separated according to their acidity.

ionic bond A bond formed when an electron passes from one atom to another. The atom that loses the electron becomes a positively charged ion, and the receiving atom becomes negatively charged, so the two ions become bonded by a strong electrostatic force. *Compare* covalent bond; hydrogen bond.

iron A MICRONUTRIENT essential for plant growth. It is found in porphyrins, and is essential for chlorophyll synthesis. Lack of iron leads to CHLOROSIS, especially in young leaves. It is also found in cytochromes, which are important components of the electron-transport chains of respiration and photosynthesis.

isoelectric focusing A technique used in ELECTROPHORESIS to separate amphoteric molecules (able to combine with either acids or bases).

isoelectric point The value of pH in a medium at which a molecule has no electric charge. Repulsion between molecules ceases, so they precipitate. This principle is used to separate mixtures of amino acids or proteins. *See* isoelectric focusing.

isoenzyme (**isozyme**) An enzyme that occurs in different structural forms within a single species. The isomeric forms all have the same molecular weight but differing structural configurations and properties, usually due to different combinations of their component subunits. They are formed by different alleles, and are a valuable tool for detecting genetic variation.

Isoetales (**quillworts**) An order of the LYCOPHYTA containing two genera – *Isoetes*, found in most parts of the world, and *Stylites*, found only in the Peruvian Andes. Quillworts are heterosporous, mainly aquatic perennials, with a dense rosette of tubular quill-like leaves arising from a short, stout rootstock.

isogamy The sexual fusion of gametes of similar size and form. It occurs in fungi and some protoctists. *Compare* anisogamy.

isolating mechanism Structural, physiological, behavioral, genetic, geographical, or other factors that restrict the interbreeding of one population with another, hence restricting gene flow. The development of isolating mechanisms promotes the formation of new varieties and species.

isoleucine A nonpolar AMINO ACID, synthesized from threonine.

isomerase An enzyme that catalyzes the conversion of a molecule from one isomeric form to another. For example, in glycolysis, glucose phosphate isomerase catalyzes the conversion of glucose 6-phosphate to fructose 6-phosphate.

isomorphism A condition seen in certain algae (e.g. *Ulva*, the sea lettuce) in which the alternating generations of the life cycle are morphologically identical (HOMOLOGOUS). *Compare* heteromorphism.

isotonic Designating a solution with an osmotic pressure or concentration equal to that of a specified other solution, usually taken to be within a cell. It therefore neither gains nor loses water by osmosis. *Compare* hypertonic; hypotonic.

isotopic dating *See* radiometric dating.

isozyme *See* isoenzyme.

IUCN *See* World Conservation Union.

Ivanovsky, Dmitri Iosifovich (1864–1920) Russian botanist. In 1862, while working on tobacco mosaic disease at the Technical Institute in St Petersburg, Ivanovsky noticed that healthy plants could be infected with the sap of diseased tobacco plants. He also noticed that the sap filtrate contained tiny crystalline particles that he believed were causative agents of the disease, but wrongly thought that they were bacteria. The Dutch bacteriologist Martinus Beijerinck repeated Ivanovsky's experiments in 1898 and showed that the crystalline particles were not bacteria, but were, in fact, a different type of organism, which he called a 'virus'.

J

Jacob–Monod model A model of regulation of gene activity in bacteria proposed by F. Jacob and J. Monod in 1960. Using the concept of the OPERON, it proposed that metabolism is regulated and coordinated by controlling messenger RNA synthesis (i.e. TRANSCRIPTION) and hence the production of enzymes. A similar but more complex system is known to operate in eukaryotes. *See* inducible enzyme.

Johannsen, Wilhelm Ludwig (1857–1927) Danish botanist and geneticist. Johannsen became interested in the *Theory of Heredity*, a work by the British anthropologist Sir Francis Galton, which stated that natural selection does not work on the offspring of self-fertilizing plants. Johannsen experimented with the Princess bean and found that if the self-fertilizing parent plants were from a mixed population, then selection did indeed work. He described the offspring of a single parent as a 'pure line' and stated that they were genetically identical. In 1905 he coined the terms 'phenotype' to describe the outward appearance of an individual and 'genotype' to describe the genetic makeup of an individual. Following the rediscovery of MENDEL'S LAWS, he published his *Elements of Heredity* in 1905 and used the word 'gene' for Mendel's factors of inheritance. His application of statistical methods to characteristics of populations proved to be a cornerstone of modern genetics.

J-shaped growth curve *See* exponential growth.

Juncaceae (rushes) A family of monocotyledonous plants, mostly herbs, with long, narrow grasslike or cylindrical leaves spirally arranged, but usually arising close to the ground. They are found in temperate and cold regions, and on tropical mountains, forming an important part of the vegetation in poorly drained areas.

junk DNA The noncoding part of DNA, i.e. the part that does not code for proteins. Its function is not yet understood, but it appears that part of it contains intructions that help control where and when the different proteins are expressed. *Compare* repetitive DNA; selfish DNA.

Jurassic The middle period of the Mesozoic era, 215–145 million years ago. The climate was warm and humid, and there was widespread expansion of forests, which gave rise to thick deposits of coal. The Jurassic has been called the 'Age of Cycads'. The seed-ferns declined, and there was a great increase in the number and diversity of cycads, palmlike gymnosperms, and conifers. Fossils of monocotyledonous pollen and stems of palms suggest that the angiosperms evolved during this period. *See also* geological time scale.

Jussieu, Antoine-Laurent de (1748–1836) French plant taxonomist. Working at the Jardin du Roi in Paris, Jussieu based his classification system on many plant characteristics rather than the small number used by LINNAEUS. Using the collections of Banks, Linnaeus, and Philibert Commesson, he established 15 classes and 100 families of plants in his classification.

K

karyogamy The fusion of two nuclei that exist within a common cytoplasm, as occurs in the formation of the zygote from two gametes. In the more advanced fungi, especially the Basidiomycota, nuclei may fuse some considerable time after fusion of hyphae of different mating strains, forming a dikaryon.

karyogram (**idiogram**) The formalized layout of the KARYOTYPE of a species, often with the chromosomes arranged in a certain numerical sequence. This usually takes the form of drawings or photographs of the individual chromosomes, often arranged in homologous pairs in numerical sequence.

karyokinesis During cell division, the process of nuclear division that precedes cytoplasmic division. There are two main types of nuclear division: MITOSIS, which results in daughter cells identical to their parents; and MEIOSIS (reduction division), which produces daughter cells which have half the number of chromosomes of their parent cells, and in which the genetic material has been recombined. *Compare* cytokinesis.

karyotype The physical appearance of the chromosome complement of a given species. A species can be characterized by its karyotype since the number, size, and shape of chromosomes vary greatly between species but are fairly constant within species. *See* karyogram.

kelp *See* Laminariales.

α-ketoglutaric acid A five-carbon dicarboxylic acid that is an intermediate in the KREBS CYCLE. It is synthesized by oxidation of isocitrate, coupled to the reduction of NAD. In the Krebs cycle it undergoes decarboxylation to form succinyl CoA. α-ketoglutarate is also involved in amino acid metabolism, especially TRANSAMINATION.

ketohexose A ketose SUGAR with six carbon atoms, e.g. fructose.

ketopentose A ketose SUGAR with five carbon atoms, e.g. ribulose.

ketose A monosaccharide SUGAR con-

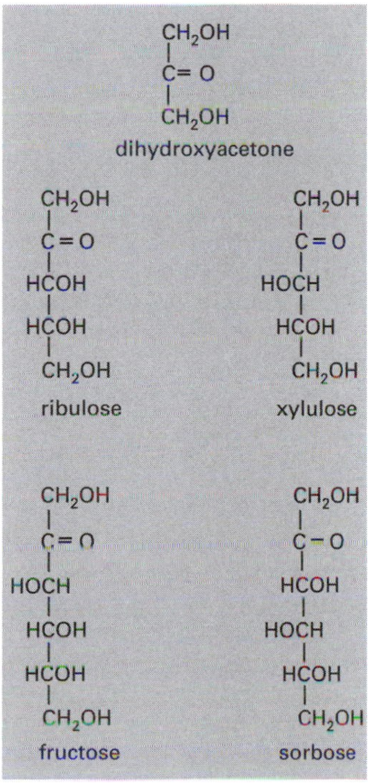

CH_2OH–C=O–CH_2OH
dihydroxyacetone

CH_2OH–C=O–HCOH–HCOH–CH_2OH
ribulose

CH_2OH–C=O–HOCH–HCOH–CH_2OH
xylulose

CH_2OH–C=O–HOCH–HCOH–HCOH–CH_2OH
fructose

CH_2OH–C=O–HCOH–HOCH–HCOH–CH_2OH
sorbose

Ketose: typical ketose sugars

taining a ketone (=CO) or potential ketone group, e.g. dihydroxyacetone.

kinase (**phosphotransferase; phosphorylase**) **1.** Any enzyme that transfers a phosphate group, usually from ATP.
2. An enzyme that activates the inactive form of other enzymes. For instance, when trypsinogen, the inactive form of trypsin, comes in contact with enterokinase, active trypsin is released.

kinetin *See* cytokinin.

kinetosome (**basal body**) A barrel-shaped body found at the base of all eukaryote UNDULIPODIA (cilic and flagella) and identical in structure to the CENTRIOLE. It is essential for formation of undulipodia.

kingdom Formerly, the highest ranking category in most classification systems. Most taxonomists today recognize the rank of domain as being higher than that of kingdom. *See* Five Kingdoms classification.

klinostat (**clinostat**) An apparatus used in tropism experiments to remove the unidirectional influence of a stimulus on a plant organ.

Kramer, Paul Jackson (1904–) American plant physiologist. Working at Duke University, Kramer made a significant contribution to the study of water-uptake mechanisms in plants. Using radioactively labeled elements he demonstrated that the region of maximum absorption of water is a few centimeters behind the root tip and that quantities of minerals are absorbed passively in the transpiration stream.

Kranz anatomy The specific arrangement of the photosynthetic tissues in the leaves of C_4 PLANTS. The cells of the bundle sheath that surround the vascular tissues are large and contain specialized elongated

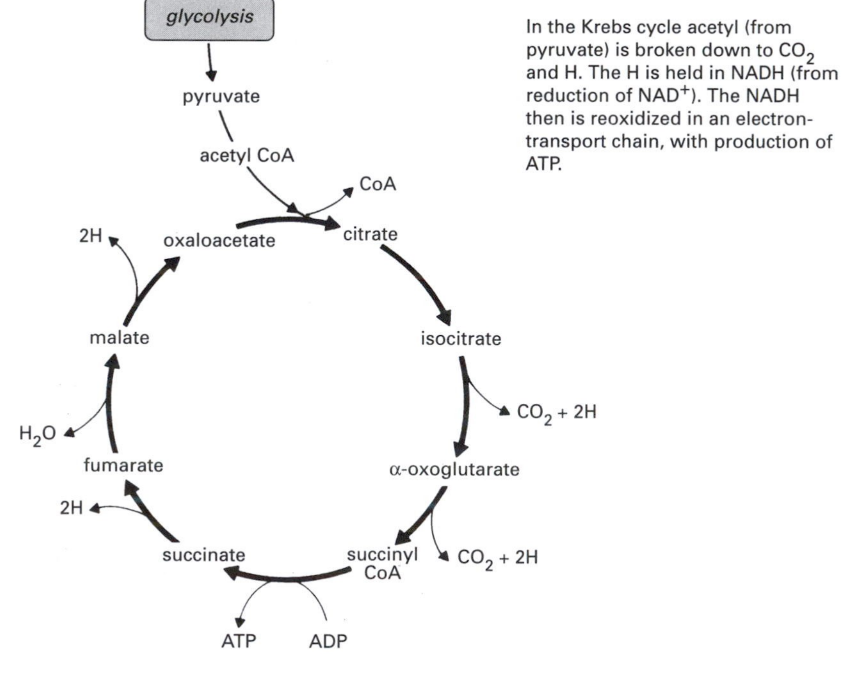

In the Krebs cycle acetyl (from pyruvate) is broken down to CO_2 and H. The H is held in NADH (from reduction of NAD^+). The NADH then is reoxidized in an electron-transport chain, with production of ATP.

Krebs cycle

chloroplasts that do not contain grana, but form starch grains. The chloroplasts of the mesophyll, by contrast, appear normal and contain grana, but do not form starch grains. In these chloroplasts carbon dioxide combines with phosphoenolpyruvate to form oxaloacetic acid, which is transported to the bundle sheath cells, where the carbon dioxide is released, then fixed by the enzyme ribulose bisphosphate carboxylase to form glycerate 3-phosphate, the first step in the Calvin cycle. *See* photosynthesis.

Krebs, Sir Hans Adolf (1900–81) German-born British biochemist best known for his elucidation of the KREBS CYCLE. He shared the 1953 Nobel Prize for Physiology or Medicine with Fritz Lipmann (1899–1986).

Krebs cycle (**citric acid cycle; TCA cycle; tricarboxylic acid cycle**) A complex cycle of reactions in which pyruvate, produced by glycolysis, is oxidized to carbon dioxide and water, with the production of large amounts of energy. It is the second stage of AEROBIC RESPIRATION, requires oxygen, and occurs in the matrix of mitochondria. 2-carbon acetate (acetyl coenzyme A), derived from pyruvate by decarboxylation, reacts with 4-carbon oxaloacetate to form 6-carbon citrate, which is then decarboxylated in a series of steps to reconstitute oxaloacetate. Some ATP is produced by direct coupling with Krebs cycle reactions, but most production is coupled to the ELECTRON-TRANSPORT CHAIN. The operation of this depends on the generation of reduced coenzymes, NADH and $FADH_2$, by the Krebs cycle. These reduced cofactors provide the energy for ATP formation in the respiratory electron-transport chain and OXIDATIVE PHOSPHORYLATION. Considering the Krebs cycle and electron-transport chain together, each pyruvate molecule yields 15 ATP molecules. Since two pyruvate molecules enter the cycle from glycolysis, 30 ATP are produced in all. The intermediates in the Krebs cycle may be used directly or indirectly in the synthesis of amino acids.

L

labeling The technique of using isotopes (usually radioactive isotopes) or other recognizable chemical groups to investigate biochemical reactions. For instance, a compound can be synthesized with one of the atoms replaced by a radioactive isotope of the element. The radioactivity can then be used to follow the course of reactions involving this compound.

Labiatae *See* Lamiaceae.

lactic acid A 3-carbon hydroxy-acid formed as a result of fermentation by certain bacteria, e.g. *Lactobacillus*.

lactic acid bacteria A group of bacteria that ferment carbohydrates in the absence of oxygen, with lactic acid always a major end product, e.g. *Lactobacillus* spp., *Streptococcus lactis*. They have a high tolerance of acid conditions. Lactic acid bacteria are involved in the formation of yoghurt, cheese, sauerkraut, and silage. They can occur as spoilage organisms and some are pathogenic causing infections.

lacuna (*pl.* **lacunae**) **1.** An intercellular cavity in plant tissues resulting from cell breakdown, tissue splitting, or organized formation. In hydrophytes a well-arranged system of these air-filled cavities may exist to provide buoyancy. A large lacuna occupies the center of many stems.
2. *See* leaf gap.
3. A depression in a lichen thallus.

lag phase *See* exponential growth.

lamella (*pl.* **lamellae**) **1.** A layer within the cytoplasm or inside an organelle, which is formed from a flattened membrane-bounded vesicle or tube. The layer thus consists of two membranes lying close together with a cavity between them. An example is the internal system of lamellae (*thylakoids*) in CHLOROPLASTS.
2. A general term applied to any thin platelike structure, e.g. the blade of a leaf.

Lamiaceae (Labiatae) A dicotyledonous family of herbs and low-growing shrubs, most of which have stems of square cross-section and simple leaves with stipules arranged in opposite pairs. There are about 6700 species, distributed worldwide but especially numerous in the Mediterranean region. The family includes many herbs, such as basil (*Hyssopus officinalis*), marjoram (*Origanum*), mint (*Mentha*), sage (*Salvia officinalis*), savory (*Satureja*), and thyme (*Thymus vulgaris*).

lamina (*pl.* **laminae**) **1.** (of angiosperms) A thin usually flat foliage *leaf blade*, commonly attached to the stem by a petiole. In most plants the leaf laminae are the main photosynthetic organs. A lamina is termed *simple* if complete, and *compound* if divided into leaflets. Internal organization shows a wide photosynthetic mesophyll layer permeated by veins and bounded by an epidermis. Leaf variation is mainly due to the structural diversity of the lamina.
2. (of algae) The bladelike part of the thallus of certain algae, notably the Phaeophyta (brown algae), e.g. kelps (*Laminaria*, *Macrocystis*, *Nereocystis*) and bladderwrack (*Fucus vesiculosus*).

Laminariales (kelps) An order of the PHAEOPHYTA (brown algae) that contains the *kelps* (e.g. *Macrocystis* and *Nereocystis*) and is distinguished by its heteromorphic ALTERNATION OF GENERATIONS in which a large sporophyte alternates with a microscopic gametophyte, and in which growth takes place in a special meristem region (*see* meristoderm) on the surface of the thallus.

laminarin The chief carbohydrate food reserve of the Phaeophyta (brown algae).

lateral meristem A MERISTEM that lies parallel to the sides of the organ in which it occurs and gives rise to secondary tissues that increase the girth of the organ. *See* cambium; secondary growth. *Compare* apical meristem.

lateral root Any ROOT that arises from the PERICYCLE of another root.

laterite A hard crust that may form on the surface of the SOIL in tropical regions with alternating wet and dry seasons.

latex A liquid found in some flowering plants contained in special cells or vessels called *laticifers* (or laticiferous vessels). It is a complex variable substance that may contain terpenes (e.g. rubber), resins, tannins, waxes, alkaloids, sugar, starch, enzymes, crystals, etc. in solution or suspension. It is often milky in appearance, as in dandelion (*Taraxacum* spp.) and lettuce (*Lactuca sativa*) but may be colorless, orange, or brown. Commercial rubber comes from the latex of the rubber plants *Ficus elastica* and *Hevea brasiliensis*. Opium comes from alkaloids found in the latex of the opium poppy (*Papaver somniferum*).

laticifers *See* latex.

Lauraceae A family of trees and shrubs (bays and laurels) usually with simple, leathery, evergreen alternate or opposite leaves without stipules. There are about 2500 species, mainly in the tropics and subtropics, especially in Amazonia and Southeast Asia.

Law of Independent Assortment *See* Mendel's laws.

Law of Segregation *See* Mendel's laws.

layering A method of vegetative propagation in which RUNNERS or STOLONS are pegged down to the soil surface. Where a node touches the soil, adventitious roots develop and a shoot arises from the LATERAL MERISTEM. If the intervening runner or stolon rots or is cut away, the daughter plants becomes independent.

leaching The removal of humus and soil nutrients in solution by water moving down the soil profile. It makes soils more acid since cations (e.g. potassium and magnesium) are replaced by hydrogen ions. Leaching leads to the formation of podsolized and lateritic soils.

leaf A flattened appendage of the stem that arises as a superficial outgrowth from the apical meristem. Leaves are arranged in a definite pattern, have buds in their axils, and show limited growth. Most foliage leaves are photosynthetic, bilaterally symmetrical, and externally differentiated into LAMINA, MIDRIB, and PETIOLE. Many variations occur. Not all foliage leaves are photosynthetic as they may be modified wholly to bud scales and spines or partly to form tendrils. See illustration overleaf.

leaf buttress A leaf primordium, appearing as a protuberance on the side of the stem apex, produced by periclinal division of the tunica and corpus, and associated with a procambium. It is the earliest stage in the development of a leaf, and forms the leaf base as the leaf grows out from it. The leaf axis arises from the leaf buttress and maintains procambial continuity in the central region. This differentiates into the vascular tissue of the developing leaf.

leaf curl Any disease of plants that causes an increase in numbers of cells that results in curling and puckering of the leaves. It may be caused by a fungus (e.g. *Taphrinia deformans* causes peach leaf curl), or by viruses (e.g. tobacco and cotton leaf curl) transmitted by insects such as whiteflies (*Bemisia*).

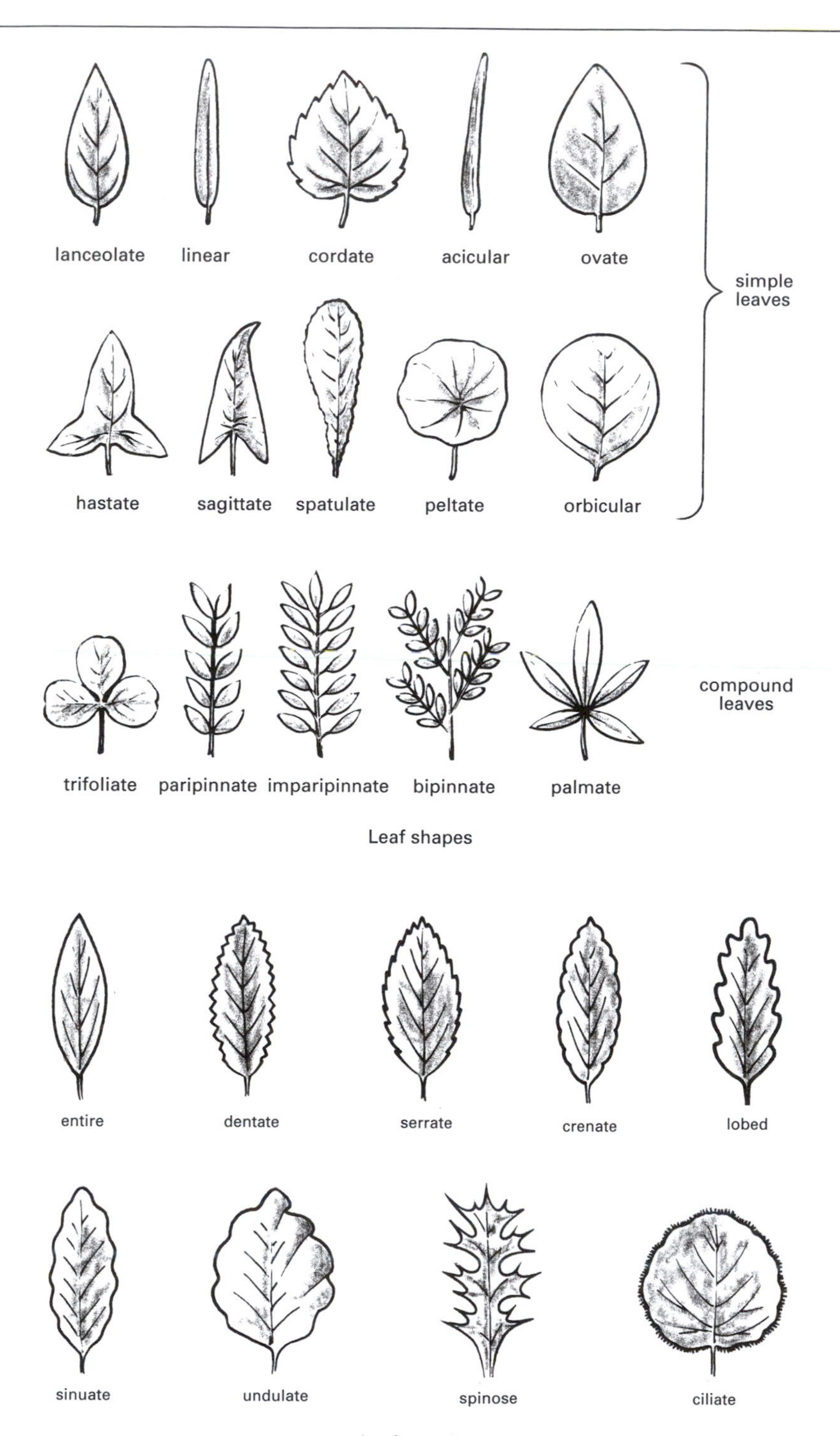

Leaf shapes

Leaf margins

leaf fall The shedding of leaves, usually associated with a fall in AUXIN concentration. In both evergreen and deciduous trees old leaves are continually shed throughout the growing season, but the remaining leaves are also shed at the onset of winter or a dry season in deciduous species. *See* abscission.

leaf gap (**lacuna**) A region of parenchyma differentiated in the stem vascular cylinder immediately above a diverging LEAF TRACE. Lateral connections insure no break in the vascular system.

leaf mosaic **1.** The nonoverlapping pattern of leaves on a plant that results mainly from the genetically programmed arrangement of leaves and branches, and in some species also by the twisting of the petioles due to EPINASTY. It ensures that the maximum amount of light is intercepted by the leaves. *See* phyllotaxis.
2. A virus disease of plants characterized by yellow and green or light- and dark-green mottling of the leaves.

leaf trace A vascular bundle or group of vascular bundles connecting the vascular systems of leaf and stem. It stretches from the leaf base to the vascular ring in the stem axis.

lecithin (**phosphatidyl choline**) One of a group of phospholipids that contain glycerol, fatty acid, phosphoric acid, and choline that are found widely in higher plants and animals, particularly as a component of cell membranes.

legume **1.** (**pod**) A dry dehiscent single- or many-seeded fruit formed from a single carpel, that dehisces mechanically by splitting, often explosively, along both sides. The opposite tensions set up by oblique fibers in the drying pericarp cause dehiscence. The valves may continue twisting and remove any remaining seeds. It is the typical fruit of the FABACEAE family, and is found in certain other families. *Compare* lomentum.
2. Any plant of the family Fabaceae.

lemma *See* bract.

lenticel *See* bark.

leptosporangiate Describing the condition, found in certain ferns (e.g. Filicales), in which the sporangium develops from a single initial cell. The wall of a leptosporangium is usually only one cell thick, and as it dries out, it flicks the spores away explosively. *Compare* eusporangiate.

leptotene *See* meiosis; prophase.

leucine An amino acid synthesized from pyruvate. It can be broken down to yield acetoacetate. *See* amino acids.

leukoplast A colorless PLASTID, i.e. one not containing chlorophyll or any other pigment. Leukoplasts are common in the cells of roots and underground stems and storage organs.

liana (**liane**) A long-stemmed woody climbing plant that grows from ground level to the canopy of trees, and often hangs down freely from branches, e.g. curare (*Strychnos toxifera*). Lianas may have stems up to 100 meters (330 feet) long, and may climb over the canopy to gain maximum sunlight. Many flower only when they reach the top of the canopy. Up to a quarter of all the woody growth in some rainforests may be lianas.

lichens Symbiotic associations between an alga or cyanobacterium (the *photobiont*) and a fungus (the *mycobiont*). They are slow-growing but can colonize areas too inhospitable for other plants. Usually the fungus is an ascomycete but occasionally it is a basidiomycete. Reproduction in lichens may be asexual by soredia (algal cells enclosed by fungal hyphae) or by sexual fungal spores, which can survive only if some algal cells are also present. Fragments of the thallus containing both phycobiont and mycobiont cells may also grow into new lichens, a form of asexual reproduction. Examples include *Peltigesa* and *Xanthoria*.

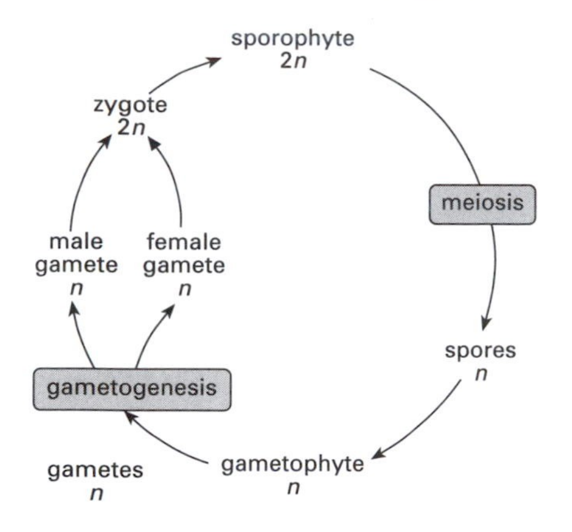

ALTERNATION OF GENERATIONS

the type of life cycle found in most plants, in which there is a definite development, however reduced, of both sporophyte and gametophyte generations

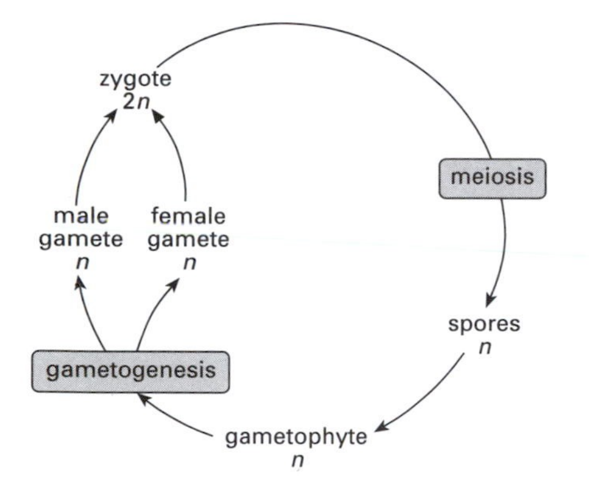

HAPLONTIC

a form of life cycle found in certain algae, e.g. the Ulotrichales, in which there is no sporophyte generation, except as the zygote

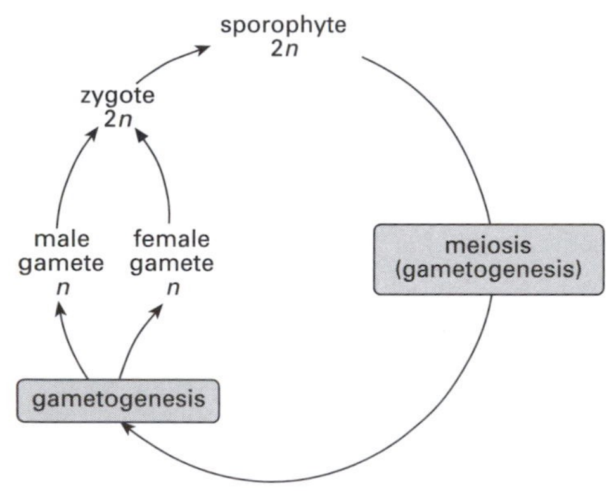

DIPLONTIC

a form of life cycle found in certain algae, e.g. the Fucales, in which there is no gametophyte generation. This cycle is also typical of animals

KEY

n diploid stage
$2n$ diploid stage

Life cycle

life cycle The sequence of developmental changes making up the span of an organism's life from the fertilization of gametes to the same stage in the subsequent generation. *See* diplobiontic; haplobiontic.

ligase An enzyme that catalyzes the bond formation between two substrates at the expense of the breakdown of ATP or some other nucleotide triphosphate. The degree of bond formation by ligases is proportional to the amount of ATP available in the cell at a particular instant.

light green A stain used in light microscopy to show up cytoplasm and cellulose. It is commonly used as a counterstain with safranin to highlight cytoplasm. *See* staining.

light microscope *See* microscope.

light reactions The light-dependent reactions of PHOTOSYNTHESIS that convert light energy into the chemical energy of NADPH and ATP, which are used in the dark reactions for carbon dioxide fixation. Light absorbed by two groups of pigments in the chloroplast – Photosystems I and II. The absorbed light energy excites chlorophyll *a* molecules, causing them to emit high-energy electrons that pass down an ELECTRON-TRANSPORT CHAIN, releasing energy to synthesize ATP (photophosphorylation) and NADPH. The light reactions take place on the thylakoid membranes of chloroplasts.

light saturation point *See* compensation point.

lignin One of the main structural materials of vascular plants. Together with cellulose it is one of the main constituents of wood, where it imparts high tensile and compressive strengths, making it ideal for support and protection. Lignified tissues include SCLERENCHYMA and XYLEM. Lignin is deposited during secondary thickening of cell walls. The degree of lignification varies from slight in protoxylem to heavy in sclerenchyma and some xylem vessels, but values of 25–30% lignin and 50% cellulose are average. It is a complex variable polymer, derived from sugars via aromatic alcohols.

ligule **1.** A scalelike outgrowth, varying in shape and size, of certain angiosperm leaves. In grasses the ligule occurs at the junction of the leaf sheath and lamina. Although membranous in most species, it may be only a fringe of hairs. Ligules are important taxonomic features in grasses.
2. A toothed, strap-shaped structure formed by the extension of one side of the corolla tube in the florets of certain species of the ASTERACEAE. The teeth indicate the number of fused petals. In some species, e.g. dandelion (*Taraxacum* spp.), all florets of a capitulum are ligulate but in others e.g. daisy (*Bellis perennis*), only the ray florets have ligules.
3. A very small tongue-shaped flap of tissue inserted on the upper surface of the leaves (microsporophylls) and sporophylls of certain clubmosses (e.g. *Selaginella*) and quillworts (*Isoetes*).

Liliaceae (lilies) A large family of monocotyledonous plants. Their taxonomy is much in dispute, with groups that were formerly sections now regarded with new evidence as being separate families. They have long, narrow parallel-veined leaves, whorled or alternate, arising at the base of the stem or along it. Most are herbs, e.g. lilies (*Lilium*) and tulips (*Tulipa*); some are succulents or dwarf shrubs, e.g. butcher's broom (*Ruscus aculeatus*), and a few are climbers (e.g. *Smilax*). Many species have perennating organs such as bulbs, corms, or rhizomes.

liming The addition of lime to a soil to reduce its acidity, improve crumb structure, or increase its calcium content. *See* flocculation.

limiting factor Any factor in the environment that governs the behavior or metabolic activity of an organism or system by being above or below a certain level. In general, a number of different factors (e.g. light intensity, temperature, carbon dioxide concentration) may be limiting, but at any one time the factor that is closest to its critical minimum is the limiting factor. For instance, the rate of photosynthesis rises with increasing light intensity as long as there is sufficient carbon dioxide available, but at high light intensities carbon dioxide may become the limiting factor instead.

limiting layer *See* meristoderm.

limnology The scientific study of freshwater and its flora and fauna. Limnology also includes studying the chemical and physical aspects of inland water. It may be divided into the study of standing water habitats, e.g. lakes and ponds, and running water habitats, e.g. rivers and streams.

linear Describing leaves that are long, narrow, and parallel-sided for most of their length.

linkage The occurrence of genes together on the same chromosome so that they tend to be inherited together and not independently. Groups of linked genes are termed *linkage groups* and the number of linkage groups of a particular organism is equal to its haploid chromosome number. Linkage groups can be broken up by CROSSING OVER at meiosis to give new combinations of genes. Two genes close together on a chromosome are more strongly linked, i.e. there is less chance of a cross over between them, than two genes further apart on the chromosome. Linked genes are symbolized Ab...Y/aB...y, indicating that Ab...Y are on one homolog while aB...y are on the other homolog. Linkage is indicated when the associated inheritance of two or more nonallelic genes is greater than would be expected from independent assortment. The genes on a single chromosome form one *linkage map*. *See* chromosome map; independent assortment.

Linnaeus, Carolus (1707–76) Swedish botanist. While working at Uppsala University he began to investigate the sexuality of plants and by 1730 he had begun to set out a classification system based on the number and arrangement of stamens and pistils. After traveling through Lapland and discovering many new species, he settled in Holland and published his *Systema Naturae*, in which he set out a classification of animals, plants, and minerals. He placed flowering plants into classes based on the number of stamens and subdivided the classes into orders based on the number of their pistils.

Linnaeus's most famous contribution to taxonomy was his system of BINOMIAL NOMENCLATURE, in which he gave plants two Latin names. The first name is the generic name and the second is the specific (or species) name.

linoleic acid A common unsaturated 18-carbon fatty acid occurring in glycerides in linseed oil, cottonseed oil, and other vegetable oils. It is an essential nutrient in the diet of mammals.

linolenic acid An unsaturated 18-carbon fatty acid occurring commonly in plants as the glyceryl ester, for example in linseed oil and poppy-seed oil. Linoleic acid is the predominant fatty acid in the chloroplast. It is an essential nutrient in the diet of mammals.

lipase Any of various enzymes that catalyze the hydrolysis of fats to fatty acids and glycerol. Lipases are important in germinating seeds, e.g. castor oil (*Ricinus communis*) where they help to break down storage fats.

lipid A collective term used to describe a group of substances in cells characterized by their solubility in organic solvents such as ether and benzene, and their absence of solubility in water.

The group is rather heterogeneous in terms of both function and structure. They encompass the following broad bands of biological roles: (1) basic structural units of cellular membranes and cytologically distinct subcellular bodies such as chloroplasts and mitochondria; (2) compartmentalizing units for metabolically active proteins localized in membranes; (3) a store of chemical energy and carbon skeletons; and (4) primary transport systems of nonpolar material through biological fluids. Plants synthesize a wide range of lipids. Long-chain fatty acids (often 16–18 carbon atoms) are esterified to glycerol for use in membranes, and PHOSPHOGLYCERIDES are particularly important components of membranes; esters of long-chain monohydric alcohols and fatty acids form the WAXES of the cuticle and the SUBERIN of the endodermis; and triacylglycerols are important storage lipids in seeds, especially in such species as the castor oil seed (*Ricinis communis*), where they are stored in oil bodies, as small lipid droplets coated with lipids and protein. The fatty acid components of plant lipids come mainly from acetyl CoA, produced in plastids (including chloroplasts) and mitochondria. GLYCOLIPIDS are synthesized in the endoplasmic reticulum, mitochondria, and plastids.

lipopolysaccharide A conjugated polysaccharide in which the noncarbohydrate part is a lipid. Lipopolysaccharides are a major component of the cell walls of Gram-negative bacteria.

lipoprotein Any conjugated protein formed by the combination of a protein with a lipid. Most membranes, including the plasma membrane, are composed mostly of lipoprotein. The lipid components are mainly phosphoglycerides and glycolipids.

litmus paper Red or blue acid–alkali indicator papers. In acids blue litmus paper turns red and in alkalis red litmus paper turns blue. Litmus solution has a pH range from 4.5–8.3.

littoral **1.** The zone of the seashore between the high and low tide mark. The term is also applied to organisms living in this zone. Since tidal ranges vary continually, the zone is often defined in terms of the upper and lower limits of certain species of organism. *Compare* benthic; sublittoral.
2. The zone between the water's edge and a depth of about six meters in a pond or lake, where light reaches the bottom sediments. Rooted hydrophytes, both emergent and submergent, are found in this zone. *Compare* profundal; sublittoral.

liverworts *See* Hepatophyta.

living fossil Modern organisms with anatomical or physiological features found elsewhere only in extinct species. An example is the maidenhair tree (*Ginkgo biloba*), discovered in Japan in the 17th century, and later in China, but found only in cultivation. Fossil members of the Ginkgophyta are common from rocks of Mesozoic age.

loam A medium-textured soil containing a mixture of large and small mineral particles. Loams are easy soils to work and combine the good properties of sandy and clay soils.

locule (**loculus**) An air-filled compartment in an ovary in which the ovules develop. The term locule can also be applied to any other cavity in an organ in which other structures develop, for example the locules of anthers in which pollen is formed. When all the carpel edges meet at the middle of a syncarpous ovary there are as many locules as carpels, and the ovary is designated bi-, tri-, quadri-, or multilocular according to number. Adjoining carpel walls form septa that separate locules. A unilocular condition occurs in monocarpellary ovaries and syncarpous ovaries lacking septa.

loculus (*pl.* **loculi**) *See* locule.

locus (*pl.* **loci**) The position of a gene on a chromosome. Alleles of the same gene occupy the equivalent locus on homologous chromosomes.

loess A fine-textured, fertile, often calcareous uniform soil comprising mainly quartz particles about 0.015–0.005 mm in diameter. Widespread in central Europe, southern Russia, China, the central USA, and Argentina, it is derived from windblown clay and silt particles originally deposited at the edge of the ice sheets at the end of the last ICE AGE. *See* chernozem.

log phase (logarithmic phase) *See* exponential growth.

lomasome An infolding of the plasma membrane found particularly in fungal hyphae and spores, and also in some algae and higher plants.

lomentum (*pl.* **lomenta**) A dry dehiscent fruit formed from a single carpel and bearing more than one seed, as in sainfoin (*Onobrychis viciifolia*). It resembles a legume or siliqua but is divided by false septa into single-seeded compartments that rupture at maturity.

long-day plant (LDP) A plant that flowers in response to a dark period shorter than a critical maximum, e.g. spinach (*Spinacea oleracea*), which flowers in summer, when days are longer and nights shorter. *See* photoperiodism; critical day length.

lumen The central space that remains, surrounded by cell walls, in a cell that has lost its living contents (e.g. in xylem elements).

lutein The commonest of the xanthophylls, an orange pigment found in green leaves and certain algae, e.g. the Rhodophyta. *See* photosynthetic pigments.

lyase An enzyme that catalyzes the separation of two parts of a molecule with the formation of a double bond in one of them. For example, fumarase catalyzes the interconversion of malic acid and fumaric acid. The reaction does not involve hydrolysis.

Lycophyta (Lycopodophyta; Lycopsida) A phylum of spore-bearing vascular plants containing about 1000 living species and 6 genera, mostly found in tropical regions. Most lycopods belong to the genera *Lycopodium* and *Selaginella*. There are five orders, three of which – the Lycopodiales, Selaginellales, and Isoetales – contain both living and fossil representatives. The remaining orders – Lepidodendrales and Pleuromeiales – are represented only by fossils. The extinct trees of the genus *Lepidodendron* were once distributed widely and contributed largely to the coal seams of the Carboniferous.

Lycopodiales An order of the LYCOPHYTA containing the homosporous clubmosses or ground pines. They are mainly tropical, some of them epiphytic, but also occur in temperate regions, especially in montane habitats. *Lycopodium* and related genera comprise the common clubmosses or (North America) ground pines.

Lycopodophyta *See* Lycophyta.

Lycopsida *See* Lycophyta.

Lysenko, Trofim Denisovich (1898–1976) Ukrainian agriculturalist. While working at the Kiev Agricultural Institute, Lysenko claimed to have invented VERNALIZATION, the cold treatment of grain to promote next-season flowering in species that would normally take two years to flower. However, this had long been a common practice in agriculture. Lysenko went further by stating that vernalized plants could pass this 'acquired characteristic' to their offspring, so that no further treatment was necessary – a return to the discredited ideas of Lamarck. In the 1930s he achieved great influence in the Soviet Union under Stalin and used this to enforce his unorthodox ideas. In particular, he rejected the chromosome theory of inheritance, acting to discredit those scientists who followed this view and even changing school textbooks and courses to suit his own beliefs. He finally fell from power in 1964.

lysine A basic amino acid synthesized from aspartic and pyruvic acids, and broken down through acetyl CoA and the Krebs cycle. Lysine is a precursor of certain alkaloids. *See* amino acids.

lysis (degeneration) The death and subsequent breakdown of a cell. *See also* autolysis; lysogeny.

lysogeny **1.** The formation of an intercellular space in plants by dissolution of cells. *Compare* schizogeny.

2. A stable phage–bacteria relationship in which lysing of the bacteria does not occur. The phage (known as a *temperate phage*) penetrates the host cell and its nucleic acid becomes integrated into the bacterial DNA. In this state the phage is termed a *prophage*; most of the viral genes are repressed and both bacteria and phage reproduce together, producing infected daughter cells. In a process called *induction*, certain environmental factors can cause the phage to leave the host DNA and resume the lytic cycle.

lysosome An organelle of plant and animal cells that contains a range of digestive enzymes whose destructive potential necessitates their separation from the rest of the cytoplasm. Lysosomes are bounded by a single membrane and have homogeneous contents that often appear uniformly gray with the electron microscope. They are usually spherical and about 0.5 μm in diameter, although lysosomal compartments may range from small Golgi vesicles to large plant vacuoles. Lysosomes may be formed directly from endoplasmic reticulum or by budding off of Golgi vesicles containing processed proteins derived from the endoplasmic reticulum. The final structure after digestion is called a *residual body*.

In ageing plant tissue AUTOLYSIS may occur, when cell components such as mitochondria or fragments of endoplasmic reticulum are broken down by lysosome enzymes, facilitating the recycling of valuable nutrients within the plant.

M

macromolecule A very large molecule, usually a polymer, having a very high molecular weight (10 000 or more atoms). Proteins and nucleic acids are examples.

macronutrient A nutrient required in more than trace amounts by an organism. It may be an organic or inorganic compound. *See* essential element. *Compare* micronutrient.

macrosclereids Elongated rod-shaped SCLEREIDS that form a close outer protective layer in the seed testas and fruit walls of some plants.

magnesium An element essential for plant and animal growth. It is contained in the chlorophyll molecule and is thus essential for photosynthesis. It is an important component of membranes, and an essential cofactor for certain phosphate-transferring enzymes, e.g. phosphohydrolase and phosphotransferase. High concentrations of magnesium ions, Mg^{2+}, are needed to maintain ribosome structure. Deficiency of magnesium may result in CHLOROSIS,, NECROSIS, stunted growth, and in some species, whitening and puckering of the leaf edges. It may be remedied by adding magnesium sulfate or magnesium oxide to the soils.

Magnoliaceae A family of primitive dicotyledons that includes *Magnolia* and the tulip trees (*Liriodendron* spp.). They are mostly trees and shrubs of warm temperate to tropical parts of east Asia and tropical North America.

maidenhair ferns *See* Adiantaceae.

male sterility A condition in which pollen production is prevented by the mutation of one or more of the genes involved. Male sterility is used by plant breeders to ensure cross-pollination takes place. It is also used in genetic engineering to ensure that genetically modified varieties do not release pollen containing modified genes.

malic acid (**2-hydroxybutanedioic acid**) A colorless crystalline 4-carbon carboxylic acid, which occurs in acid fruits such as grapes and gooseberries. In biological processes malate ion is an important intermediate in the KREBS CYCLE.

Malpighi, Marcello (1628–1694) Italian histologist. Often described as the father of microscopy, Malpighi, a philosopher and professor of medicine, initiated the detailed study of plant and animal tissues. He is best known for his investigations into mammalian blood and also studied many body organs, including the skin. He described the stomata on the abaxial surfaces of leaves, the annual rings of woody plants, and also made a distinction between monocotyledonous and dicotyledonous plants.

maltose A sugar found in germinating cereal seeds. It is a disaccharide composed of two glucose units linked by an α-1,4–GLYCOSIDIC BOND. Maltose is an important intermediate in the enzyme hydrolysis of starch. It is further hydrolyzed to glucose.

manganese A metallic MICRONUTRIENT needed by plants for growth. Manganese ions are cofactors in various enzymes, e.g. kinases, IAA oxidase, and certain enzymes involved in the biosynthetic pathway for chlorophyll. Manganese is required for the light reactions of Photosystem II (*see* photosynthesis). Deficiency of manganese may cause dwarfing, mottling of the upper leaves, CHLOROSIS between the leaf veins, 'blight' of sugar cane, and 'gray speck' of oats.

mangrove Any of a range of trees and shrubs that form forests and dense thickets along muddy coasts and estuaries and in salt marshes, some of which have prop roots. Adventitious roots arch downward to the mud and send up new trunks. The prop roots trap sediment, building up the land and aiding SUCCESSION. Many mangrove species have PNEUMATOPHORES, or 'breathing knees', that rise up out of the mud. They have abundant lenticels, through which air diffuses to supply the waterlogged roots below the mud.

mannitol A soluble sugar alcohol (carbohydrate) found widely in plants and forming a characteristic food reserve of the Phaeophyta (brown algae). It is also the main soluble sugar in fungi and lichens. Mannitol is a hexahydric alcohol, i.e. each of the six carbon atoms has an alcohol (hydroxyl) group attached. It may be derived from mannose or fructose.

mannose A simple sugar found in many polysaccharides. It is an aldohexose, isomeric with glucose. In some plants, such as members of the family Fabaceae (legumes). It is the main unit of storage polysaccharides, forming polymers called mannans. It is also a component of some hemicelluloses. Reduction of mannose yields mannitol, a major sugar in fungi, lichens, and some algae.

mantle *See* mycorrhiza.

maquis A kind of Mediterranean scrub found on poor soils in regions with a pronounced dry season. It is made up of thickets of evergreen sclerophyllous (having thick leathery leaves) or spiny bushes and shrubs up to about 3 meters (9 ft) tall, e.g. gorse (*Ulex*), broom (*Cytisus*, *Genista*), laurels (*Laurus*), and members of the Ericaceae (heath family), interspersed with aromatic herbs and smaller shrubs, such as thyme (*Thymus*), *Cistus*, and myrtle (*Myrtus*), and scattered small trees such as olive (*Olea europaea*) and figs (*Ficus*). Similar vegetation in California is called CHAPARRAL.

Marattiaceae (**giant ferns**) A family of large tropical EUSPORANGIATE ferns, sometimes called the giant ferns, that dates back to the Carboniferous period. They have stout erect stems and large compound fronds, in some species, e.g. *Angiopteris evecta*, up to 4.5 m (15 ft) long.

Marchantiales An order of thallose liverworts with flat, ribbonlike, dichotomously branching thalli that lie close to the ground. *Marchantia*, a common species of gardens, riverbanks, and wet moorland, has umbrellalike, separate male and female reproductive structures on long upright stalks.

marker gene A gene of known location and function that can is used to establish the relative positions and functions of other genes. During gene transfer, a marker gene may be linked to the transferred gene to determine whether or not the transfer has been successful. *See* chromosome map; genetic engineering.

marsh *See* swamp.

Marsileaceae A family of highly specialized heterosporous water ferns. The leaves are borne on creeping rhizomes, and may be cylindrical and pointed, as in *Pilularia*, or divided into two or four leaflets, as in *Marsilea* and *Regnelidium*.

mass flow A hypothesis put forward by Münch (1930) to explain the mechanism of phloem transport. The movement of substances is believed to be the result of changes in osmotic pressure. Thus in an actively photosynthesizing region (source) where sugars are being produced the osmotic pressure is high and water is taken in. Conversely, in regions (sinks) where photosynthetic products are being used up or converted to storage compounds there is a lowering of osmotic pressure and water is lost. A system is then set up in which there is mass flow of water and dissolved sugars from source to sink.

medulla 1. *See* pith.
2. In fungal fruiting bodies, the layer of longitudinal hyphae.
3. In lichens, the layer of loosely organized hyphae lying below the cortex and algal layer.
4. In multicellular algal thalli, the inner layers of the thallus, comprising nonpigmented cells that often form a storage tissue.

medullary ray A linear group of living parenchyma cells in the secondary xylem of a woody plant. They extend from the vascular cambium for various distances into the wood.

megaphyll (**macrophyll**) A foliage LEAF with a branched system of veins in the blade and with the leaf trace leaving a leaf gap in the stele. It is typical of ferns and seed-bearing plants. The large often pinnately divided megaphyll of ferns, often termed a *frond*, contrasts with the generally much smaller leaves (MICROPHYLLS) of clubmosses and horsetails. The scale- or needlelike leaves of conifers have characteristics of both types since they have a single vein like a microphyll but leave a leaf trace like a megaphyll.

megasporangium (*pl.* **megasporangia**) A sporangium that produces megaspores. In *Selaginella* the megasporangium is borne in the axis of a sporophyll located in a strobilus. Usually all the spore mother cells degenerate except one, which forms a tetrad of cells. One or more of these may develop into MEGASPORES, which, when shed, develop into the female gametophyte. The ovule of seed plants may be considered equivalent to the megasporangium of the pteridophytes. *Compare* microsporangium. *See* heterospory.

megaspore The larger of the two types of spores in heterosporous pteridophytes and seed plants that produces the female gametophyte. In some pteridophytes the megaspores are released from the sporophyte, in others they are retained and fertilized on the sporophyte; in the seed plants they are retained. The megaspores of pteridophytes and gymnosperms produce a female gametophyte called a PROTHALLUS that forms two or more archegonia each containing a haploid female gamete. In angiosperms the megaspore becomes the embryo sac, lacking an obvious prothallus but containing, most commonly, eight nuclei, one of which is organized as the female gamete. Other angiosperms have four or sixteen or more nuclei in the embryo sac. *Compare* microspore.

megasporophyll A leaf, modified leaf, or leaflike structure that bears the megasporangium. Simple megasporophylls include the fertile ligulate leaves of heterosporous lycopods (e.g. *Selaginella*). They are usually grouped in a STROBILUS that may also contain MICROSPOROPHYLLS or vegetative leaves. The carpel of angiosperms and the ovuliferous scale of gymnosperms are modified megasporophylls.

meiosis The process of cell division leading to the production of daughter nuclei with half the genetic complement of the parent cell. Cells formed by meiosis give rise to gametes and fertilization restores the correct chromosome complement.

Meiosis consists of two divisions during which the chromosomes replicate only once. Like mitosis the stages PROPHASE, METAPHASE, and ANAPHASE can be recognized. However during prophase homologous chromosomes attract each other and become paired forming bivalents. At the end of prophase genetic material may be exchanged between the chromatids of homologous chromosomes. Meiosis also differs from mitosis in that after anaphase, instead of nuclear membranes forming, there is a second division, which may be divided into metaphase II and anaphase II. The second division ends with the formation of four haploid nuclei, which develop into gametes. *Compare* mitosis. *See* alternation of generations; telophase.

membrane A structure consisting mainly of lipid and protein (lipoprotein) surrounding all living cells as the *plasma membrane*, or *plasmalemma*, and also found surrounding organelles within cells. Membranes function as selectively permeable barriers, controlling passage of substances between the cell and its organelles, and their internal environment, either actively or passively. Membranes are typically 7.5–10 nm in thickness with two regular layers of lipid molecules (a *bilayer*) containing various types of protein molecules. Some proteins penetrate through the membrane, while others are associated with one side; some float freely over the surface while others remain stationary. Some are enzyme controlling, for instance, in the active transport of molecules or ions through the membrane. Larger molecules or particles can pass from one side of a membrane by ENDOCYTOSIS or EXOCYTOSIS.

The lipids are mostly phospholipids. These are polar molecules: one end (the phosphate end) is *hydrophilic* (water-loving) and faces outward, while the other end (consisting of two fatty acid tails) is *hydrophobic* (water-hating) and faces inward. Short chains of sugars may be associated with the proteins or lipids forming glycoproteins and glycolipids. The particular types of carbohydrates, lipids, and proteins determine the characteristics of the membrane, affecting, for example, cell–cell recognition (as in embryonic development and immune mechanisms), permeability, and hormone recognition. Membranes may contain efficient arrangements of molecules involved in certain metabolic processes, e.g. electron transport and phosphorylation (ATP production) in mitochondria and chloroplasts. *See* osmosis; freeze fracturing; plasma membrane.

membrane potential The potential difference that exists across a membrane as a result of the action of PROTON PUMPS, which maintain an imbalance of positive and negative ions on either side of the membrane. Such pumps are usually coupled to ATP: energy is required to maintain the potential difference.

Mendel, Gregor Johann (1822–84) Austrian plant geneticist. Having developed an interest in horticulture as a child growing up on a peasant farm, Mendel was able to carry out many plant breeding experiments after entering an Augustinian monastery at Brünn (now Brno in the Czech Republic). He used the pea plant in his experiments and selected seven characteristics of the plant to observe. First, he isolated his selected pure-bred parent plants and carried out his crosses by transferring pollen from one to another. He then collected and counted the seeds and grew them on, patiently observing and carefully recording the characteristics and their frequency in the next and often in successive generations. Mendel concluded from his results that each character is controlled by two 'factors' in the somatic cells, but each gamete can pass on only one of these fac-

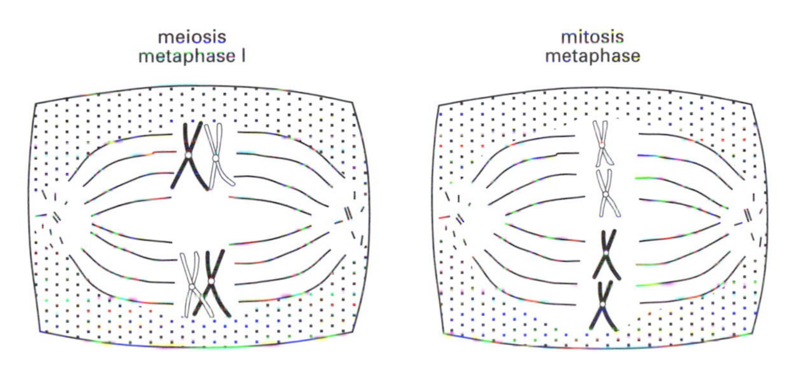

Meiosis: the differences in chromosome behavior in meiosis and mitosis

tors to the next generation. He also discovered that the factors were distributed at random in the gametes and that each pair of factors segregates independently of one another (*see* Mendelism; Mendel's laws). Mendel's 'factors' of inheritance are, in fact, genes and he is recognized as being the first person to provide genetics with a mathematical basis.

Mendel published his results in a local scientific journal and his paper was widely circulated, but unfortunately ignored. When Mendel became abbot of the monastery, he had to abandon his scientific studies. It was not until 1900, some 16 years after his death, that his work was rediscovered by DE VRIES and others and its accuracy and ingenuity appreciated.

Mendelian ratio The ratio of contrasting alleles present in the offspring of a cross involving genes that behave according to MENDEL'S LAWS.

Mendelism The theory of inheritance according to which characteristics are determined by particulate 'factors', or genes, that are transmitted by the germ cells. It is the basis of classical genetics, and is founded on the work of MENDEL in the 1860s. *See* Mendel's laws.

Mendel's laws Two laws formulated by MENDEL to explain the pattern of inheritance he observed in plant crosses. The first law, the Law of Segregation, states that any character exists as two factors, both of which are found in the somatic cells but only one of which is passed on to any one gamete. The second law, the Law of Independent Assortment, states that the distribution of such factors to the gametes is random; if a number of pairs of factors is considered, each pair segregates independently.

Today Mendel's 'characters' are termed genes and their different forms (factors) are called alleles. It is known that a diploid cell contains two alleles of a given gene, each of which is located on one of a pair of homologous chromosomes. Only one homolog of each pair is passed on to a gamete. Thus the Law of Segregation still holds true. Mendel envisaged his factors as discrete particles but it is now known that they are grouped together on chromosomes. The Law of Independent Assortment therefore applies only to unlinked genes. *See* linkage.

meristem A distinct region of actively or potentially actively dividing cells primarily concerned with growth. Numerous meristems occur in plants. In active meristems separation occurs between the cell that remains meristematic (initial) and the cell ultimately being differentiated. Two basic meristematic groups are the primary apical meristems at root and shoot apices, and the secondary lateral meristems, which include vascular and cork cambia. *See also* intercalary meristem; secondary growth.

meristoderm (limiting layer) The outermost cellular layer of the thallus of certain Phaeophyta (brown algae). It consists of small densely packed rectangular cells containing brown pigmented plastids and covered by a mucilaginous layer to prevent desiccation. The meristoderm maintains its meristematic activity and assists the outer cortical layers in adding to the thickness of the thallus and replacing tissues worn away by tidal action.

mesarch Denoting a stele or part of a stele in which the protoxylem is surrounded by metaxylem. *Compare* centrarch; endarch; exarch.

mesocarp *See* pericarp.

mesophilic Designating microorganisms with an optimum temperature for growth between 25–45° C. *Compare* psychrophilic; thermophilic.

mesophyll Specialized tissue located between the epidermal layers of the leaf and composed mainly of photosynthetic cells (CHLORENCHYMA). Veins, supported by sclerenchyma and collenchyma, are embedded in the mesophyll. *Palisade mesophyll* consists of cylindrical cells, at right angles to the upper epidermis, with many chloroplasts and small intercellular spaces.

It is the main photosynthesizing layer in the plant. *Spongy mesophyll*, adjacent to the lower epidermis, comprises interconnecting irregularly shaped cells with few chloroplasts and large intercellular spaces that communicate with the atmosphere through stomata allowing gas exchange between the cells and the atmosphere. The distribution of mesophyll tissue varies in different leaves depending on the environment in which the plant lives.

mesophyte A plant that is adapted to grow under adequate conditions of water supply and has no particular adaptations to withstand environmental extremes. In drought conditions wilting is soon apparent as the plants have no special mechanisms to conserve water. Most angiosperms are mesophytes. *Compare* hydrophyte; xerophyte.

Mesozoic The middle era in the most recent (Phanerozoic) eon of the geological time scale, dating from about 230–66 million years ago. It is divided into three main periods: the Triassic, Jurassic, and Cretaceous. During the Triassic period seed-ferns predominated, but as the climate warmed the cycads came to dominate, and the Mesozoic is sometimes called 'The Age of Cycads'. In the Jurassic period, tropical and temperate regions were dominated by forests of palmlike cycads and conifers, and the dawn redwoods made their first appearance. Ginkgoes were also widespread. The first angiosperms arose in the Cretaceous period, and rapidly diversified, becoming the dominant vegetation by the end of the Mesozoic as the climate cooled and dried. *See also* geological time scale

messenger RNA (mRNA) The form of RNA that transfers the information necessary for protein synthesis from the DNA in the nucleus to the ribosomes in the cytoplasm. One strand of the double helix of DNA acts as a template along which complementary RNA nucleotides become aligned and joined together by the enzyme mRNA polymerase. These form a polynucleotide identical to the other DNA strand, except that the thymine bases are replaced by uracil. This polynucleotide is called *heterogeneous nuclear RNA* (hnRNA) and it contains both coding and noncoding sequences (*see* exon; intron). In the nucleolus POST-TRANSCRIPTIONAL PROCESSING occurs: adenine polynucleotide (ply A) is added to one end and a guanine derivative to the other, and (in eukaryotic cells) the introns are then removed to produce mRNA (sometimes called mature mRNA). The whole process is termed *transcription*. The new mRNA molecule thus has a copy of the genetic code, which directs the formation of proteins in the ribosomes. *Compare* transfer RNA.

metabolism The chemical reactions that take place in cells. The molecules taking part in these reactions are termed *metabolites*. It is metabolic reactions, particularly those producing energy, that keep cells alive. Metabolic reactions characteristically occur in small steps, comprising a *metabolic pathway*. Metabolic reactions involve the breaking down of molecules to provide energy (CATABOLISM) and the building up of more complex molecules and structures from simpler molecules (ANABOLISM).

metabolite *See* metabolism.

metaphase The stage in mitosis and meiosis when the chromosomes become aligned along the equator of the nuclear spindle. Metaphase follows prophase and precedes anaphase.

metaphloem The primary PHLOEM formed from the PROCAMBIUM after the PROTOPHLOEM. It is found behind the zone of elongation below the meristem, and is more durable than the protophloem. In plants showing no secondary thickening the metaphloem is responsible for the transport of most of the organic materials in the plant, but this function is taken over by the secondary phloem in regions where secondary tissues have differentiated.

metaxylem The primary XYLEM elements that are differentiated from the PROCAMBIUM after the PROTOXYLEM. They are found some distance behind the apical meristem beyond the zone of elongation. The secondary cell walls show reticulate and scalariform thickening and are thus inextensible.

methionine A nonpolar sulfur-containing AMINO ACID synthesized from aspartate. Methionine can be broken down to succinyl CoA, which is further oxidized in the Krebs cycle. The ATP-activated derivative of methionine, S-adenosyl methionine, donates its methyl group to many different molecules in various metabolic pathways.

methylene blue A blue dye used in light microscopy to stain bacterial protoplasm. It is also used to stain nuclei blue. *See* staining.

microbiology The study of microscopic organisms (e.g. bacteria and viruses), including their interactions with other organisms and with the environment.

microbody A common organelle of plant and animal cells, bounded by a single membrane, spherical, and usually about 0.0–1.0 μm in diameter. Microbodies originate from the ENDOPLASMIC RETICULUM and contain oxidative enzymes such as catalase, which decomposes hydrogen peroxide, a toxic waste product of the activities of other enzymes in the microbody.

GLYOXYSOMES contain enzymes of the glyoxylate cycle, transaminases, and enzymes associated with β-oxidation of fatty acids. They play a major role in conversion of lipids to sucrose in fatty or oily seedling tissues, e.g. the endosperm of castor-oil seeds. PEROXISOMES occur in certain animal tissues, and are numerous in photosynthetic cells of plant leaves, where they are concerned with glycolate metabolism in photorespiration and contain high levels of glycolate oxidase and other associated enzymes. Glycolate comes from chloroplasts and products such as glycine are passed to mitochondria. Hence these three organelles often appear close together. *See* photorespiration.

microevolution EVOLUTION on a scale within a species, which results from the action of natural selection on the genetic variation between individuals of a given population.

microfibril A ribbonlike structure at least 60 nm long and about 45 nm in diameter, made of parallel chains of cellulose molecules linked by hydrogen bonds. Microfibrils are found in the CELL WALLS of plants and some algae. In secondary cell walls they are packed more closely, leaving less room for water between them.

microfilament A minute filament, about 0.4–0.7 nm wide, found in eukaryotic cells and having roles in cell motion and shape. Microfilaments are made of two helically twisted strands of globular subunits of the protein ACTIN, almost identical to the actin of muscle. They can undergo rapid extension or shortening by subunit assembly or disassembly, or form complex three-dimensional networks. Their occasional association with myosinlike protein (as in muscle) suggests they may also be contractile. They often occur in sheets or bundles just below the plasma membrane and at the interface of moving and stationary cytoplasm. They are involved in CYTOKINESIS, CYCLOSIS, cell movements, e.g. ameboid movement, and movement of subcellular components, e.g. pinocytotic vesicles.

micrograph A photograph taken with the aid of a microscope. *Photomicrographs* and *electron micrographs* are produced using optical microscopes and electron microscopes respectively.

micrometer[1] In microscopy, a device for measuring the size of an object under the microscope. An eyepiece micrometer (*graticule*) of glass or transparent film, with a scale etched or printed on it, is placed in the eyepiece so that both the object to be measured and the scale are in focus. The scale of the graticule changes at different magnifications and it must therefore be calibrated against a stage micrometer, which is a glass slide with a scale etched into it, and placed on the microscope stage.

micrometer[2] (**micron**) Symbol: μm A unit of length equal to 10^{-6} meter (one millionth of a meter). It is often used in measurements of cell diameter, sizes of bacteria, etc. Formerly, it was called the *micron.*

micron *See* micrometer.

micronutrient (**trace element**) A nutrient required in trace amounts by an organism. For example, a plant can obtain sufficient of the essential trace element manganese from a solution containing 0.5 parts per million of manganese. Many act as enzyme cofactors or are components of pigments. They include boron, cobalt, copper, manganese, molybdenum, and zinc. *See* deficiency disease.

microphyll A foliage leaf that has a single unbranched vein running from base to apex and no leaf gap associated with the leaf trace: the stele remains entire when a leaf trace branches off to a leaf. Microphylls are the typical leaves of clubmosses and horsetails. *Compare* megaphyll.

micropropagation The production of large numbers of plants under laboratory conditions. It usually involves the use of TISSUE CULTURE to produce large numbers of genetically identical plants (clones), but can involve seeds, e.g. of orchids. It is used to produce disease-free plants, and also to multiply genetically engineered plants, and to increase rare plant populations for conservation purposes.

micropyle A pore leading to the nucellus formed by incomplete integument growth around the apex of the ovule. Pollen tubes usually pass through it prior to fertilization. In most seeds the micropyle forms a small hole in the testa through which water is absorbed, but in some seeds it is closed.

microscope An instrument designed to magnify objects and thus increase the resolution with which one can view them. *Resolution* is the ability to distinguish between two separate adjacent objects. Radiation (light or electrons) is focused either through the specimen by a *condenser lens* or onto it. The resulting image is magnified by further lenses. Since radiation must pass through the specimen in transmission microscopy, it is usual to cut larger specimens into thin slices of material (sections) with a microtome. Biological material has little contrast and is therefore often stained. If very thin sections are required the material is preserved and embedded in a supporting medium.

The *light microscope* (optical microscope) uses light as a source of radiation. With a *compound microscope* the image is magnified by two lenses, an *objective lens* near the specimen, and an *eyepiece*, where the image is viewed, at the opposite end of a tube. Its maximum magnification is limited by the wavelength of light: it can distinguish between two points only 0.3 μm apart. Its main use is the observation of thin sections of tissues and organs, and the study of fine detail of microorganisms. For viewing relatively large specimens and for dissection, the lower resolution stereoscopic binocular microscope is used: it uses incident light and has two eyepieces, giving a three-dimensional image. Various modifications of the light source and background illumination allow different aspects of cell structure to be observed, e.g. bright-field illumination, DARK-GROUND ILLUMINATION, PHASE CONTRAST MICROSCOPY, interference microscopy, and CONFOCAL MICROSCOPY. Much greater resolution became possible with the introduction of the *electron microscope*, which uses electrons as a source of radiation, because electrons have much shorter wavelengths than light. However, only dead material can be observed because the specimen must be in a vacuum and electrons eventually heat and destroy the material. The electron microscope uses electromagnetic lenses to focus a parallel beam of electrons onto an object and produce an image of the resulting electron scattering projected onto a photographic plate or fluorescent screen. The process takes place in a vacuum to minimize unwanted electron scattering. Since the wavelength of an electron beam is 0.005 nm, an electron microscope has about 300 times greater resolution than a light microscope. In theory, this microscope can distinguish between objects 0.0025 nm apart, but in practice, resolutions of less than about 1 nm are rarely attained. Electron microscopes are of two main types, the *transmission electron microscope* (*TEM*) and the *scanning electron microscope* (*SEM*). The former produces an image by passing electrons through the specimen. With the scanning microscope electrons scan the surfaces of specimens rather as a screen is scanned in a TV tube, allowing surfaces of objects to be seen with greater depth of field and giving a three-dimensional appearance to the image. The surface of entire objects can be scanned, and material to be viewed is often given an electron-reflecting coating. Scanning microscopes cannot operate at such high magnifications as transmission microscopes, but can still resolve to at least 2 nm. Of growing importance is the *scanning probe microscope* (atomic force microscope), in which a highly sensitive probe mounted on a cantilever is moved to and fro over a surface, and responds to the minute intermolecular distance between the probe and the surface, which causes the probe to be deflected from the surface. The deflection of the probe is measured with a laser beam, providing a topographical map of the surface without penetrating it. The probe may be dragged over the surface of the object, tap the object at intervals, or travel just above the surface. The scanning probe microscope can achieve resolutions of less than 1 nm in x-, y-, and z-directions, and can produce images of individual molecules such as myosin and DNA. In non-touching mode it provides a tool for the study of real-time changes in the surface without interfering with processes causing the changes, e.g. the change in structure of starch as it is digested by an enzyme. Modifications of the technique can give further information; it is used, for example, to measure the volume of chromosomes.

microsomes Fragments of endoplasmic reticulum and Golgi apparatus in the form of vesicles formed during homogenization of cells and isolated by high-speed centrifugation. Microsomes from rough endoplasmic reticulum are coated with ribosomes and can carry out protein synthesis in the test tube.

microsporangium (*pl.* **microsporangia**) In heterosporous plants, the sporangium that produces the MICROSPORES, located on the microsporophyll. The microsporangium wall splits to disperse the mature microspores. Microsporangia are found in some pteridophytes (e.g. *Selaginella*, water ferns) and are represented by the pollen sacs in gymnosperms and angiosperms. *Compare* megasporangium.

microspore The smaller of the two types of spores produced in large numbers by seed plants and heterosporous pteridophytes. In pteridophytes, microspores develop into the male gametophyte generation, but in gymnosperms and angiosperms, the microspores (POLLEN grains) develop into a very reduced gametophyte represented by the pollen tube and the vegetative nucleus (pollen-tube nucleus) and the generative nuclei. The latter are the male gametic nuclei. *Compare* megaspore.

microsporophyll A leaf or modified leaf on which the microsporangium is borne. Simple microsporophylls include the scales found in the male cones of gymnosperms and the fertile photosynthetic leaves of clubmosses, usually grouped in a strobilus. In gymnosperms the male scale is the microsporophyll, while in angiosperms the stamen is a highly modified microsporophyll. *Compare* megasporophyll.

microtome An instrument for cutting thin sections (slices a few micrometers thick) of biological material for microscopic examination. The specimen is usually embedded in wax for support and cut by a steel knife. Alternatively it is frozen and a *freezing microtome*, which keeps the specimen frozen while cutting, is used (cryo-microtomy). For electron microscopy, extremely thin (20–100 nm) sections can be cut by an *ultramicrotome*. Here the specimen is embedded in resin or plastic for support and mounted in an arm that advances slowly, moving up and down, toward a glass or diamond knife. As sections are cut they float off on to the surface of water contained in a trough behind the knife.

microtubule A thin cylindrical unbranched tube of variable length found in eukaryotic cells, either singly or in groups. Its walls are made of the globular protein *tubulin*. Microtubules have a skeletal role, helping cells to maintain their shape. They occur freely in the cell, and also form part of the structure of the SPINDLE during cell division, bringing about chromosome movement. Microtubules also help to orientate materials and structures in the cell, e.g. cellulose fibrils during the formation of plant cell walls, and may be involved in the transport of materials with and between cells, e.g. phloem transport. They are also part of the centrioles found in many fungal and algal cells, and the basal bodies and undulipodia of many motile unicells.

middle lamella In mature plant tissues, the material cementing the walls of adjacent cells. It is derived from the first-formed layer of the primary cell wall, which is laid down by the PHRAGMOPLAST (cell plate), and is composed mainly of calcium and magnesium pectates.

midrib 1. The thick vein running down the middle of the leaf from the petiole to the leaf tip of a plant.
2. Any thickened linear structure running down the center of an algal thallus, such as a seaweed, e.g. *Fucus* spp.

mildew A fungal disease of plants in which the hyphae appear on the surface of the plant.

Mississippian In North America, the Lower Carboniferous period, dating from 345 to 310 million years ago. *See* Carboniferous; Pennsylvanian.

mitochondrial DNA (**mt-DNA**) The DNA found in mitochondria. It is a circular molecule not associated with histones or other proteins and independent of nuclear DNA. It codes for specific RNA components of ribosomes, and also for certain respiratory enzymes that occur in the mitochondria and are synthesized on mitochondrial ribosomes. Plant mt-DNA evolves very slowly (unlike animal mt-DNA), and can break apart and recombine. It is usually transmitted from females to their offspring in the cytoplasm of the ova, but mitochondria do not usually pass into male gametes.

mitochondrion (*pl.* **mitochondria**) An organelle of all plant and animal cells chiefly associated with aerobic respiration. It is surrounded by two membranes separated by an intermembrane space; the inner membrane forms fingerlike processes called *cristae*, which project into the gel-like *matrix*. Mitochondria are typically sausage-shaped, but may assume a variety of forms, including irregular branching shapes. The diameter is always about 0.5–1.0 μm, and the length averages 2 μm. Mitochondria contain the enzymes and cofactors of AEROBIC RESPIRATION and therefore are most numerous in active cells (up to several thousand per cell). The reactions of the KREBS CYCLE take place in the matrix and those of electron transport coupled to oxidative phosphorylation (i.e. the respiratory chain) on the inner membrane. Within the membrane the components of the respiratory chain are highly organized. The matrix is also involved in amino acid metabolism via Krebs cycle acids and transaminase enzymes, and in fatty acid oxidation. Like chloroplasts, mitochondria contain their own ribosomes and DNA, which is circular in form, like that of bacteria, and reproduce by binary fission. It is believed that mitochondria and chloroplastsmay be the descendants of once independent organisms that early in evolution invaded eukaryotic cells, leading to an extreme form of symbiosis. *See* endosymbiont theory. *See also* electron-transport chain.

mitosis (**karyokinesis**) The ordered process by which the cell nucleus and cytoplasm divide in two during the division of body (i.e. nongermline) cells. The chromosomes replicate prior to mitosis to form two sister chromatids, which are then separated during mitosis in such a way that each daughter cell inherits a genetic complement identical to that of the parent cell. Although mitosis is a continuous process it is divided into four phases; prophase, metaphase, anaphase, and telophase. Telophase may be followed by cytokinesis. *Compare* meiosis; amitosis; endomitosis.

mitosporic fungi *See* Fungi Anamorphici.

molecular systematics A branch of biology that compares functionally equivalent macromolecules from different organisms as a basis for classification. Sequences of amino acids in proteins (e.g. enzymes) or of nucleotides in nucleic acids (e.g. ribosomal RNA) are determined using automated techniques, and compared statistically using sophisticated computer programs. Essentially, how closely two organisms are related in evolutionary terms is reflected in the degree of similarity of their macromolecules.

molybdenum A MICRONUTRIENT (trace element) needed for plant growth. It forms part of the enzymes nitrate reductase, which is involved in nitrate reduction, and nitrogenase, the key enzyme in nitrogen fixation. Deficiency of molybdenum may lead to CHLOROSIS between the leaf veins.

Monera An alternative name for the kingdom BACTERIA (Prokaryotae).

monochasial cyme (**monochasium**) *See* inflorescence.

Monocotyledonae A class of the flowering plants (phylum Anthophyta) deriving their name from single cotyledon in the embryo (though some dicotyledons have this, too). They are usually herbaceous plants and most do not show secondary growth. Examples of monocotyledons are the grasses (Poaceae) and lilies (Liliaceae).

monoculture The growing of a single species or variety of crop over a large area. The economic reasons for this are the ability to grow crops of uniform height for harvesting, and economies of scale in the harvesting process. The major disadvantage is increased susceptibility to pests and pathogens, especially if a single variety is grown.

monoecious Denoting plants in which the male and female reproductive organs are borne on separate structures on the same individual plant. Monoecious flowering plants bear separate unisexual male and female flowers, e.g. maize and many temperate trees such as birch, hazel, and oak. Many monoecious plants are wind-pollinated. *Compare* dioecious; hermaphrodite.

monohybrid An indiviudal heterozygous at one locus (i.e. for a single gene), obtained by crossing homozygous parents with different alleles at a given locus; for example, Mendel's cross between tall (TT) and dwarf (tt) garden peas to give a tall monohybrid (Tt). When a monohybrid is selfed, dominant and recessive phenotypes appear in the offspring in the ratio of 3:1 (the *monohybrid ratio*). *Compare* dihybrid.

monoploid *See* haploid.

monopodial Describing the system of branching in plants in which the main axis of the stem (the monopodium) continues to grow indefinitely by the terminal bud. Monopodial growth is also called indefinite or racemose branching and is typical of the formation of a racemose inflorescence. *Compare* sympodial.

monosaccharide A SUGAR that cannot be hydrolyzed to simpler carbohydrates of smaller carbon content. Glucose and fructose are examples. Monosaccharides may be classified according to the number of carbon atoms they contain: 5-carbon pentoses (e.g. ribose); 6-carbon hexoses (e.g. glucose, fructose). The general formula of a monosaccharide is $(CH_2O)_n$.

monosomy *See* aneuploidy.

monotrichous Describing BACTERIA that possess one flagellum, e.g. *Vibrio.*

Moraceae A family of trees, shrubs, lianas, and herbs, found mainly in tropical and subtropical regions. There are about 12 000 species. The Moraceae contains many important food plants, such as edible figs (*Ficus carica*), breadfruit (*Artocarpus artilis*), jackfruit (*A. heterophyllus*), and mulberry (*Morus*).

morel *See* Pezizales.

morphogenesis The development of form and structure.

morphology The study of the form of organisms. The term may be used synonymously with 'anatomy' although generally the study of external form is termed 'morphology' while the study of internal structures is termed 'anatomy'.

mosaic **1.** A hybrid organism whose cells differ genetically, although they have all arisen from a single zygote.
2. A pattern of light and dark green patches on the leaves of a plant, usually due to viral infection.
3. *See* leaf mosaic.

mosquito ferns *See* Azollaceae.

mosses In older classifications, a class of bryophytes containing leafy plants with multicellular rhizoids. In the FIVE KINGDOMS CLASSIFICATION, mosses comprise the phylum BRYOPHYTA. They differ from liverworts in their greater differentiation of the gametophyte and also their complex mechanisms of capsule dehiscence, with no formation of elaters in the capsule. Orders include the Bryales (e.g. *Funaria*, *Polytrichum*, *Mnium*) and the Sphagnales (e.g. *Sphagnum*).

mucilage Any of various gumlike carbohydrates that swell when wet, becoming slimy and jellylike. *See gum.*

mucopolysaccharide *See* polysaccharide.

Mucorales **(pin molds)** An order of the ZYGOMYCOTA, containing fungi that form a dense mycelium bearing sporangia or conidia. The tips of the spore-bearing hyphae often become pigmented, hence the name pin molds (e.g. *Mucor mucedo*). Most are saprobes in soil, dung, and other organic debris. Many cause spoilage of stored food. A few are parasites on other fungi, plants, or animals.

multicellular Consisting of many cells.

multifactorial inheritance *See* continuous variation.

multiple allelism The existence of a series of alleles (three or more) for one gene. Only two alleles of the series can be present in a diploid cell. Dominance relationships within an allelic series are often complicated. Multiple alleles are common in the incompatibility systems of plants, such as *Sinapis*, poppy (*Papaver*), and clover (*Trifolium*). The numbers of such alleles are often considerable, e.g. Brussels sprouts (*Brassica oleracea* var. *bullata*) has up to 40 incompatibility alleles, or *S alleles*.

multiple fruit **(composite fruit)** A type of pseudocarpic ('false') fruit that incorporates parts of the inflorescence. Examples are figs (*Ficus*) and pineapples (*Ananus comosus*).

mutagen Any physical or chemical agent that induces MUTATION or increases the rate of spontaneous mutation. Chemical mutagens include compounds that react with nucleotides, e.g. nitrous acid, proflavin; base analogues, which are incorporated in place of normal bases during DNA replication; and acridines, which cause base-pair deletions or additions. Physical mutagens include short-wave radiation such as ultraviolet light, x-rays, and cosmic rays; and ionizing radiation, such as α and β particles.

mutation Mutations are the ultimate source of all genetic variation. They are inherited only if they occur in the cells that give rise to the gametes; somatic mutations may give rise to chimaeras and cancers. GENE MUTATIONS alter only a single gene, resulting in new allelic forms of the gene and hence new variations upon which natural selection can act. Most mutations are deleterious but are often retained in the population because they also tend to be recessive and can thus be carried in the genotype without affecting the viability of the organism. The natural rate of mutation is low, and varies with different gene loci. The mutation frequency can be increased by mutagens. *See also* chromosome mutation; mutagen; polyploid.

mutualism The close relationship between two or more species of organisms in which all benefit from the association. There are two types of mutualism: *obligatory mutualism*, in which one cannot survive without the other, for example the algal/fungal partnership found in lichens; and *facultative* or *nonobligatory mutualism*, in which one or both species can survive independently, for example, in the association between the bacterium *Rhizobium* and members of the family Fabaceae, which leads to the formation of root nodules in which nitrogen fixation occurs, the bacterium can live independently of the plant; in lichens often the alga can live independently of the fungus. In many mutualistic associations, the relationship can change from a parasitism to a mutualism or vice versa. *Compare* symbiosis; commensalism; amensalism.

mycelium (*pl.* **mycelia**) A filamentous mass comprising the vegetative body of a fungus, each filament being called a hypha. The mycelium often forms a loose mesh as in *Mucor*, or hyphae may anastomose to form a netlike structure, as in Ascomycota and Basidiomycota. *See* hypha.

mycoplasmas (**PPLO; pleuropneumonia-like organisms**) A group of extremely small bacteria that naturally lack a rigid cell wall. They often measure less than 200 nm in diameter and their cells are delicate and plastic. Mycoplasmas are implicated in certain plant diseases, such as witches broom of alfalfa, maize stunt, and yellows diseases. They can kill tissue cultures and cause serious diseases in humans and animals, e.g. pleuropneumonia.

mycorrhiza (*pl.* **mycorrhizae**) The association between the hyphae of a fungus and the roots of a higher plant. Two main types of mycorrhiza exist, *ectomycorrhizae* (*ectotrophic mycorrhizae*), in which the fungus forms a *mantle* around the smaller roots, as in many temperate and boreal trees, and *endomycorrhizae* (*endotrophic mycorrhizae*), in which the fungus grows around and within the cortex cells of the roots, as in many herbaceous plants. Orchids and heathers have further specialized types of mycorrhizae, in which there is often a mantle and cell penetration. In ectomycorrhizae the fungus, which is usually a member of the Agaricales, benefits by obtaining carbohydrates and possibly B-group vitamins from the roots. The trees benefit in that mycorrhizal roots absorb nutrients more efficiently than uninfected roots, and it is common forestry practice to insure the appropriate fungus is applied when planting seedling trees. In endomycorrhizae the fungus is generally a species of *Rhizoctonia* and again both partners benefit nutritionally from the relationship. In a special form of endomycorrhiza, the *vesiculo-arbuscular mycorrhiza* (*arbuscular mycorrhiza*), the fungus lives between the cells of the cortex and projections of its hyphae actually penetrate the cortical cells. Sometimes the projections are finely branched hyphae called *arbuscules*.

myosin A contractile protein found in cells, which is involved in the movement of cytoplasm, for example in the cytoplasmic streaming of plant cells, the ameboid movement of some protoctists and fungus gametes, and the movement of slime mold plasmodia. Myosin is thought to interact with the protein ACTIN to bring about movement.

Myrtaceae A family of trees, shrubs and creepers found in most parts of the tropics and warm temperate regions, especially Australia. The family includes many important forest trees such as *Eucalyptus* species, especially in Australia, sources of timber, tannin and oils. Some species produce edible fruits, e.g. guava (*Psidium guajava*), or spices, e.g. allspice (*Pimenta dioica*) and cloves (*Eugenia caryophyllus*).

myxobacteria (**gliding bacteria; slime bacteria**) A group of bacteria in which individual cells are typically rod-shaped and covered in a slime made from polysaccharides, but which may congregate to form gliding swarms or various upright reproductive structures under certain conditions.

Myxomycota The phylum of protoctists that contains the plasmodial slime molds. They occur in two forms: as individual amebae that feed on bacteria by engulfing them with pseudopodia; and as a multinucleate haploid or diploid mass of protoplasm called a *plasmodium* that can move around and is often pigmented.

N

NAD (nicotinamide adenine dinucleotide) A pyridine-based nucleotide that acts as a coenzyme to oxido-reductase enzymes in electron-transfer reactions, such as the oxidation of citrate to α-ketoglutarate in the KREBS CYCLE. The reduced form is written as NADH, and may act as a reducing agent or be reoxidized in the respiratory chain coupled to ATP formation. Its main importance is as an electron acceptor in oxidation reactions.

NADP (nicotinamide adenine dinucleotide phosphate) A pyridine-based nucleotide that acts as a COENZYME to oxido-reductase enzymes. Unlike NAD, it usually acts as an electron donor in enzyme-catalyzed reduction reactions.

Naegeli, Karl Wilhelm von (1817–91) Naegeli was educated at Zurich, Geneva, and Berlin, where he studied philosophy before resuming his botanical work. He produced an accurate account of cell division (1842) while studying pollen formation and even observed chromosomes in dividing nuclei, calling them 'transitory cytoblasts'. His observations disproved SCHLEIDEN'S theory that new cells were formed by budding off the nuclei of existing cells. Working on apical growth in plants in 1845, he concluded that there were distinct regions of cell division (meristems). Naegeli also laid the foundations of cell ultrastructure when he studied starch grains in cells and formulated his micellar theory. Sadly, his own entrenched views on heredity and evolution, which arose from his studies on the hawkweed (*Hieracium*), led him to reject MENDEL's very important work on pea plants.

nano- Symbol: n A prefix denoting one thousand-millionth, or 10^{-9}. For example, 1 nanometer (nm) = 10^{-9} meter. *See* SI units.

nastic movements (nasties) Movements of plant parts in which direction of movement is independent of the direction of the stimulus that induces them. With *photonasty* the stimulus is light. For instance, at constant temperature *Crocus* and tulip (*Tulipa*) flowers open in the light and close in the dark because of slight growth movements. Opening is caused by *epinasty* – i.e. greater growth of the upper surface of a plant organ – and closed by *hyponasty* – i.e. greater growth of the lower surface of a plant organ. Similarly, at constant light intensity crocus and tulip flowers show *thermonasty*; i.e. they open in warm air and close in cool air. Such day–night rhythms are examples of *nyctinasty* and may also occur in leaves, as in *Oxalis*. Nongrowth nastic movements also occur and are more rapid. The 'sensitive plant' (*Mimosa pudica*) rapidly closes its leaflets upwards and the petioles droop in response to touch, shock (*seismonasty*), or injury. This plant, and many other legumes, shows nongrowth nyctinastic movements. Movement is the result of osmotic changes in special swollen groups of cells (*pulvini*) at the bases of the moving structures. When the stimulus is contact, the movement is *haptonastic* as in the closure movements of insectivorous plants. The closure of the two halves of the Venus fly-trap leaf (*Dionaea muscipula*) is a nongrowth haptonastic movement caused by a loss of turgor in the cells along the midrib following stimulation of the sensitive hairs on the leaf. *See also* taxis; tropism.

nasties *See* nastic movements.

natural selection The process, which DARWIN called the 'struggle for survival', by which organisms less adapted to their environment tend to perish, and better-adapted organisms tend to survive. There has been much debate about the level at which natural selection acts: traditionally it is regarded as the action of the environment on individual organisms that determines which genotypes survive and reproduce, but there are strong arguments for natural selection acting at the level of the gene (the 'selfish gene' hypothesis) and some authors argue that it happens in the long term at higher levels such as population or species. In the end the less successful types will die out. According to DARWINISM, natural selection acting on a varied population results in evolution.

necrosis The death of a cell or group of cells in a living plant, especially where the dead area is discolored. Necrosis is often caused by fungal infections.

necrotrophic Describing a parasitic organism that obtains its nutrients from dead cells and tissues of its host.

nectar A sugar-containing fluid secreted by the nectaries in plants. It is mainly produced in flowers and attracts insects or other animals, who feed on it and may unintentionally transfer pollen from one flower to another, thus facilitating cross-pollination. Floral nectar contains variable proportions of different monosaccharide and disaccharide sugars and, frequently, other substances such as amino acids and volatile compounds as well. Some plants, particularly trees, produce nectar on stems and leaves. This may be collected by insects such as ants that can protect the plant from insect herbivores.

nectary A patch of glandular epidermal cells usually on the receptacle or the petals, but sometimes in other parts of the flowers or in specialized organs. They produce a sugary liquid (nectar) that attracts insects.

negative feedback *See* feedback loop.

negative staining A method of preparation of material for electron microscopy used for studying three-dimensional and surface features, notably of viruses, macromolecules (e.g. enzyme complexes), and the cristae of mitochondria. A stain is used that is not taken up by certain components of the specimen. Usually the stain covers the background and penetrates surface features of the specimen, but leaves the specimen itself unstained. For example, nigrosin or Indian ink is used to make bacteria visible. In electron microscopy a similar technique mixes the specimen material with electron-dense negative stains, such as phosphotungstic acid (PTA) or potassium molybdate, which is not taken up by proteinaceous material in the specimen. The electron-transparent proteinaceous material shows up as dark areas against a light background on the resulting electron micrograph. *See* staining.

neo-Darwinism DARWIN's theory of evolution through NATURAL SELECTION, modified and expanded by genetic studies arising from the work of MENDEL and his successors. This fusion of Darwin's theory of natural selection with Mendel's genetics was called the *neo-Darwinian synthesis* in the 1930s. The inclusion of genetics in evolutionary studies answered many questions that Darwin's theory raised but could not adequately explain because of lack of knowledge at the time it was formulated. Notably, genetics has revealed the source of variation on which natural selection operates, namely mutations of genes and chromosomes, and provided mathematical models of how alleles fluctuate in natural populations, thereby quantifying the process of evolution. More recent discoveries in molecular biology have added to our understanding of the causes of variation and the nature of evolution at the molecular level.

neritic *See* sublittoral.

neutral theory The theory that most evolutionary changes are caused by random GENETIC DRIFT (random changes in the frequencies of alleles in a population) rather than by NATURAL SELECTION. Genetic MUTATIONS arise at random. Many are neutral – they do not affect the fitness or survival of the carrier. They therefore survive in the genome and changes in their frequencies are due to the breeding system and chance rather than to natural selection. *See* gene frequency; variation.

niacin *See* nicotinic acid.

niche *See* ecological niche.

nicotinic acid (**niacin**) A carboxylic acid that is one of the water-soluble B-group of VITAMINS. It also forms part of the alkaloid nicotine, hence its name. It is sythesized from various precursors, but especially from tryptophan. As nicotinamide, it is a constituent of two coenzymes, NAD and NADP, which operate as hydrogen and electron-transfer agents and play a vital role in metabolism. In humans, deficiency of nicotinic acid causes pellagra. Diets based on maize, which is particularly poor in nicotinic acid, tend to lead to an increase in pellagra in the population.

Nidulariales The bird's nest fungi – an order of the BASIDIOMYCOTA in which the basidioma is nestlike, usually cup-shaped (e.g. splash cups, *Cyathus striatus*) or funnel-shaped (e.g. white-egg bird's nest, *Crucibulum laeve*), containing several separate spore cases (the 'eggs') that are exposed when the covering membrane dies back. The spores are dispersed by rain splash. Bird's nest fungi may be found on the ground, or on wood or dung.

ninhydrin A reagent used to test for the presence of proteins and amino acids. A colorless aqueous solution turns blue in the presence of α-amino acids in solution. When dissolved in an organic solvent it is used as a developer to color amino acids on chromatograms. If a chromatogram treated with ninhydrin is heated strongly the amino acids appear as purple spots that can be identified by measuring the Rf value. Ninhydrin is carcinogenic. *See also* paper chromatography.

nitrification The oxidation of ammonia to nitrite, and/or nitrite to nitrate, carried out by certain *nitrifying bacteria* in the soil. The chemosynthetic bacteria *Nitrosomonas* and *Nitrobacter* carry out the first and second stages respectively of this conversion. The process is important in the NITROGEN CYCLE since many plants assimilate nitrate as their source of nitrogen. *Compare* denitrification.

nitrifying bacteria *See* nitrification.

nitrogen An essential element found in all amino acids and therefore in all proteins, and in various other important organic compounds, e.g. nucleic acids. Gaseous nitrogen forms about 80% of the atmosphere but is unavailable in this form except to a few nitrogen-fixing bacteria. Nitrogen is incorporated into plants as the nitrate ion, NO_3^- or, especially in acid soils, the ammonium anion NH_4^+, absorbed in solution from the soil by roots. Plants suffering from nitrogen deficiency tend to develop CHLOROSIS and become etiolated, the effects being seen first in the oldest parts. *See also* nitrogen cycle.

nitrogenase *See* dinitrogenase.

nitrogen cycle The circulation of nitrogen between organisms and the environment. Atmospheric gaseous nitrogen can be used directly only by certain nitrogen-fixing bacteria (e.g. *Clostridium*, *Nostoc*, *Rhizobium*). They convert nitrogen to ammonia, nitrites, and nitrates, which are released into the soil by excretion and decay. Some are free-living, while others form symbiotic associations with plants (*see* nitrogen fixation). Another method by which atmospheric nitrogen is fixed is by lightning, which causes nitrogen and oxygen to combine. The oxides so produced dissolve in rain to form nitrous and nitric acids; in the soil these acids combine with mineral salts to form nitrites and nitrates. When plants and animals die, the organic nitrogen they contain is converted back into nitrate in the process termed *nitrification*. Apart from uptake by plants, nitrate may also be lost from the soil by *denitrification* and by leaching. The use of nitrogen fertilizers in agriculture and the emission of nitrous oxides in car exhaust fumes have influenced the nitrogen cycle and contributed to urban air pollution and acid precipitation. *See* acid rain.

nitrogen fixation The formation of nitrogenous compounds from atmospheric nitrogen. In nature this may be achieved by electric discharge in the atmosphere or by the activities of certain microorganisms. For example, symbiotic bacteria of the genus *Rhizobium* are associated with leguminous plants (FABACEAE), causing the root cortex to form *root nodules* that house the bacteria. These bacteria contain the nitrogenase enzyme that catalyzes the fixation of molecular nitrogen to ammonium ions, which the plant can assimilate. In return the legume supplies the bacteria with carbohydrate.

node The point of leaf insertion on a stem. At the apex of the stem the nodes are very close together but become separated in older regions of the stem by *intercalary growth*, which forms the internodes. In certain monocotyledons such as those forming bulbs, the nodes are very closely spaced on a condensed stem.

noncompetitive inhibition A form of enzyme inhibition in which the inhibitor binds to the enzyme at a site other than the active site, altering the conformation of the enzyme and affecting its ability to bind with the substrate.

noncyclic photophosphorylation *See* photosynthesis.

nondisjunction The failure of homologous chromosomes to move to separate poles during anaphase I of meiosis, both homologs going to a single pole. This results in two of the four gametes formed at telophase missing a chromosome (i.e. being $n - 1$). If these fuse with normal haploid (n) gametes then the resulting zygote is *monosomic* (i.e. $2n - 1$). The other two gametes formed at telophase have an extra chromosome (i.e. are $n + 1$) and give a *trisomic* zygote (i.e. $2n + 1$) on fusion with a normal gamete. If two gametes deficient for the same chromosome fuse then *nullisomy* ($2n - 2$) will result, which is almost always lethal, and if two gametes with the same extra chromosomes fuse, *tetrasomy* ($2n + 2$) results. All these abnormal chromosome conditions are collectively referred to as *aneuploidy*.

nonsense triplet A triplet of bases (CODON) in DNA or RNA that does not code for an amino acid. Some nonsense codons (stop or termination codons) specify the termination of polypeptide synthesis during TRANSLATION. *See* genetic code.

nucellus (*pl.* **nucelli**) The parenchymatous tissue core of an ovule, enclosing the megaspore or egg cell (or in angiosperms the EMBRYO SAC). The pollen tube gains entry to the nucellus through a gap in the surrounding integuments called the MICROPYLE. In some angiosperm species the nucellus persists as the perisperm, providing nourishment for the developing embryo.

nuclear membrane (**nuclear envelope**) *See* nucleus.

nuclease *See* endonuclease; exonuclease.

nucleic acid hybridization (**DNA hybridization**) The pairing of a single-stranded DNA or RNA molecule with another such strand, forming a DNA–DNA or RNA–DNA hybrid. In order to achieve hybridization the base sequences of the strands must be complementary. This phenomenon is exploited in many techniques, notably in GENE PROBES, which are designed to bind to particular complementary base sequences among a mass of DNA fragments.

nucleic acids Organic acids whose molecules consist of chains of alternating sugar and phosphate units, with nitrogenous bases attached to the sugar units. Essentially they are polymers of NUCLEOTIDES formed by condensation reactions between the component nucleotides, which are linked by phosphodiester bonds. There are two forms of nucleic acids, DNA (deoxyribonucleic acid) and RNA (ribonucleic acid). In DNA the sugar is deoxyribose; in RNA it is ribose. Nucleic acids occur in the cells of all organisms, and may be single- or double-stranded.

nucleolar organizer *See* nucleolus.

nucleolus (*pl.* **nucleoli**) A more or less spherical structure found in nuclei of eukaryote cells, and easily visible with a light microscope, staining densely with basic dyes. There may be one to several per nucleus. The nucleolus is the site of manufacture of ribosomal subunits and is thus most conspicuous in cells making large quantities of protein. Nucleoli disappear during cell division. The nucleolus synthesizes ribosomal RNA (rRNA) and is made of RNA (about 10%) and protein. It forms around particular loci of one or more chromosomes called *nucleolar organizers*. These loci contain numerous tandem repeats of the genes coding for ribosomal RNA. TRANSCRIPTION of this code requires a specific RNA polymerase found only in the nucleolus.

nucleoplasm The material of the NUCLEUS, especially the material inside the nuclear envelope.

nucleoprotein A compound consisting of a protein associated with a nucleic acid. Examples of nucleoproteins are the chromosomes, made up of DNA, some RNA, and histones (proteins); and the ribosomes (ribonucleoproteins), consisting of ribosomal RNA and proteins.

nucleoside A molecule consisting of a PURINE or PYRIMIDINE base linked to a sugar, either ribose or deoxyribose. Adenosine, cytidine, guanosine, thymidine, and uridine are common nucleosides.

nucleosome *See* chromosome.

nucleotide The compound formed by condensation of a nitrogenous base (a PURINE or PYRIMIDINE) with a sugar (ribose or deoxyribose) and phosphoric acid. ATP is a *mononucleotide* (consisting of a single nucleotide), the coenzymes NAD and FAD are *dinucleotides* (consisting of two linked nucleotides), and the nucleic acids are *polynucleotides* (consisting of chains of many linked nucleotides).

nucleus (*pl.* **nuclei**) An organelle of eukaryote cells containing the genetic information (DNA) and hence controlling the cell's activities. It is the largest organelle, typically spherical and bounded by a double membrane, the *nuclear envelope* or *nuclear membrane*, which is perforated by many pores (*nuclear pores*) that allow exchange of materials with the cytoplasm. The outer nuclear membrane is an extension of the endoplasmic reticulum. In the nondividing (interphase) nucleus the genetic material is irregularly dispersed as CHROMATIN; during nuclear division (mitosis or meiosis) this condenses into densely staining CHROMOSOMES, and the nuclear envelope disappears, as do the nucleoli that are normally present. *See* nucleolus.

nullisomy *See* aneuploidy.

nut A dry indehiscent fruit resembling an achene but derived from more than one carpel. It has a hard woody pericarp and characteristically a single seed. Cupules, supporting the nuts, may be distinctive of the species. Examples are the fruits of the beech (*Fagus*), hazel (*Corylus*), oak (*Quercus*), and sweet chestnut (*Castanea*). *See* achene.

nutation *See* circumnutation.

nyctinasty (**nyctinastic movements**) The opening and closing of plant organs, especially flowers and leaves, in response to daily changes in temperature and light. *See* nastic movements.

O

obligate Describing an organism that requires specific environmental conditions for its survival and cannot adopt an alternative mode of living. For example, an obligate aerobe is an organism that can grow only under aerobic conditions and cannot survive in anaerobic conditions. *Compare* facultative.

octadecanoic acid *See* stearic acid.

offset A short runner arising from an axillary bud near the base of a stem and growing horizontally above ground away from the parent, seen in daisies (*Bellis*) and houseleeks (*Sempervivum*). It stores no food but turns up at the end and produces a new plant from the apical bud. Like stolons and suckers, offsets are a means of vegetative propagation in angiosperms.

oil 1. A TRIACYLGLYCEROL that is liquid at room temperature. The commonest fatty acids in oils are oleic and linolenic acids, which are unsaturated. Storage oils, synthesized in the endoplasmic reticulum, make up as much as 60% of the dry weight of certain seeds, such as the castor bean (*Ricinus communis*), corn (*Zea mays*), flax (*Linum usitatissimum*), soybean (*Glycine max*), sunflower (*Helianthus*), safflower (*Carthamus tinctorius*), and mustards (family Brassicaceae). Oils can be stored in *oil bodies* (*elaiosomes*) – oil-secreting structures that attract ants to disperse the seeds. Oil may also make up part of the fleshy part of a fruit, as in olive, the oil palm, and coconut palm.
2. *Essential oils*. Any of a number of volatile oils secreted by aromatic plants that are the source of characteristic odors or tastes. They are stored as droplets in plant glands. Most essential oils are terpenoids or derivatives of benzene. Some flowers, such as the vine *Dalechampia* (Euphorbiaceae) and some orchids, produce oils that are collected by pollinating bees. Some repel insects or grazing animals, while others (allelochemicals) deter encroaching neighboring plants. Resinous oils that appear at the site of wounds help to prevent loss of sap and protect against the entry of pathogens and parasites.

oil immersion A microscopic technique using special high-powered objective lenses. A drop of immersion oil (e.g. cedarwood oil) is placed on the coverslip of a microscope slide and the objective lens carefully lowered into it. The oil has the same refractive index as the lens glass and increases the resolving power obtainable by letting a wider angle of rays enter the objective lens.

oleic acid An 18-carbon unsaturated fatty acid occurring as the glyceride in oils and fats. Oleic acid occurs naturally in larger quantities than any other fatty acid in plant cells except in chloroplasts, where linolenic acid is the dominant fatty acid. In many organisms oleic acid can be synthesized directly from stearic acid and further enzymatic paths exist for conversion to linoleic acid and linolenic acid. This pathway does not occur in humans and the higher animals so plant sources are an essential dietary element.

Oligocene The epoch of the Tertiary period 38–25 million years ago. The world's climate ranged from temperate to subtropical, and the epoch was marked by an expansion of grasslands at the expense of forests, a change that promoted the evolution of large herbivores. Around the great

Tethys Sea there were tropical swamps, which gave rise to extensive deposits of lignite in Germany and neighboring countries. The angiosperms were evolving and diversifying rapidly, and overtook the gymnosperms in their abundance. Flowers and insects were evolving together, and complex coevolution was taking place as pollination mechanisms diversified. *See also* geological time scale.

oligosaccharide *See* polysaccharide.

oligotrophic Describing lakes and ponds that are low in nutrients and consequently low in productivity. With little decaying vegetation, the oxygen content of the water is adequate for fish and other animal life. *Compare* eutrophic.

one gene–one enzyme hypothesis The theory that each gene controls the synthesis of one enzyme, which was advanced following studies of nutritional mutants of fungi. Thus by regulating the production of enzymes, genes control the biosynthetic reactions catalyzed by enzymes and ultimately the character of the organism. Genes also code for proteins, or polypeptides that form proteins, other than enzymes, so the idea is perhaps more accurately expressed as the *one gene–one polypeptide hypothesis*.

ontogeny The course of development of an organism from fertilized egg to sexual maturity and the production of gametes and the next generation. Occasionally ontogeny is used to describe the development of an individual structure.

oogamy Sexual reproduction involving the fusion of two dissimilar gametes. The male gamete is usually motile and smaller than the female gamete, which is usually nonmotile, contains a food store, and may be retained by the parent. The term is generally restricted to descriptions of plants, particularly those that produce female gametes in oogonia. It is an extreme form of anisogamy. *Compare* isogamy. *See* anisogamy.

oogonium (*pl.* **oogonia**) The female reproductive organ of certain algae and fungi, often distinctly different in shape and size from the male reproductive organ (the antheridium). This initially unicellular gametangium contains one or more large nonmotile haploid eggs called oospheres. These may be liberated prior to fertilization, e.g. *Fucus*, or remain within the oogonium, e.g. *Pythium*.

Oomycete *See* Oomycota.

Oomycota (**Oomycete**) A phylum of funguslike protoctists that includes the water molds, downy mildews, and white rusts. Most oomycetes live in fresh water or soil. They are mainly parasites or saprophytes, and feed by extending hyphae into the tissues of their host, releasing digestive enzymes and absorbing nutrients through the hyphal walls. Some are important crop pests, such as potato blight (*Phytophthora infestans*), damping off fungus (*Pythium*), mildew of grapes (*Plasmopara viticola*), and *Saprolegnia parasitica*, a parasite of aquarium fish.

oosphere A large nonmotile female gamete or egg cell. In lower plants oospheres may be produced inside an oogonium; they may be retained inside the oogonium and fertilized there, or released for fertilization in the water. In angiosperms the oogonium is enclosed in the EMBRYO SAC and protected by the integuments of the ovule.

oospore A diploid zygote of certain protoctists, produced by fertilization of the female gamete (the oosphere), by the male gamete, the antherozoid. The zygote may form a thick coat and go through a resting period before germination. The term 'oospore' distinguishes a zygote produced oogamously from a zygote produced isogamously or anisogamously and called a zygospore. *See* zygote.

OP *See* osmotic pressure.

operator gene A bacterial regulatory gene that contains the genetic code for starting TRANSCRIPTION of one or more structural genes. The structural genes together with the relevant operator genes are linked together in a functional unit called an OPERON. The operator is under the control of a regulator gene, which codes for a small protein called a *repressor molecule*, which binds to the operator gene and prevents it from initiating transcription. If production of the repressor ceases or if another molecule, such as the substrate of an enzyme, binds with it, preventing it from binding to the operator gene, TRANSCRIPTION can proceed.

operculum (*pl.* **opercula**) **1.** The circular lid of the capsule of many mosses. It covers the peristome and may be forcibly blown off the capsule by pressure developing in the lower portion of the capsule, or by the swelling of cells of the annulus.
2. In the pollen grains of many species, a lid covering the aperture, which is displaced by the emerging pollen tube.
3. A cap covering the ostiole of the perithecium in certain ascomycetes.

operon A genetic unit found in prokaryotes and comprising a group of closely linked genes acting together and coding for the various enzymes of a particular biochemical pathway. They function as a unit, because either all are transcribed or none. At one end is an *operator*, which contains a site called the PROMOTER. For transcription of the *structural genes* (the genes that actually code for the enzymes) to proceed, mRNA polymerase must first bind with the promoter. Whether or not this happens is controlled by another gene outside the operon, the *regulator gene.* The regulator gene produces a *regulator protein* that binds with the operator, renders it inoperative, and so prevents enzyme production. The presence of a suitable substrate prevents this binding, and so enzyme production can commence. An example is the *lac operon* in the bacterium *Escherichia coli*, which is involved in the metabolism of lactose. In the absence of an *inducer* molecule (in this case, the substrate lactose or one of its derivatives), no transcription occurs. But if lactose binds to the regulator protein, the regulator can no longer bind to the operator, so transcription can proceed and enzyme synthesis is initiated. *See* Jacob–Monod model; repressor.

Ophioglossales (**moonworts and adder's-tongue ferns**) An order of eusporangiate ferns (*see* Filicinophyta) characterized by fronds that are divided into a distinct sterile green blade and a fertile spike bearing sporangia sunken into its sides. Unlike all other ferns, the fronds are not rolled up as they push up through the ground. They are found in temperate and tropical regions.

opiate An ALKALOID derived from the opium poppy (*Papaver somniferum*) in a latex that oozes from cut seed capsules. Opiates are highly addictive drugs that resemble endorphins, chemicals naturally produced in the mammalian body that suppress pain and enhance mood by occupying certain receptor sites on nerve cells that affect transmission of impulses. Opiates occupy the same receptor sites. They are the source of codeine, morphine, heroin, and other drugs, many of which have medicinal uses in the treatment of severe pain.

Orchidaceae The orchid family comprising monocotyledonous perennial herbs. The family includes terrestrial forms, epiphytes, and full saprophytes with no green parts (e.g bird's-nest orchid (*Neottia nidus-avis*) and coralroot, (*Corallorhiza* spp.)). It is the largest family of flowering plants containing over 25,000 species. See illustration overleaf.

order A collection of similar families. The Latin names of plant orders generally end in *ales* (e.g. Liliales). Orders may be divided into suborders. Similar orders constitute a class. *See* International Code of Botanical Nomenclature.

Ordovician The second oldest period of the Paleozoic era, some 510–440 million years ago. Algae were abundant in the oceans, especially mat-forming algae and algae in symbiotic relationships in the corals of the reefs dominating the many

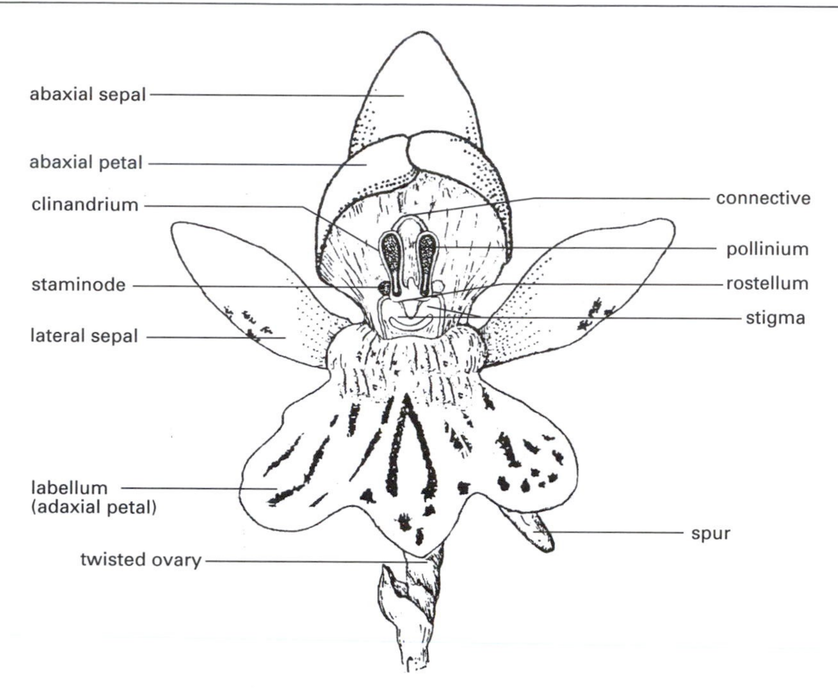

Orchidaceae: front view of an orchid flower

shallow, warm seas of the time. Some algae may also have been terrestrial, forming mats on wet ground. *See also* geological time scale.

organ A part of an organism that is made up of a number of different tissues specialized to carry out a particular function. Examples include the stem, leaf, flower, and root.

organ culture The maintenance or growth of living organs, usually embryonic, *in vitro*. Mature plant organs, notably roots, may be cultured indefinitely in suitable media. Leaves, embryos, meristems, and ovules have also been successfully cultured.

organelle A discrete membrane-enclosed subcellular structure with a particular function. The largest organelle is the nucleus; other examples are CHLOROPLASTS, MITOCHONDRIA, VACUOLES, and RIBOSOMES. Organelles allow division of labor within the cell.

organochlorines Organic compounds containing chlorides, used in pesticides such as DDT, and in other chemicals used to control plant pests, which are not readily broken down and persist in tissues of animals, affecting breeding success and causing other problems. They are banned from use in most countries, but are still used in some, especially where insect-borne diseases such as malaria are prevalent.

origin of life The development of living organisms from inorganic matter. Geological evidence strongly suggests that life originated on earth about 4600 million years ago. The earliest evidence of life is bacterialike fossils in rocks over 3500 million years old. The basic components of organic matter – water, methane, ammonia, and related compounds – were abundant in the atmosphere, which had much greater reducing properties than today's atmosphere. Until recently, it was widely accepted that life evolved in warm lagoons or hot springs, where energy from the sun (cosmic rays) and lightning storms caused simple molecules to recombine into increasingly complex organic molecules that eventually showed the characteristics of living organisms. However, recent discoveries suggest that life may have evolved deep underground in fissures in hot rocks – a habitat in which today vast numbers of thermophilic bacteria live. Life probably had its origins in organic self-replicating molecules that consumed the chemicals around them to duplicate themselves. These molecules were probably a form of RNA.

ornithine A nonprotein AMINO ACID that occurs in the cell walls of certain bacteria. It is an important intermediate in the synthesis of the amino acid arginine, and the synthesis of many alkaloids.

orthogenesis An early theory of evolution in which evolutionary change was envisaged as occurring in a definite direction and along a predetermined route, irrespective of natural selection.

orthotropism *See* tropism.

orthotropous (atropous) *See* ovule.

osmium tetroxide A stain used in electron microscopy because it contains the heavy metal osmium. It is fat-soluble, and unsatured fats will reduce it to black osmium dioxide. It therefore stains lipids, including membranes, particularly intensely. It also acts as a fixative.

osmometer An instrument that is used to measure OSMOTIC PRESSURE (WATER POTENTIAL). In its simplest form, a porous pot is made selectively peremeable by impregnating its pores with copper ferrocyanide. The pot is filled with an osmotically active solution, and a piston placed on the solution. The pot is then immersed in a container of pure solvent. Solvent entering the porous pot by osmosis lifts the piston until the force of the water entering the pot equals the hydrostatic pressure created by the weight of the piston.

osmosis The movement of solvent from a dilute solution to a more concentrated solution through a *semipermeable membrane* (selectively permeable membrane) – one that allows the passage of some kinds of molecule and not others. For example, if a concentrated sugar solution (in water) is separated from a dilute sugar solution by a selectively permeable membrane, water molecules can pass through from the dilute solution to the concentrated one by diffusion.

Osmosis between two solutions will continue until they have the same concentration. If a certain solution is separated from pure water by a membrane, osmosis also occurs. The pressure necessary to stop this osmosis is called the OSMOTIC PRESSURE (OP) of the solution. The more concentrated a solution, the higher its osmotic pressure. Osmosis is a very important feature of both plant and animal biology. Cell membranes act as *differentially permeable membranes* and osmosis can occur into or out of the cell. It is necessary for an animal to have a mechanism of osmoregulation to stop the cells bursting or shrinking. In the case of plants, the cell walls are slightly 'elastic' – the concentration in the cell can be higher than that of the surroundings, and osmosis is prevented by the pressure exerted by the cell walls (*wall pressure*). Where the solvent is water, physiologists now describe the tendency for water to move in and out of cells in terms of WATER POTENTIAL.

osmotic potential *See* water potential.

osmotic pressure (OP) The force (pressure) that must be applied to a solution to prevent the passage of water or another pure solvent through a selectively permeable membrane separating the solvent from the solution (i.e. to prevent OSMOSIS occurring). Osmotic pressure rises as the concentration of the solution rises. The osmotic pressure is equal to the diffusion pressure deficit plus the turgor pressure. This term is becoming obsolete, as it ignores capillary and imbibitional forces that are significant in certain situations. A better concept is that of WATER POTENTIAL.

Osmundales (**royal ferns**) An order of ferns often large, retaining some primitive features, whose whose sporangia combine eusporangiate and leptosporangiate characters (*see* Filicinophyta).

ostiole A small pore through which spores or gametes are released from the reproductive bodies of certain fungi and algae. Examples include the CONCEPTACLES of wracks (*Fucus* spp.).

outbreeding Breeding between individuals that are not closely related. In plants the term is often used to mean cross-fertilization, and various methods exist to promote it, e.g. stamens maturing before pistils, SELF-INCOMPATIBILITY. The most extreme form – crossing between species – usually results in sterile offspring and there are various mechanisms to discourage it. Outbreeding increases heterozygosity, giving more adaptable and more vigorous populations. *Compare* inbreeding. *See* heterostyly; protandry; protogyny.

ovary The swollen base of the CARPEL in the gynoecium of plants, containing one or more ovules, each attached by a *funiculus*. The gynoecium of angiosperms may consist of more than one carpel that fuses in certain species forming a complex ovary. Where the fused carpel walls remain intact, this is a *multilocular ovary*; where they break down it becomes a *unilocular ovary*. If the ovary lies below the other floral organs in a flower, it is said to be an *inferior ovary*, as in *Rosa*. In such cases, the ovary usually appears sunk into and fused with a cup-shaped receptacle. In a floral formula, an inferior ovary is denoted by a line above the carpel (gynoecium) symbol and number. If the petals, sepals, etc. are inserted below it, it is a *superior ovary*, as in buttercup (*Ranunculus*). In a floral formula a superior ovary is denoted by a line below the carpel number. After fertilization, the ovary wall becomes the *pericarp* of the fruit enclosing seeds in its central hollow.

ovate Up to four times as long as broad, tapering at both ends, with the broadest part below the middle.

ovule Part of the female reproductive organs in seed plants. It consists of the nucellus, which contains the embryo sac, surrounded by the integuments. The ovule is surrounded by a cuticle, and is thought to be derived from the megasporangium of its nonseed-bearing ancestors. After fertilization the ovule develops into the seed. In angiosperms the ovule is contained within an ovary, and attached to the placental tissue of the ovary by a stalk, the *funiculus*. It may be orientated in different ways being upright, inverted, or sometimes horizontal. Where the developing ovule has turned through 180°, so that the micropylar end has folded over and lies close to the base of the funiculus, it is termed *anatropous*. This is the most common position of the ovule in flowering plants. Where the ovule has developed vertically so that the micropylar end is directly over the funiculus, it is termed *orthotropous*, e.g. *Polygonum*.

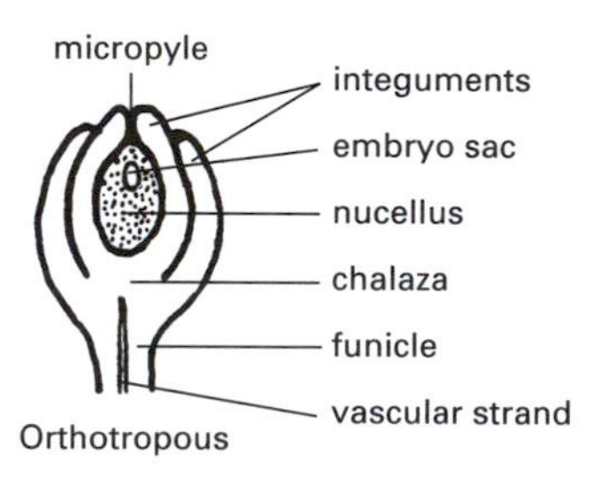

Ovule: types of orientation

Where the funiculus appears to be attached half way between the chalaza and the micropyle, and the micropylar end is turned through 90° relative to the orthotropous condition so that the ovule is horizontal, it is termed *campylotropous*, e.g. mallow (*Malva sylvestris*).

In gymnosperms ovules are larger but are not contained within an ovary. They are borne on ovuliferous scales. Gymnosperm seeds are thus naked while angiosperm seeds are contained within a fruit, which develops from the ovary wall.

ovuliferous scale The megasporophyll found in the axis of the bract scale in the female strobili of the Coniferophyta. It is a large woody structure and bears ovules and later seeds, on its upper surface. It is thus homologous with the carpel of the Anthophyta.

ovum (*pl.* **ova**) (**egg cell**) An unfertilized egg cell.

oxalic acid (**ethanedioic acid**) A dicarboxylic acid, that occurs in various plants, e.g. in the leaves of rhubarb (*Rheum*), dumbcanes (*Dieffenbachia*), wood sorrel (*Oxalis acetosella*), and the garden oxalis (*Oxalis* spp.). It plays a role in the removal of excess cations, such as sodium, potassium, and calcium, which are incorporated into oxalates.

oxaloacetic acid (**OAA**) A water-soluble carboxylic acid, structurally related to fumaric acid and maleic acid. Oxaloacetic acid is an intermediate of the KREBS CYCLE, and in the Hatch–Slack pathway of C_4 PLANTS. It is produced from L-malate in an NAD-requiring reaction and itself is a step toward the formation of citric acid in a reaction involving pyruvate and coenzyme A, part of a series of reactions that are used to replace Krebs cycle intermediates removed for use in biosynthesis. When these reactions run in reverse order, they form a pathway for glucose synthesis.

oxidase An enzyme that catalyzes a reaction involving the oxidation of a substrate using molecular oxygen as electron acceptor, e.g. cytochrome c oxidase in the respiratory ELECTRON-TRANSPORT CHAIN. Many oxidases are flavin-linked dehydrogenases in which the reduced COENZYME can be reoxidized by molecular oxygen to form water and hydrogen peroxide. *See* dehydrogenase; oxidative phosphorylation.

β-oxidation The main pathway for the breakdown of fatty acids to acetyl CoA. β-oxidation involves the removal of carbon acids from fatty acid chains two at a time, forming one molecule of acetyl CoA each time. The acetyl CoA may enter the Krebs cycle and be further oxidized, or it may enter the glyoxylate cycle and be converted to sugar. β-oxidation is a major energy-liberating process in germinating oil-rich seeds, where the associated enzymes are located in the glyoxysome together with those of the glyoxylate cycle.

oxidative phosphorylation The production of ATP from phosphate and ADP in aerobic respiration. Oxidative phosphorylation occurs in mitochondria, the energy being provided by steps in the electron-transport chain.

oxygen An element essential to living organisms both as a constituent of carbohydrates, fats, proteins, and their derivatives, and in aerobic respiration. It enters plants in both carbon dioxide and water, the oxygen from water being released in gaseous form as a byproduct of photosynthesis. Plants are the main, if not the only, source of gaseous oxygen and as such are essential in maintaining oxygen levels in the air for aerobic organisms.

ozone layer A layer consisting of ozone (O_3) molecules scattered through the stratosphere roughly 15–50 km above the earth's surface. The ozone is formed by the action of solar radiation on oxygen molecules. The ozone layer absorbs about 99% of the harmful ultraviolet radiation entering the earth's atmosphere, and hence provides a shield for living organisms without which life on earth would be impossible. This absorption of ultraviolet radiation has two consequences: it warms

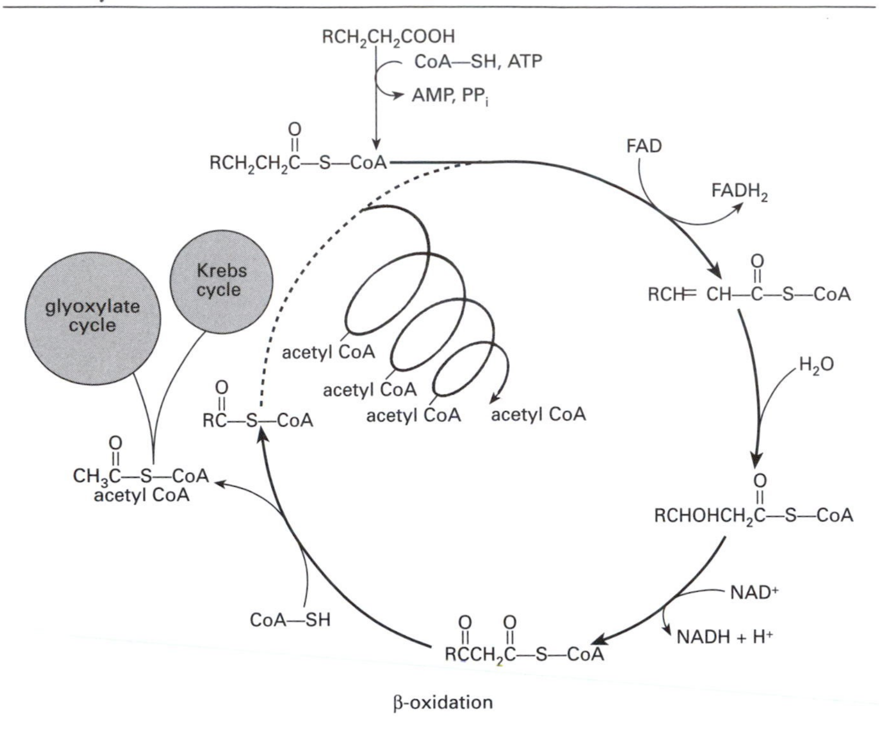

β-oxidation

the stratosphere, helping to maintain the temperature of the Earth's surface; and it stabilizes the stratosphere, preventing it from mixing with other layers, so reducing winds in the lower atmosphere.

In the early 1980s it was discovered that the ozone layer was becoming thinner, with seasonal 'holes' appearing over Antarctica and elsewhere. This depletion is believed to be caused by increasing atmospheric concentrations of chlorine compounds, especially chlorofluorocarbons (CFCs) – very stable chemicals used as refrigerants and aerosol propellants – which disrupt the delicate balance between ozone production and breakdown in the atmosphere. A single atom of chlorine or fluorine can destroy about 10^5 molecules of ozone. Other ozone-destroying chemicals include nitrogen oxides from aircraft and halons – gases widely used in industry and in certain kinds of fire extinguishers. Such findings led to the introduction in 1987 of restrictions on the use of CFCs, and a complete ban proposed to take effect by 2000. However, even with a ban in force it may take many years before the ozone layer is stabilized.

P

P680 *See* photosynthesis.

P700 *See* photosynthesis.

pachytene In MEIOSIS, the stage in mid-prophase I that is characterized by the contraction of paired homologous chromosomes. At this point each chromosome consists of a pair of chromatids and the two associated chromosomes are termed a tetrad.

pairing *See* synapsis.

palea *See* bract.

paleobotany *See* paleontology.

Paleocene The oldest epoch of the Tertiary, 65–55 million years ago. It was a period of warm humid climate with large areas of temperate and subtropical forests, but only limited grasslands, as the grasses were of very recent origin. During the Paleocene there was a great deal of coevolution of plants and animals, the plants evolving adaptations to animal pollination and fruit and seed dispersal. Bilateral flowers evolved, e.g. those of the pea family (Fabaceae) and orchids (Orchidaceae). In the latter part of the Paleocene, large angiosperm fruits and seeds evolved as fruit- and seed-eating mammals also evolved and diversified. *See also* geological time scale.

paleoecology The investigation of prehistoric ecology as revealed by studying fossils and ancient human artefacts, pollen (*see* palynology) samples, and the mineral deposits in which such structures occur.

paleontology The study of extinct organisms, including their fossil remains, and impressions left by them. Sometimes the subject is divided into *paleobotany*, the study of fossil plants, *paleozoology*, the study of fossil animals, and PALEOECOLOGY.

Paleozoic The first and oldest era in which multicellular life became abundant, about 590–230 million years ago. It is divided into six main periods: the Cambrian, Ordovician, Silurian, Devonian, Carboniferous, and Permian. Beginning with aquatic invertebrates and algae, the era ended with the invasion of land by tree ferns and reptiles. *See also* geological time scale.

palisade mesophyll *See* mesophyll.

palmella (*pl.* **palmellae**) A stage formed under certain conditions in various unicellular algae in which, after division, the daughter cells remain within the envelope of the parent cell and are thus rendered immobile. The cells may continue dividing giving a multicellular mass, which is contained in a gelatinous matrix. Palmelloid forms may develop undulipodia and revert to normal mobile cells at any time.

palmitic acid (**hexadecanoic acid**) A 16-carbon saturated fatty acid occurring widely in fats and oils of animal and vegetable origin. It is the commonest saturated fatty acid in plants, and an intermediate in the synthesis of oleic and stearic acids. *See also* carboxylic acid; oleic acid.

palynology The study of living and fossil pollen. It is mainly used as a means of obtaining information on the composition and extent of past floras, and may give reliable quantitative information on the vegetative cover many thousands of years ago. *See also* pollen analysis.

panicle *See* inflorescence.

pantothenic acid (**vitamin B_5**) The precursor of coenzyme A, pantothenic acid is essential for several fundamental reactions in metabolism. It is one of the water-soluble B group of VITAMINS, and is synthesized by plants and bacteria but not by vertebrates, so is an essential vitamin in their diet.

paper chromatography A chromatographic method by which minute amounts of material can be analyzed. An absorbent paper strip with a drop of test material at the bottom is dipped into the solvent (which rises up the paper by capillary action) and removed when the solvent front almost reaches the top of the strip.

papilla (*pl.* **papillae**) A short sometimes cone-shaped projection from a cell, usually an extension of the outer wall of an epidermal cell. Papillae are often found on petals, giving them a velvety appearance. *See* hair.

pappus (*pl.* **pappi**) The ring of hairs, scales, or teeth that makes up the calyx in flowers of the ASTERACEAE. It persists on the fruit and serves to aid the wind dispersal of seeds. An example is the cypsela of the dandelion (*Taraxacum*) in which the pappus remains attached by a long thin stalk and acts as a parachute.

parallel evolution (**parallelism**) The development of similar features in closely related organisms as a result of strong selection in the same direction. This may occur between species of the same genus that are widely separated geographically but which live in similar environments. There are few examples of this phenomenon and some authorities deny its existence. *Compare* convergent evolution.

paramylum A starchlike polysaccharide composed of β(1-3)-linked units of glucose, which is the assimilation product of euglenas (Eustigmatophyta).

paraphysis (*pl.* **paraphyses**) A sterile unbranched (usually) multicellular hair found in large numbers between the reproductive organs of certain algae and bryophytes. Club-shaped hyphae in the hymenial layer of certain basidiomycete fungi are also termed paraphyses.

parasitism An association between two organisms in which one, the *parasite*, benefits at the expense of the other, the *host*. The tolerance of the host varies from being almost unaffected to serious illness and often death. An *obligate* parasite such as potato blight (*Phytophthora infestans*) can live only in association with a host, whereas a *facultative* parasite such as the damping-off fungus (*Pythium*) can exist in other ways, for example as a saprophyte.

paratonic movements Movements of plants or plant parts in response to external stimuli. They may be divided into mechanical movements (e.g. the hygroscopic movements of dead cells) and movements caused by the stimulation of sensitive cells (e.g. tropisms and nastic movements). Taxes are also paratonic movements. *Compare* autonomic movements.

parenchyma Tissue made up of living thin-walled cells that are not differentiated for any specific function, but in which important metabolic processes are carried out. The leaf mesophyll and the stem medulla and cortex consist of parenchyma. The vascular tissue is also interspersed with parenchyma; for instance the medullary rays of secondary vascular tissue. Parenchyma often forms a basic ground tissue in which other tissues are embedded. It is generally considered to be the tissue from which other types of tissues evolved.

parthenocarpy The development of fruit in unfertilized flowers, resulting in seedless fruits. It may occur naturally, as in the banana (*Musa*), which is triploid. It may also be induced artificially by the application of auxin, as in commercial tomato growing.

parthenogenesis Development of unfertilized ovules to form new individuals. It occurs regularly in certain plants, e.g. dandelion (*Taraxacum*), in which meiosis is defective, resulting in a diploid egg. In hawkweeds (*Hieracium*) a cell of the nucellus takes the place of the megaspore. *See also* apomixis.

partial dominance *See* incomplete dominance.

passage cells Endodermal cells that have CASPARIAN STRIPS but remain otherwise unthickened after deposits of lignin and cellulose have been laid down elsewhere in the endodermis. Passage cells are found opposite the protoxylem and allow transport of water and solutes from the cortex to the stele.

pathogen Any organism that is capable of causing disease or a toxic response in another organism.

PCR *See* polymerase chain reaction.

peat Partially decomposed plant material that accumulates in waterlogged anaerobic conditions in temperate humid climates, often forming a layer several meters deep. Peat varies from a light spongy material (*Sphagnum* moss) to a dense brown humidified material in the lower layers.

pectic substances Polysaccharides that, together with hemicelluloses, form the matrix of plant cell walls. They serve to cement the cellulose fibers together. Fruits are a rich source, especially apples (*Malus*), pears (*Pyrus communis*), and citrus fruits. *See also* middle lamella.

pedicel *See* inflorescence.

peduncle *See* inflorescence.

pelagic Inhabiting the open upper waters rather than the bed of a sea or ocean. *Compare* benthic. *See also* photic zone; plankton.

pellicle (**periplast**) A thin flexible transparent outer protective covering of some unicellular algae that lack a cellulose cell wall, e.g. *Euglena*. It is made of protein, and maintains the shape of the body.

Pennsylvanian The US name for the late Carboniferous period in the geological time scale, 320 to 286 million years ago. Rocks that originated during this period are widespread in the state of Pennsylvania, USA.

pentose A SUGAR that has five carbon atoms in its molecules. Examples include ribose, deoxyribose, and arabinose. Pentoses are synthesized in the Calvin cycle, derived from hexoses in the PENTOSE PHOSPHATE PATHWAY, and also made by decarboxylation of nucleoside diphosphate sugars. Polymers of pentoses are called *pentosans*.

pentose phosphate pathway (**hexose monophosphate shunt**) A pathway of glucose breakdown in which pentoses are produced, in addition to reducing power (NADPH) for many synthetic reactions. It is an alternative to GLYCOLYSIS, and produces sugars containing 3 to 7 carbon atoms. Glucose 6-phosphate is decarboyxlated and converted to a 5-carbon sugar, with the formation of NADPH. This 5-C sugar than enters a sequence of reactions, until eventually glucose 6-phosphate is reformed.

PEP *See* phosphoenolpyruvate.

pepo A succulent fruit resembling a berry. The hard outer ring orginates either from the epicarp or from noncarpellary tissues of the plant. In members of the Cucurbitaceae, such as *Cucurbita* species (e.g. cucumber, squash, and pumpkin), the rind is formed from the receptacle, while the flesh enclosing the seeds is derived from the ovary wall.

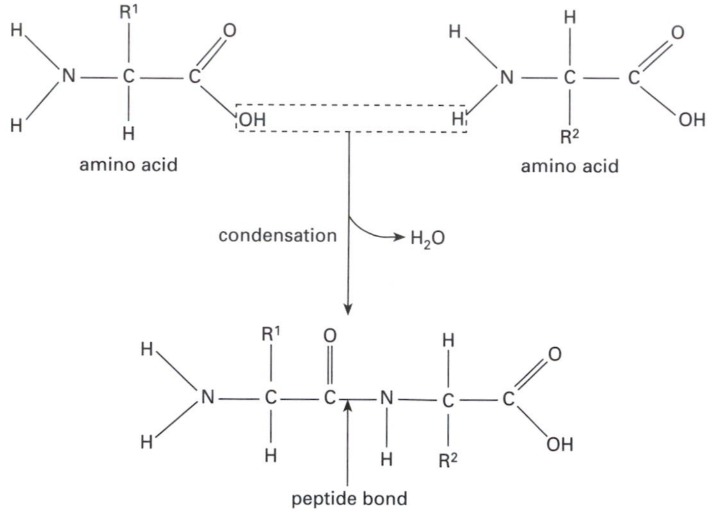

Peptide bond

peptidase An enzyme that is responsible for catalyzing the hydrolysis of certain peptide bonds. The peptidases help break down peptides into amino acids.

peptide A molecule consisting of a number of amino acid molecules linked together. Peptides can be regarded as formed by a CONDENSATION REACTION in which the carbonyl group of one amino acid reacts with the amino group of another amino acid with the elimination of water. This link between amino acids is called a *peptide bond*. According to the number of amino acids linked together they are called *di-*, *tri-*, *oligo-*, or *polypeptides*. In general, peptides have an amino group at one end of the chain and a carboxyl group at the other. They can be produced by the partial hydrolysis of proteins.

peptide bond *See* peptide.

peptidoglycan The principal component of the cell walls of most BACTERIA (but not of Archaea), consisting of polysaccharide molecules cross-linked by short peptides chains to form a rigid framework.

perennating organ In biennial or perennial plants, a storage organ that enables the plant to survive an adverse season, such as winter or a dry season (*perennation*). Most such organs are underground, e.g. rhizomes, corms, bulbs, swollen taproots, and tubers. Perennating organs are also used by many plants in asexual reproduction (vegetative growth). Buds developing in the axils of scale leaves on CORMS, BULBS, and stem TUBERS may become detached, put down ADVENTITIOUS roots, and grow into independent plants. Similarly, new shoots arise from buds on RHIZOMES, and the intervening segments of rhizome may break or rot away.

perennation *See* perennating organ.

perennial A plant that may live for several years. Perennials may reproduce in their first growing season or may have to attain a certain age before seed production commences. *Herbaceous perennials* die back each year and survive until the next growing season as tubers (e.g. *Dahlia*), bulbs (e.g. daffodil, *Narcissus*), rhizomes (e.g. *Iris*), etc. *Woody perennials*, such as trees and shrubs, persist above ground throughout the year but may show adaptations (e.g. leaf fall) to survive unfavorable seasons.

perforation plates The remains of the cross walls between the vessel elements in the xylem vessels of angiosperms, the Gnetophyta, and some ferns, which leave an opening allowing water to move freely through the vessel. The cross wall may have disintegrated completely so that the vessel is effectively one long cylinder (e.g. oak) or parts of the cross wall may remain as bars across the cavity of the vessel, e.g. alder (*Alnus*). *Compare* sieve plate.

perianth The part of the flower that encircles the stamens and carpels. It usually consists of two whorls of leaflike structures that are differentiated into the sepals (calyx) and petals (corolla). In many monocotyledons there is no such distinction, and the individual perianth units are called *tepals*, e.g. tulip (*Tulipa*). In some flowers, especially those that are wind pollinated, the perianth is reduced (e.g. grasses) or absent (e.g. willow).

periblem *See* ground meristem; histogen theory.

pericarp The ovary wall that becomes the wall of the FRUIT as the fruit develops. Depending on how the pericarp tissues differentiate, the resulting fruit may be dry (papery or leathery) or succulent. The outer layer is the *exocarp*, often a tough skin. The middle layer is the *mesocarp*, whose texture varies with different fruits, e.g. it is juicy in drupes like the plum (*Prunus domestica*), and hard in almonds (*Prunus dulcis*). The mesocarp may be protective or aid in dispersal. The *endocarp* is the innermost layer and forms the stony covering of the seed in a drupe, but in other fruits (e.g. the berry), it is indistinguishable from the mesocarp. If the pericarp opens when ripe, the fruit is said to be DEHISCENT; if it remains closed, it is INDEHISCENT.

periclinal Describing a line of cell division parallel to the surface of the organ, resulting in the formation of periclinal walls between daughter cells and an increase in girth of the organ. *Compare* anticlinal.

pericycle The parenchymatous layer of cells that lies within the endodermis, forming the outermost part of the stele. It remains meristematic in most roots and gives rise to the lateral roots. In most dicotyledonous roots showing secondary growth, the pericycle is also involved in the origination of the vascular cambium and the phellogen.

periderm The secondary tissue that arises from the activity of the cork cambium. It forms the outer layer of the stem and root in plants with secondary growth, having a protective role. It is made up of the PHELLOGEN (cork cambium), which gives rise to the PHELLEM (cork) and PHELLODERM (secondary cortex). Periderm may also develop at the site of a wound, forming a barrier to pathogens.

perigyny The arrangement in flowers in which the perianth and androecium are inserted on the receptacle around the gynoecium rather than above or below it. The receptacle is extended to a flat or saucer-shaped organ and carries the gynoecium in the middle and the other floral parts around the edge. The OVARY is technically superior even though in extreme forms of perigyny in which the receptacle is cup-shaped, e.g. the wild rose (*Rosa* sp.), the floral parts may be inserted at a level above the gynoecium. In perigynous flowers the ovary is not fused with the receptacle. *Compare* epigyny; hypogyny.

periplast *See* pellicle.

perisperm The nutritive tissue in the seeds of many Caryophyllaceae (the carnation family). It is a diploid tissue derived from the nucellus rather than the embryosac. Such seeds therefore differ from other angiosperm seeds that have either cotyledons or endosperm tissue acting as the food store.

peristome A single or double ring of teeth around the opening of the capsule in mosses that is involved in spore dispersal. It is revealed when the operculum falls off or is removed. The teeth absorb moisture from the atmosphere, and the coiled fibers of which they are made change length and twist and bend when subjected to humidity changes, either opening and closing the capsule to release the spores or scattering them through jerky movements.

perithecium (*pl.* **perithecia**) *See* ascoma.

permanent stain *See* staining.

permanent wilting point The point at which soil has dried to the extent that plants can no longer remove the remaining water held on the soil particles, and begin to wilt and will not recover, even if moved to a cool, dark place, unless more water is added to the soil. At permanent wilting point the water potential of the soil is equal to or lower than the water potential of the plant.

Permian The most recent period of the Paleozoic era, some 280–250 million years ago. As the climate warmed and dried during the Permian, the ferns, seed ferns, and lycophytes that dominated Permian marshes and swamps were joined by the evolving coniferophytes and some other gymnosperms, whose evolution was probably driven by the need to cope with drier environments. These groups gradually began to replace the older floras dominated by nonseed-bearing plants. *See also* geological time scale.

Peronosporales An order of the OOMYCOTA characterized by sexual reproduction (usually) by zoospores and sexual reproduction by zoospores produced from oospores. The Peronosporales contains saprobes and many harmful plant parasites. Most are parasites found in damp terrestrial habitats; they include downy mildews (*Peronospora* spp.), late blight of potatoes (*Phytophtherainfestans*), and damping-off fungus (*Pythium* spp.).

peroxidase An enzyme that catalyzes the oxidation of certain organic molecules using hydrogen peroxide as electron acceptor.

peroxisome *See* microbody.

petal One of the usually brightly colored parts of the flower, which together make up the corolla. The petals are pigmented and often scented to attract insects. They are reduced or absent in most wind-pollinated flowers. Petals are thought to be modified leaves, with a much simplified internal structure and vascular system, having only one vascular bundle. They may derive from sterile stamens. *Compare* sepal.

petiole The stalk that attaches the leaf blade to the stem. It is similar to the stem except that it is asymmetrical in cross-section with the vascular and strengthening tissues arranged in a V shape rather than a circle. Leaves that lack a petiole are termed *sessile.*

Pezizales (**cup fungi; morels**) An order of ascomycete fungi in which the asci (*see* ascus) are borne in a HYMENIUM that lines the surface of a cup-shaped fruiting body (apothecium). The spores are discharged forcibly into the air. Most are saprobes living in soil, dung, or rotting wood. A few species are involved in mycorrhizae.

pH A measure of the acidity or alkalinity of a solution on a scale 0–14. A neutral solution (such as pure water) has a pH of 7. Acid solutions have a pH below 7; alkaline solutions have a pH above 7. The lower the pH value, the higher the concentration of H^+. The pH is given by $\log_{10}(1/[H^+])$, where $[H^+]$ is the hydrogen ion concentration in moles per liter. Thus one integer on the pH scale represents a tenfold difference in the concentration of H^+.

Phaeophyta (**brown algae**) A phylum of protoctists comprising mainly marine algae, notably the macroscopic thallose seaweeds that inhabit the intertidal zones. They contain the pigments chlorophyll *a* and *c*, β-carotene, and the xanthophylls (including fucoxanthin), which give the algae their characteristic brown color.

phage *See* bacteriophage.

phagocytosis The taking up of food particles by a cell using protrusions of the outer part of the cell, generated by cytoplasmic flow. These protrusions surround the particle, enclosing it in a membrane-bound vesicle and so introducing it into the cell as a food vacuole. Lysosomes then congregate around the food vacuole, fusing with it and releasing hydrolytic enzymes to digest the particle. *See* endocytosis.

Phallales (stinkhorns) An order of basidiomycetes in which the BASIDIOMA has a two-layered wall enclosing a slimy, gelatinous layer. *See* Basidiomycota.

phanerogam In early classifications, any plant whose reproductive organs are flowers or cones. *Compare* cryptogam.

phanerophyte A perennial plant with persistent shoots and buds on upright stems well above soil level. Phanerophytes include herbs, shrubs, trees, and climbing plants. *See* Raunkiaer's plant classification.

Phanerozoic *See* Precambrian.

phase contrast microscopy A form of light microscopy that amplifies the difference in the phase of light after passing through a specimen to give greater contrast in the image. It is based on the fact that light is diffracted (changes phase) according to the refractive index or thickness of the material through which it passes. The phase contrast microscope uses a special condenser, which produces a particular pattern of light, usually a hollow cone. This pattern is matched by a groove in a glass plate (the *retardation plate*) in the condenser. Light that is not diffracted passes through the holes in the plate. Light that has been diffracted by the specimen passes through the thicker part of the plate, so is slowed down even further. This technique can be used to study real-time events such as cell division.

phellem (cork) A protective layer of radically arranged cells produced to the outside of the PHELLOGEN (cork cambium). The cork replaces the epidermis in certain woody plants, forming an impervious layer broken only by lenticels. The older cork cells are dead, suberized, and frequently only air-filled, although lignin, fatty acids, and tannins often accumulate. The cork oak, *Quercus suber*, of Southern Europe, forms a very thick layer and is the main source of commercial corks.

phelloderm The inner layer of the PERIDERM. During secondary growth, the meristematic PHELLOGEN (cork cambium) gives rise to parenchyma tissue, the periderm, which forms the secondary cortex. The cells of the phelloderm arise in radial columns from the phellogen, and this distinguishes them from the primary cortex.

phellogen (cork cambium; periderm cambium) A subepidermal layer of cells forming a lateral meristem that arises following the onset of secondary growth. The cells of the phellogen (cork cambium) give rise externally to the PHELLEM (cork) and internally to the PHELLODERM.

phelloid A thin-walled unsuberized cork cell.

phenolics (phenols) Compounds that contain a benzene ring or other aromatic ring substituted with one or more hydroxyl groups, e.g. phenol (C_6H_5OH). Phenolics are similar to alcohols but are more soluble in water. Many have sharp, spicy odors. It is thought that most phenolics are secreted by plants to deter herbivores as many are enzyme inhibitors and disrupt digestion. Plant phenolics include the FLAVONOIDS, LIGNIN, and TANNINS.

phenolphthalein An acid–alkali indicator that is colorless in acids and red in alkalis. It has a pH range from 8.4–10.0 and is a frequently used indicator for the detection of pH change in acid–alkali titrations. Phenolphthalein, together with borax, is used to test for saccharide derivatives, e.g. glycerol. Such derivatives turn the solution from red to colorless but on boiling the red color returns.

phenotype The observable characteristics of an organism, which are determined by the interaction of the GENOTYPE with the environment. Many genes present in the genotype do not show their effects in the phenotype because they are masked by dominant alleles. Genotypically identical organisms may have very different phenotypes in different environments, an effect particularly noticeable in plants grown in various habitats.

phenylalanine An aromatic AMINO ACID derived from phosphoenolpyruvate and erythrose 4-phosphate. It is a precursor of many aromatic compounds and of alkaloids such as curare and morphine. Phenylalanine can be deaminated to cinnamic acid, required in the synthesis of lignin.

phloem Plant vascular tissue in which food is transported from areas where it is made to where it is needed or stored. It consists of SIEVE ELEMENTS and COMPANION CELLS. The PROTOPHLOEM and METAPHLOEM are primary tissues and derived from the procambium, while the secondary phloem is formed from the VASCULAR CAMBIUM. There are also fibers (sclerenchyma), and parenchymatous packing tissue in the phloem. *See also* mass flow.

phloroglucinol A temporary stain that dyes lignin magenta red in thin sections of plant tissue. It is usually acidified with hydrochloric acid before use. *See* staining.

phosphatide A glycerophospholipid. *See* lipid.

phosphodiester bond A covalent bond linking a sugar and a phosphate group by means of an oxygen bridge, for example, the bonds linking sugars and phosphate groups in the backbone of NUCLEIC ACIDS. Phosphodiester bonds are formed by condensation reactions and may be broken by hydrolysis.

phosphoenolpyruvate (PEP) A 4-carbon phosphorylated organic acid that is the main substrate for carbon dioxide fixation in C_4 PLANTS, being carboxylated to oxaloacetic acid in a reaction mediated by the enzyme *phosphoenolpyruvate carboxylase* (*PEP carboxylase*). It is also an important intermediate of GLYCOLYSIS, being the immediate precursor of pyruvic acid. In the GLYOXYLATE CYCLE for the synthesis of glucose it is generated from oxaloacetate.

phosphoglyceric acid *See* glycerate 3-phosphate.

phosphoglyceride (phospholipid) A complex lipid similar to acylglycerides (*see* acylglycerol), but with a backbone of glycerol 3-phosphate rather than glycerol. Phosphoglycerides are major consitituents of the membranes of cells and organelles, with the exception of chloroplasts, in which GLYCOLIPIDS are more important. They have both hydrophilic and hydrophobic groups, so are reasonably soluble in both water and lipids.

phospholipid *See* lipid.

phosphorescence In general usage the term is applied to the emission of 'cold light' – light produced without a high temperature. The light comes from excited atoms produced directly in the reaction – not from the heat produced. It is thus an example of *chemiluminescence*. In biological systems, the light emitted is often light that has been previously absorbed, but which is re-emitted at a different wavelength. *Phosphorescent* seas are due to the bioluminescence of dinoflagllates (Dinomastigota), in particular *Noctiluca* species.

phosphorus One of the essential elements in living organisms, a macronutrient needed for plant growth. Phospholipids are important in cell membrane structure, and phosphates are necessary for the formation of the sugar–phosphate backbone of nucleic acids. Phosphates are also necessary for the formation of high-energy bonds in compounds such as ATP. Phosphorus has other important roles in living tissues, being a component of certain coenzymes. The phosphate ion, PO_4^{3-}, is an important buffer in cell solutions. Plants take up phosphorus from the soil water as the PO_4^{3-} ion. Phosphorus deficiency results in stunted roots, bluish, bronzed, or purple leaves, and poor germination, seed set, and fruit ripening.

phosphorylase *See* kinase; phosphorylation.

phosphorylation The addition of a phosphate group to a compound. Phosphorylation reactions are catalyzed by phosphorylase enzymes. Phosphorylation is the first step in many metabolic pathways, the phosphate bond serving as an energy source in a later stage of the pathway. For example, glycolysis starts with the phosphorylation of glucose and fructose 1,6-bisphosphate. Phosphorylation is also needed for the activation or deactivation of some enzymes. *See also* oxidative phosphorylation; photophosphorylation.

phosphotransferase *See* kinase.

photic zone The surface layer of an ocean or lake that is penetrated by sunlight and in which the phytoplankton flourish. Red and yellow wavelengths of light penetrate to about 50 m while blue and violet light may reach 200 m. The diatoms, which are the main components of phytoplankton, may be found down to 80 m. Most of the plankton lives in the top 10 m. Beyond 200 m the water is perpetually dark. The depth at which the rate of fixation of carbon dioxide by photosynthesis is equal to the rate at which it is lost by respiration is called the *compensation depth*. The part of the photic zone above this depth is termed the *euphotic zone*, the part below the compensation depth is the *dysphotic zone*. The depth of the euphotic zone depends on the turbidity of the water. *See* plankton. *Compare* aphotic zone.

photoautotrophism *See* autotrophism; phototrophism.

photoheterotrophism *See* heterotrophism; phototrophism.

photolysis Chemical breakdown caused by light. In PHOTOSYNTHESIS the process is important in providing hydrogen donors by the splitting of water, as follows:

$$2H_2O \rightarrow O_2 + 4H^+ + 4e^-$$

In the CHLOROPLAST photolysis of water occurs in the oxygen- releasing complex linked to Photosystem II. The protons are released into the lumen of the thylakoid. The high-energy electrons go to replace electrons displaced from the reaction center of Photosystem II by the absorption of light energy. The electrons are the energized by light energy and pass to the electron-transport chain of noncyclic photophosphorylation, which generates ATP and reduced NADP.

photomicrograph *See* micrograph.

photonasty (**photonastic movements**) A nastic movement in response to light. Flowers such as the marigold (*Calendula officinalis*) open on exposure to light. Care must be taken not to confuse photonasty with nyctinasty, in which day–night opening/closing of flowers is governed by an internal CIRCADIAN RHYTHM. *See* nastic movements.

photo-oxidation The oxidation of a substance as a result of the absorption of light energy. Very high light intensity can cause oxidation (bleaching) of photosynthetic pigments such as chlorophylls, inactivating them.

photoperiodism The response of an organism to changes in day length (*photoperiod*). In plants, leaf fall and flowering are common responses to seasonal changes in day length.

Plants are classified as *short-day plants* (*SDPs*), e.g. *Chrysanthemum*, or *long-day plants* (*LDPs*), e.g. cucumber (*Cucumis sativus*), according to whether they flower in response to short or long days. *Day-neutral plants*, e.g. pea (*Pisum sativum*), have no photoperiodic requirement. In fact, the critical factor is not the length of the day, but the length of the dark period, since flowering of SDPs is inhibited by even a brief flash of red light in the dark period (a PHYTOCHROME response), and an artificial cycle of long days and long nights inhibits flowering in LDPs. Thus, it is the interaction between light and dark periods that in some way affects flowering through the mediation of phytochrome. Phytochrome is a protein that can absorb red light, which converts it to a form (P_{FR}) that absorbs far-red light. Absorption of far-red light causes the phytochrome to revert to the red light-absorbing form (P_R). The P_{FR} form of phytochrome inhibits flowering in SDPs (P_{FR} slowly disappears during long nights) and promotes flowering in LDPs (P_{FR} remains at high levels in short nights). The light stimulus is perceived by the leaves and in some unknown way transmitted to the floral apices. *See* circadian rhythm; critical day length; thermoperiodism; vernalization.

photophosphorylation (**photosynthetic phosphorylation**) The conversion of ADP to ATP using light-induced electron transport. *See* photosynthesis.

photoreceptor Any light-sensitive organ or organelle. The paraflagellar body, a swelling on the undulipodium of certain species of *Euglena*, is an example.

photorespiration (**C_2 cycle**) A light-

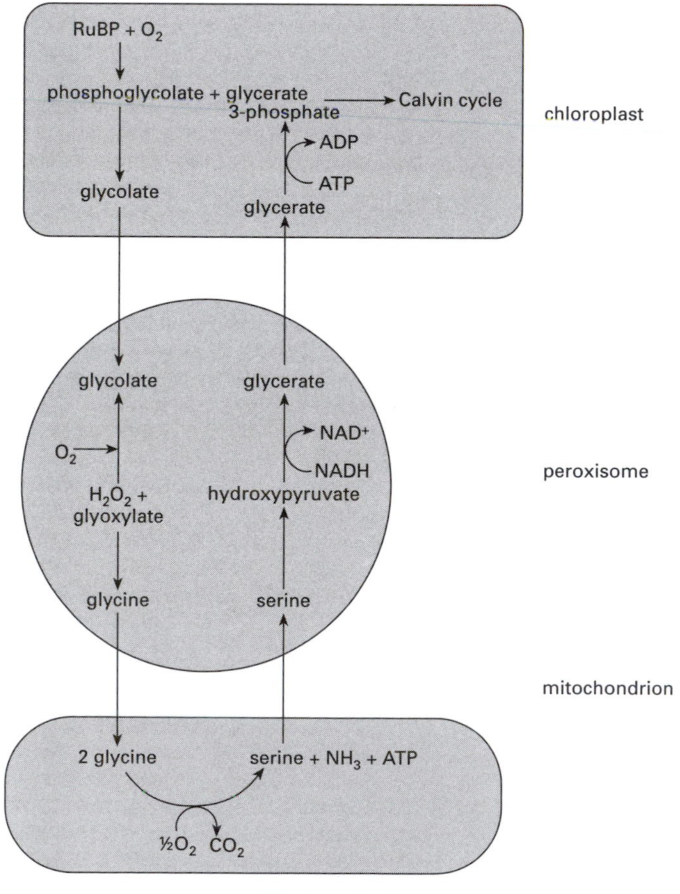

Photorespiration

dependent metabolic process of most green plants that resembles true (or 'dark') respiration only in that it uses oxygen and produces carbon dioxide. Carbon dioxide production during photorespiration may be up to five times greater than in dark respiration. Photorespiration takes place in PEROXISOMES rather than mitochondria, and it is not coupled to oxidative phosphorylation. It wastes carbon dioxide and energy, using more ATP than it produces. Its substrate is glycolate produced in C_3 plants by the oxidation of ribulose bisphosphate to glycerate 3-phosphate and phosphoglycolic acid. During photorespiration the enzyme glycolate oxidase catalyzes the oxidation, by molecular oxygen, of glycolic acid to glyoxylic acid and hydrogen peroxide in the peroxisomes. The glyoxylate is converted to glycine, and carbon dioxide is released when the glycine molecules are converted to serine in the mitochondria. The term photorespiration is being replaced by the C_2 cycle (oxidative photosynthetic carbon cycle), to better reflect the fact that unlike respiration it uses rather than releases energy.

It is estimated that in C_3 plants 50% of the potential yield of photosynthesis can be lost through photorespiration in warm weather. It is therefore economically important and ways of inhibiting the process are being investigated.

photosynthesis The synthesis of organic compounds using light energy absorbed by chlorophyll. With the exception of a small group of bacteria, organisms photosynthesize from inorganic materials. All green plants photosynthesize, as well as algae and certain bacteria. In green plants, photosynthesis takes place in chloroplasts, mainly in leaves and stems. Directly or indirectly, photosynthesis is the source of carbon and energy for all except chemoautotrophic organisms. The mechanism is complex and involves two sets of stages: *light reactions* followed by *dark reactions*. The overall reaction in green plants can be summarized by the equation:

$$CO_2 + 4H_2O \rightarrow [CH_2O] + 3H_2O + O_2$$

The light reactions of photosynthesis involve reaction centers in the chloroplasts. Each reaction center lies in a flat sheet of 200-400 pigment molecules in the thylakoid membrane, the antenna complex, and contains a pair of chlorophyll *a* molecules. The whole unit is called a *photosystem*. There are two kinds of photosystems in green plants, *Photosystems I and II*. Both are usually involved in the light reactions of photosynthesis. Light energy absorbed by the pigments of the antenna complex is passed to the reaction-center chlorophyll molecules, from which it passes along an electron-transport chain. Photosystem II contains a type of chlorophyll *a* (P680) that shows maximum light absorption at a wavelength of 684 nm. When activated by light, a pair of electrons are excited and leave Photosystem II, being replaced by electrons from the photolysis of water:

$$2H_2O \rightarrow O_2 + 4H^+ + 4e^-$$

Molecular oxygen is released, and the protons pass into the lumen of the thylakoid. The electrons pass through an ELECTRON-TRANSPORT CHAIN in the thylakoid membrane, being accepted first by plastoquinone, then by the cytochrome *b/f* complex (a proton pump that pumps H^+ into the thylakoid lumen), and on via plastocyanin to Photosystem I, which contains chlorophyll *a* (P700) with a maximum light absorption at 700 nm. Here the electrons are boosted by more absorbed light energy to an even higher energy level and pass through a second electron-transport chain involving the protein ferredoxin, reducing NADP to NADPH + H^+ for use in other reactions, such as various syntheses. This sequence, which involves both photosystems, is termed *noncyclic electron flow*. However, it is possible for electron flow to involve only Photosystem I, and the ferredoxin/cytochrome *b/f/* plastocyain electron- transport chain, electrons returning to Photosystem I to be reenergized. This *cyclic electron flow* does not produce NADPH. It simply generates a proton gradient across the thylakoid membrane. This is used to drive ATP synthesis (*cyclic photophosphorylation*). ATP synthesis associated with the noncyclic electron-transport chain is termed *noncyclic photophosphorylation*. The dark reactions of photo-

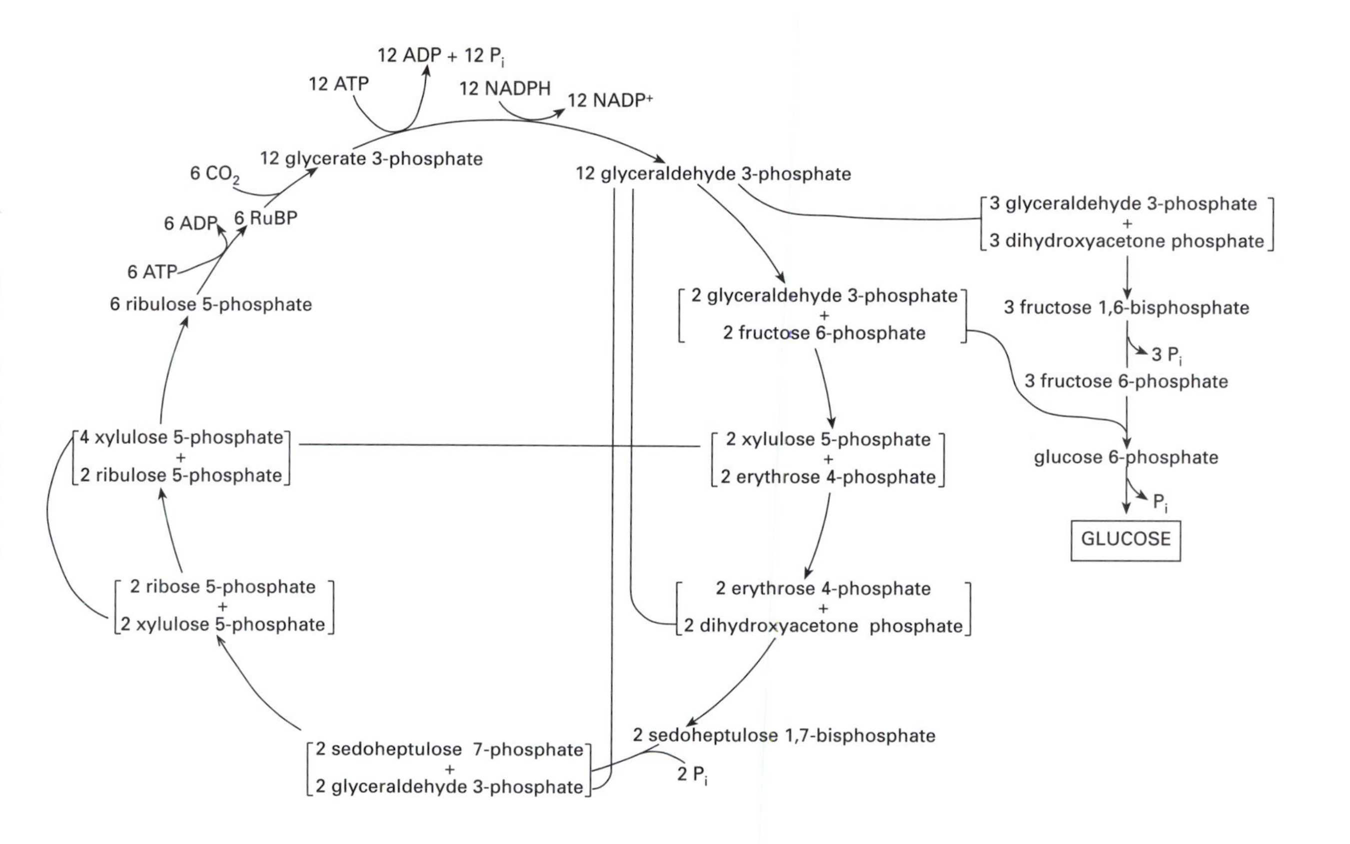

Photosynthesis: the Calvin cycle

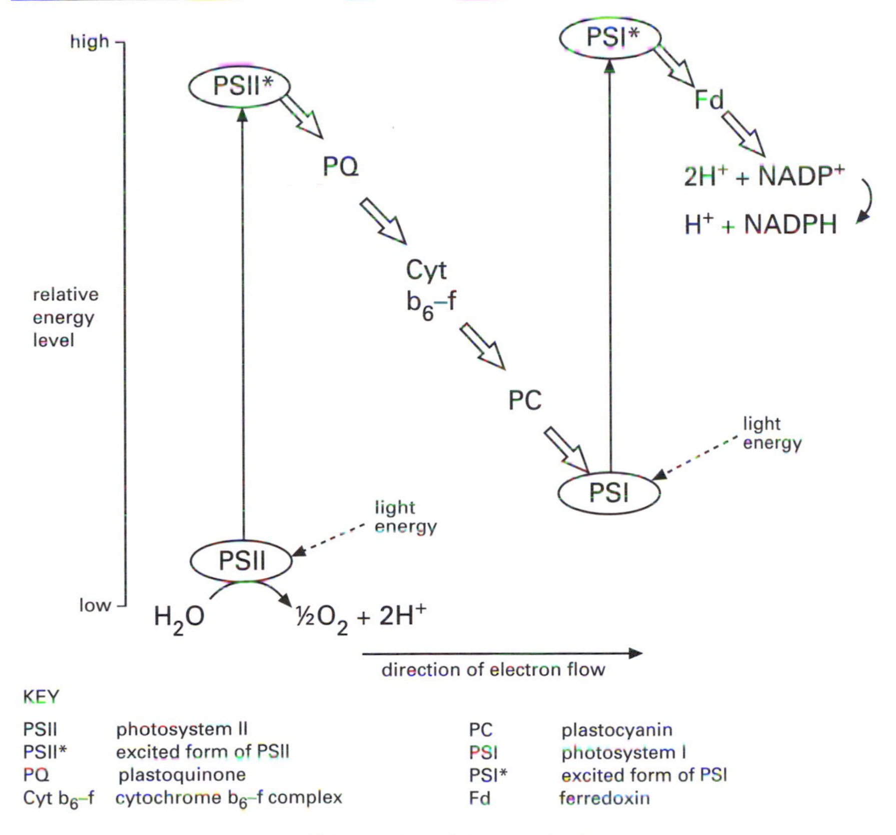

Light reaction of photosynthesis

synthesis take place in solution in the chloroplast stroma. Here, ATP and NADPH from the light reactions are used to reduce carbon dioxide to carbohydrate. Carbon dioxide is 'fixed' by combination with the 5-carbon sugar ribulose bisphosphate (RuBP), forming two molecules of phosphoglyceric acid (PGA). This reaction is catalyzed by the enzyme *ribulose bisphosphate carboxylase* (*rubisco*). In a series of reactions using NADPH and ATP from the light reactions, PGA is converted to a succession of 3-, 4-, 5-, 6-, and 7- carbon sugar phosphates in a series of reactions collectively termed the *Calvin cycle*. These products are then used in the synthesis of carbohydrates, fats, proteins, and other compounds, and RuBP is regenerated. *See* C_3 plant; C_4 plant.

photosynthetic bacteria A group of bacteria able to photosynthesize through possession of chlorophyll pigments. They include the Cyanobacteria, green sulfur bacteria (Chlorobia), green nonsulfur bacteria (Chloroflexa), purple sulfur bacteria, and purple nonsulfur bacteria (both in the phylum Proteobacteria). The photosynthetic bacteria fix carbon using the Calvin cycle, but use a variety of hydrogen sources.

photosynthetic pigments Pigments that absorb the light energy required in photosynthesis. They are located in the chloroplasts of plants and algae, whereas in most photosynthetic bacteria they are located in thylakoid membranes, typically distributed around the cell periphery. All photosynthetic organisms contain chlorophylls and carotenoids; some also contain phycobilins. Chlorophyll *a* is the *primary pig-*

ment in plants, algae, and cyanobacteria, and bacteriochlorophyll is the pigment used by many photosynthetic bacteria, since energy absorbed by this is used directly to drive the light reactions of photosynthesis. The other pigments (chlorophylls *b*, *c*, and *d*, and the carotenoids and phycobilins) are *accessory pigments* that pass the energy they absorb on to chlorophyll *a*. They broaden the spectrum of light used in photosynthesis. *See* absorption spectrum.

photosystem *See* photosynthesis.

phototaxis (**phototactic movement**) A TAXIS in response to light. Many motile algae are positively phototactic, e.g. *Volvox*. *Chlamydomonas* cells are also positively phototactic, swimming toward light to increase their photosynthesis rate, but if the light intensity becomes too great, they show negative phototaxis, swimming away to protect their pigments. Chloroplasts also show phototaxis, and take up specific orientations in relation to the direction of the incident light.

phototrophism A type of nutrition in which the source of energy for synthesis of organic requirements is light. Most phototrophic organisms are autotrophic (i.e. show *photoautotrophism*); these comprise the green plants, Cyanobacteria, and some photosynthetic bacteria (the purple and green sulfur-bacteria).

phototropism (**phototropic movement**) A directional growth movement of part of a plant in response to light. The phenomenon is clearly shown by the growth of shoots and coleoptiles toward light (*positive phototropism*). The stimulus is perceived in the region just behind the shoot tip. If light falls on only one side of the apex then auxins produced in the apex tend to diffuse toward the shaded side. Thus more auxin diffuses down the stem from the shaded side of the tip. This results in greater elongation of cells on the shaded side thus causing the stem to bend toward the light source. Most roots are *aphototropic* (light-insensitive) but some (e.g. the adventitious roots of climbers such as ivy) grow away from light (*negative phototropism*). *See also* tropism.

phragmoplast A barrel-shaped body appearing in dividing plant cells during late anaphase and telophase between the two separating groups of chromosomes. It consists of MICROTUBULES associated with the SPINDLE, together with Golgi apparatus, endoplasmic reticulum, and ribosomes, and transports vesicles that coalesce to form the early cell plate.

phycobilins (**phycobiliproteins**) A group of accessory photosynthetic pigments found in Cyanobacteria and Rhodophyta (red algae). They comprise the blue *phycocyanins*, which absorb extra orange and red light, and the red *phycoerythrins*, which absorb green light, enabling red algae to grow at depth in the sea.

phycocyanin *See* phycobilins.

phycoerythrin *See* phycobilins.

phylloclade *See* cladode.

phyllode An expanded flattened petiole that acts as the photosynthetic organ if the lamina is missing or very reduced. Phyllodes are seen in various Australian species of *Acacia*. *Compare* cladode.

phyllotaxis (**phyllotaxy**) The arrangement of leaves on a stem. There may be one, two, or several leaves at each node. When there are three or more leaves forming a circle around the node the arrangement is said to be *whorled*, e.g. bedstraws (*Galium* spp.). When leaves arise singly the arrangement may be *spiral*, e.g. poplar (*Populus* spp.), or *alternate*, e.g. hazel (*Corylus* spp.), and when they arise in pairs the arrangement is termed *opposite*, e.g. lilac (*Syringa* spp.). Opposite phyllotaxis may be either *distichous* or *decussate* depending on whether the leaf pairs are in the same plane up the stem or arise alternately at right angles to each other.

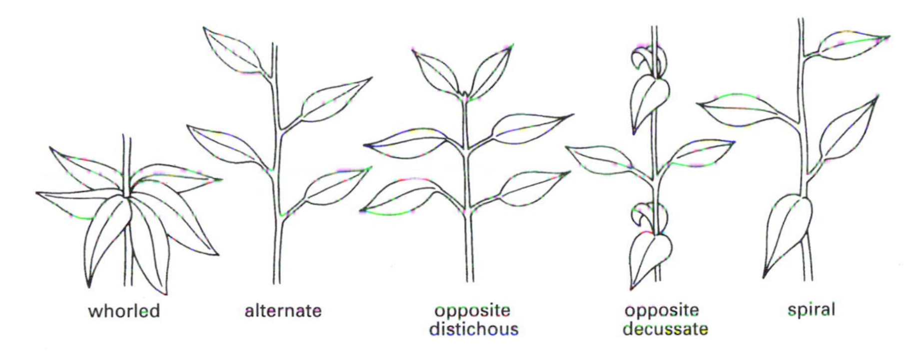

plans showing angles of divergence in various leaf arrangements

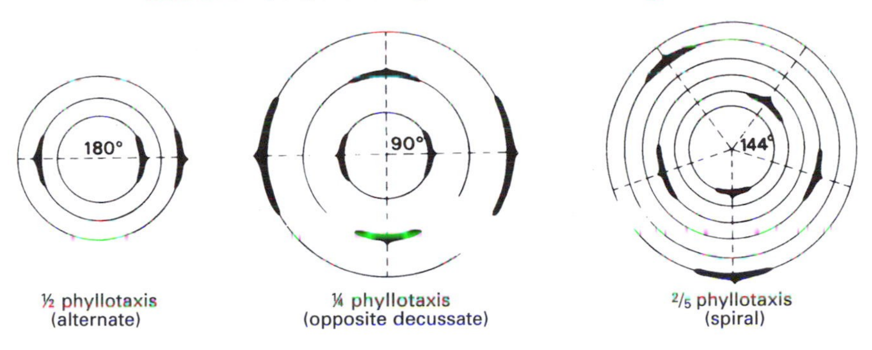

Phyllotaxis

phylogenetic Relating to the study of evolutionary relationships between and within taxonomic levels, in particular the relationships between groups of organisms based on their past evolutionary history.

phylogeny The evolutionary history of groups of organisms. *See* phylogenetic.

phylum (*pl.* **phyla**) One of the major groups into which a kingdom of organisms is classified. Phyla may be divided into *subphyla.* The names of plant phyla end in *-phyta*, e.g. Anthophyta, Lycophyta. In some plant classifications (especially older ones) the term 'division' is used instead of phylum.

physiological drought A situation in which plants are unable to take up sufficient water even though there is water in the soil: they are effectively in a drought situation. It occurs when the concentration of solutes in the soil water is equal to or higher than that in the root cells, so water is unable to enter the root by OSMOSIS. Physiological drought occurs in salt marshes and other brackish coastal habitats, and in deserts where there are high concentrations of salts in the upper layers of the soil. It also occurs in cold weather because the permeability of the endodermal cells decreases rapidly below 5°C, so the root resists the entry of water (*root resistance*).

physiological race The existence of physiologically distinct but morphologically identical races within a species. Such physiological races are important in host–pathogen studies, particularly in planning programs to breed for crop resistance.

physiology The way in which organisms or parts of organisms function. *Compare* morphology.

phytoalexin A chemical produced by a plant that inhibits the growth of a pathogenic fungus. For example, when attacked by the fungus *Ceratostomella*, the sweet potato (*Ipomoea batatus*) concentrates the terpenoid ipomeamarone in its tissues, and this concentration has been shown to be related to the degree to which it resists the attack.

phytochrome A proteinaceous pigment found in low concentrations in most plant organs, particularly meristems and dark-grown seedlings. It exists in two interconvertible forms. P_R (or P_{660}) has an absorption peak at 660 nm (red light) and P_{FR} (or P_{730}) at 730 nm (far-red light). Natural white light favors formation of P_{FR}, the physiologically active form, while far-red light encourages reversion to P_R if given soon after the red light. Light intensities required for conversion are very low and it occurs within seconds. In some species, P_{FR} may revert to P_R in the dark, a process inhibited by low temperatures. *See* circadian rhythm; photoperiodism.

phytohormone *See* hormone.

phytoplankton *See* plankton.

pico- Symbol: p A prefix denoting one million-millionth, or 10^{-12}. For example, 1 picogram (pg) = 10^{-12} gram. *See* SI units.

pileus (*pl.* **pilei**) The cap of the mature mushroom or toadstool (sporophore) in certain basidiomycete fungi (e.g. *Agaricus*). Its undersurface comprises gills or pores lined with basidia. The upper surface may be fleshy or leathery, and may retain the flaky remains of the universal veil.

pili (*sing.* **pilus**) (**fimbriae** *sing.* **fimbria**) Fine, straight, hairlike protein structures emerging from the walls of certain bacteria. They confer the property of 'stickiness' whereby bacteria tend to adhere to one another. They are hollow tubes and may number from one to several hundred.

piliferous layer The region of the root epidermis that gives rise to the root hairs. It is located just behind the zone of elongation and is the main absorptive area of the root.

Pinaceae A family of conifers that bear separate male and female cones, and in which the seeds are borne between the cone scales and do not have arils. Most species are evergreen trees or shrubs, but the larches or tamaracks (*Larix* spp.) are deciduous. One of the oldest living trees is a bristlecone pine (*Pinus aristata*), thought to be about 4900 years old. The family includes cedars (*Cedrus*), firs (*Abies*), hemlocks (*Tsuga*), and larches and pines (*Pinus*).

pin mold *See* Mucorales.

pinna (*pl.* **pinnae**) *See* pinnate.

pinnate Describing a compound leaf with leaflets (*pinnae*) arranged in two rows, one on each side of the midrib (flowering plants, e.g. *Vicia*) or of a central stalk (ferns, e.g. *Dryopteris*). The pinnae may themselves be divided into leaflets, which are called *pinnules*.

pinnule *See* pinnate.

pinocytosis *See* endocytosis.

pioneer species A species that colonizes a new physical environment, for example land exposed by retreating glaciers, felling of forest, or bare mud on coasts or estuaries. Pioneer species are the first stage in plant succession.

pistil In angiosperms, the seed-containing structure. In an apocarpous gynoecium it corresponds to the carpel, while in a syncarpous gynoecium it is made up of two or more fused carpels.

pit A gap in the secondary cell wall that enables communication between thickened cells, e.g. tracheids. According to whether or not the secondary wall forms a lip over the pit, pits are described as *bordered* or *simple*, respectively. If a pit occurs singly it is termed a *blind pit*. Usually pits occur in pairs so that the only barrier separating adjacent cells is the middle lamella and the respective primary cell walls. The primary cell wall adjacent to the cavity is called the pit membrane.

pitcher plant Any carnivorous plant with pitcher-shaped leaves. Insects and other small animals fall into the pitchers, which are usually filled with rainwater or plant digestive enzymes, and drown; the plant obtains nitrates and perhaps other nutrients from their decomposing bodies.

pith (**medulla**) The central region of the stem and, in some herbaceous plants, root that is normally composed of parenchymatous tissue. It occurs to the inside of the stele.

pitted thickening The most extensive form of thickening found in XYLEM vessels and tracheids, where the secondary cell wall is continuous apart from small areas called PITS. The pits may be arranged in an alternate or opposite pattern. If they are elongated horizontally and parallel to each other, this is described as *scalariform thickening*.

placenta (*pl.* **placentae**) **1.** The region of tissue occurring on the inner surface of the ovary wall of the carpels of flowering plants where the ovules develop. The arrangement of ovules within the ovary (*see* placentation) depends on whether there are one or many carpels and whether the carpels are free or fused.
2. A central swelling on the lower surface of fern pinnules linking clusters of sporangia with the main frond. *See also* sorus.

placentation The position of the ovule-bearing placentae in angiosperm seeds. Placentation varies according to whether there are one or many carpels. If there is only one ovule in the ovary, it may be attached at the base of the ovary wall (*basal placentation*), or at the apex (*apical placentation*). If there is more than one ovule, and there is only one carpel in the ovary, as in pea (*Pisum sativum*), placentation is along the ventral suture and is termed *marginal placentation*. Marginal placentation also occurs in apocarpous polycarpellary ovaries of LEGUMES but various types of placentation are seen in syncarpous ovaries. Where the carpels are fused to give a unilocular ovary, the ovules may be found along the two placentae at each line of fusion giving *parietal placentation*, as in violet (*Viola* spp.) and passionflower (*Passiflora*). If instead the ovules are borne on a central column, placentation is termed *central placentation*; a modification of this is *free-central placentation*, in which the column does not extend to the top of the ovary, as in primrose (*Primula vulgaris*). Where other carpels to give a multilocular ovary, the marginal placentae of the carpels may fuse in the center of the ovary, giving *axile placentation*, as in tulip (*Tulipa* spp.).

plagiotropism *See* tropism.

plankton A varied collection of aquatic organisms that drift freely, not being attached to any substrate and not possessing any organs for locomotion. The most important components of the plant plankton (*phytoplankton*) are the diatoms upon which the planktonic animals (*zooplankton*) feed. Other members of the phytoplankton include other microscopic algae and cyanobacteria.The plankton form the basis of the food chain in the sea. *Compare* benthic.

plant An organism that can (usually) make its own food by taking in simple inorganic substances and building these into complex molecules by a process termed photosynthesis. This process uses light energy, absorbed by a green pigment called chlorophyll, which is found in all plants but no animals. There are a few exceptions, in the form of certain parasitic plants. Most plants have cellulose cell walls, and have starch as a storage polysaccharide, whereas animals have no cell walls, store glycogen, and do not have plastids or chlorophyll; and fungi usually lack cellulose and also do not form plastids. Plants lack motility, with the exception of the gametes of many species. One major characteristic that distinguishes plants from other plantlike organisms, such as algae or fungi, is the possession of an embryo that is retained and nourished by maternal tissue. Fungi and algae lack embryos and develop from spores. Plants also differ from fungi in having a regular alternation of diploid and haploid generations. Plants also differ from animals by generally responding to stimuli very slowly, the response often taking a matter of days and occurring only if the stimulus is prolonged. *Compare* algae; Fungi; Protoctista.

plant hormone *See* hormone.

plasmagel (**ectoplasm**) The gel-like region of cytoplasm located in a thin layer just beneath the plasma membrane, particularly of unicells that move in an ameboid fashion, and cells that exhibit cyclosis.

plasmagene A gene contained in a structure outside the nucleus, e.g. the genes in mitochondria and chloroplasts. *See* cytoplasmic inheritance.

plasmalemma (*pl.* **plasmalemmae**) *See* plasma membrane.

plasma membrane (**cell membrane; plasmalemma**) The membrane that surrounds all living cells. It ranges from 7.5–10 nm in thickness and consists of phosphoglyceride molecules (about 40%) and proteins (about 60%). The membrane backbone consists of phosphoglyceride molecules arranged in two rows with their hydrophilic polar heads facing outwards and the hydrophobic hydrocarbon tails in the center. This lipid layer is penetrated by globular proteins, which in some places extend from one side of the membrane to the other. The whole structure has fluidlike properties, as the molecules are able to move laterally – the so-called *fluid-mosaic model*. The membrane is probably being constantly renewed. The plasma membrane is selectively permeable, and controls which substances can enter and leave the cell.

plasmasol (**endoplasm**) The sol-like form of cytoplasm, located inside the PLASMAGEL. It is free-flowing and contains the cell organelles.

plasmid An extrachromosomal genetic element found within bacterial cells that replicates independently of the chromosomal DNA. Plasmids typically consist of circular double-stranded DNA molecules of molecular weight 10^6–10^8. They carry a variety of genes, including those for antibiotic resistance (*R factors*), toxin production, and enzyme formation, and may be advantageous to the cell. Some plasmids, called *episomes*, can reversibly insert themselves into the bacterial DNA, replicating with it.

plasmodesma (*pl.* **plasmodesmata**) A fine strand of cytoplasm that serves to connect the protoplasm of adjacent plant cells, allowing material to pass from one cell to its neighbor.

plasmodial slime molds *See* Myxomycota.

Plasmodiophora A phylum of protoctists that are obligate endoparasites of plants, fungi, and algae. Most plasmodiophorans cause little damage to the host plant, but a few cause serious diseases, e.g. *Plasmodiophora brassicae* causes clubroot of cabbage and related brassicas.

plasmodium (*pl.* **plasmodia**) *See* Myxomycota.

plasmogamy Fusion of protoplasm, usually referring to the fusion of cytoplasm but not nuclei. It usually occurs between gametes or protoplasts. Plasmogamy in the absence of karyogamy (fusion of nuclei) occurs between fungal mycelia of different strains of some ascomycetes and basidiomycetes to form a heterokaryon.

plasmolysis Loss of water from a walled cell (e.g. of a plant or bacterium) due to osmosis to the point at which the protoplast shrinks away from the cell wall. The point at which this is about to happen is called *incipient plasmolysis*. Here the cell wall is not being stretched; i.e. the cell has lost its turgidity or become *flaccid* (wall pressure is zero). Wilting of herbaceous plants occurs at this point.

plastid An organelle enclosed by two membranes (the envelope) that is found in plants and certain protoctists (e.g. algae), and develops from a proplastid. Various types exist, but all contain DNA and ribosomes. Plastids range from 10 μm to less than 1.0 μm in diameter. They include chloroplasts, chromoplasts, amyloplasts, etc., and are often interconvertible.

plastid inheritance *See* plastogene.

plastocyanin An electron carrier in PHOTOSYNTHESIS, involved in the transfer of electrons from photosystem II to photosystem I.

plastogene A gene present in a self-replicating plastid. Inheritance studies have shown that plastogenes control leaf color in some plants, e.g. yellow-leaved *Primula sinensis*, is an example of *plastid inheritance*. *See* cytoplasmic inheritance; plasmagene.

pleiomorphism The occurrence of different morphological stages during the life of an organism. Examples are the different spore forms of the rust fungi. *Compare* polymorphism.

Pleistocene The first epoch of the Quaternary period, which started with a glaciation about two million years ago and ended with the last glaciation about 10 000 years ago. Several ICE AGES drove many organisms toward the equator while others became extinct. For example, tundra covered temperate parts of the USA and Central Europe where today deciduous forests are the natural vegetation. The boreal forests extended south almost to northern Louisana and the Mediterranean. Fossil pollen deposits provide evidence for the past distribution of plants. Some migrating plants found the tops of mountains further south resembled the conditions they had left, and survived in isolated areas (*refugia*). When the climate warmed again, these became centers of diversity. The cycle of southward then northward migration led to the mixing of plants from both cooler and warmer regions, leading to new patterns of competition and reorganization of plant communities. Modern man (*Homo sapiens*) evolved during this period, and probably caused further disruption of plant commununties by his use of fire and the start of forest clearance. *See also* geological time scale; pollen analysis.

plerome *See* histogen theory.

pleuropneumonia-like organisms *See* mycoplasmas.

Pliocene The epoch of the Tertiary period, about 7 to 2 million years ago, which followed the Miocene. In the Pliocene the hominids, such as *Australopithecus* and *Homo*, became clearly distinguishable from the apes. Grasses became more abundant, perhaps because the climate was becoming drier. Modern conifers were spreading, and leptosporangiate ferns underwent adaptive radiation. *See also* geological time scale.

plumule The shoot apex and first rudimentary leaves in the mature embryo and the seedling. In seedlings showing epigeal germination, the plumule is taken above ground between the cotyledons. When germination is hypogeal, only the plumule emerges from the soil; it has a hooked tip to protect it as it pushes up through the soil. *Compare* radicle.

pneumatophore (**aerophore**; **breathing root**) A specialized negatively geotropic root produced by certain aquatic vascular plants (especially trees such as mangroves, e.g. *Sonneratia*) that protrudes well above soil level. The aerial part is covered with LENTICELS through which gases can diffuse to and from the highly developed system of intercellular airspaces. Pneumatophores are an adaptation to waterlogged, poorly aerated soils, allowing roots to exchange gases.

Poaceae (**Gramineae, grasses**) The grasses, a family of monocotyledons with 9500 species. Most are annual or perennial herbs, although some are woody (but do not have secondary thickening), e.g. bamboos. Grasslands are the dominant vegetation on the prairies, steppes and savannas, and in times past supported vast herds of grazing mammals. Today these areas are also important are for growing cereal grasses, such as barley (*Hordeum vulgare*), maize (*Zea mays*), millet (*Panicum miliaceum),* oats (*Avena sativa*), rye (*Secale cereale*), sorghum (*Sorghum*), and wheat (*Triticum*). Rice (*Oryza sativa*) and sugar cane (*Saccharum officinarum*) are also grasses.

pod *See* legume.

Podocarpaceae A family of evergreen conifers. It includes one of the oldest living organisms, a huon pine, *Dacrydium franklinii*, from Tasmania, possibly 10 000 years old.

podsol (**podzol**) The type of SOIL found under heathland and coniferous forests in temperate and boreal climates where there is heavy rainfall and long cool winters, such as the coniferous forests of northern North America and Eurasia. It is strongly acid and often deficient in nutrients, especially iron compounds and lime, as a result of leaching by the heavy rain or snow-melt in spring. Beneath the humus layer lies a leached horizon (*A horizon*) composed mainly of quartz sand. In the clay-rich *B horizon* below the leached layer iron compounds accumulate, staining the layer brown and forming an impermeable *hard pan* that prevents drainage. Where this layer is sufficiently impermeable to maintain the A horizon in a waterlogged condition, the soil is termed a *gley podsol.*

point mutation *See* gene mutation.

polarity A lack of symmetry: a definite orientation that arises during the differentiation of a cell, tissue, or organ. For example, in the development of the plant embryo from the zygote, one end becomes the radicle, the other gives rise to cotyledons.

polarizing microscope A light MICROSCOPE that uses polarized light to illuminate the specimen. The orientation of molecules in certain crystalline substances affects the transmission of polarized light.

polar nuclei The, usually two, haploid nuclei found midway along the EMBRYO SAC, forming a binucleate *central cell.* They may fuse to form the diploid *definitive nucleus.* The endosperm is formed from the fusion of one or both polar nuclei (occasionally several) with one of the male gametes from the pollen tube.

pollen The MICROSPORES of seed plants, produced in large numbers in the pollen sacs, or stamens. They are formed by meiosis of somatic pollen mother cells. Pollen grains are rich in mitochondria, Golgi apparatus, and endoplasmic reticulum. In many gymnosperms the nuclei divide within the pollen grain to produce a small number of cells (2–40 depending on the group) representing the male prothallus. In angiosperms and some conifers (e.g. *Cupressus*) the pollen grains contain just two nuclei: a GENERATIVE NUCLEUS and a VEGETATIVE NUCLEUS (tube nucleus).When a pollen grain germinates, a pollen tube pushes its way out through an aperture in the pollen grain wall, and the nuclei migrate into the tube. The generative nucleus divides to produce two sperm cells consisting of only nucleus and cytoplasm, one of which will fertilize the ovum, while the other will fuse with the POLAR NUCLEI to form the endosperm. The vegetative nucleus degenerates after fertilization. Pollen grains are highly variable in size and shape ranging from 5μ to over 200μ in diameter and from circular to tetrahedral or dumbbell shaped. The wall of the pollen grain consists of two layers, a highly resistant outer *exine*, which may be intricately sculpted and is punctured by apertures,

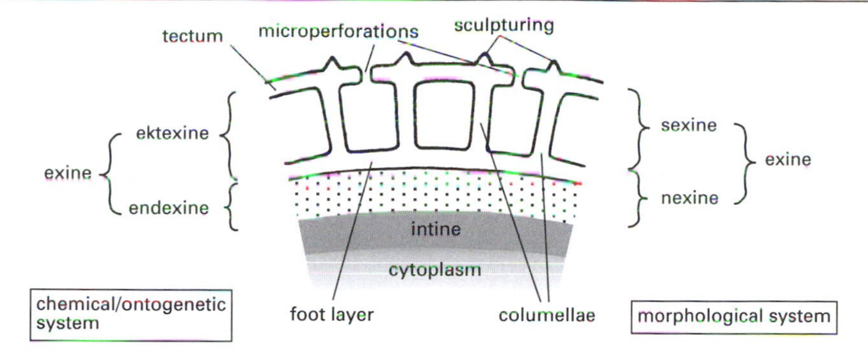

Pollen: section of a pollen grain showing the exine

and an inner *intine*, which sometimes protrudes through these apertures. The size and shape of the pollen grains and the number and orientation of the apertures and sculpturing is characteristic of particular families, genera, or sometimes species. Pollen grains are adapted according to the method of pollination, those carried by insects often being sticky or barbed, while wind-pollinated plants generally produce smooth light pollen. *See also* pollen analysis.

pollen analysis A means of obtaining information on the composition and extent of past floras and environments by examining fossil pollen grains, spores, and similar structures in peat and sedimentary deposits. The outer wall (*exine*) of the pollen grain is often diagnostic (*see* pollen) and very resistant to decay. *See also* palynology.

pollen culture *See* anther culture.

pollen mother cell (PMC) In angiosperms and gymnosperms, a somatic cell that gives rise to four haploid pollen grains by meiosis.

pollen sac A chamber in which the pollen is formed in the angiosperms and conifers.

pollen tube A filamentous outgrowth of the pollen grain that in most seed plants grows out through a pore in the exine and transports the male gametes to the ovule. Germination of the pollen grain to give the pollen tube usually takes place only when the pollen is compatible with the stigma tissue. The pollen tube may grow down the stylar canal, or it may secrete enzymes that digest the tissues of the style, creating a path for the tube. It usually enters the ovule through the micropyle, but in some species, such as beech (*Fagus*) it may enter through the chalaza (*chalazogamy*).

pollination The transfer of pollen from the anther to the stigma in angiosperms, and from the sporangiophores to the micropyle in gymnosperms. If the pollen is compatible (i.e. of the right type) then the pollen grains germinate, producing a pollen tube that grows down the style carrying the pollen nuclei to the ovule. Plants may be self-pollinating, e.g. barley (*Hordeum vulgare*), thus ensuring that seed will be set, even in the absence of other members of the same species. However, *self-pollination* also leads to homozygosity and, potentially, less adaptable plants in the long term; thus, in many plant species mechanisms exist to prevent it and promote CROSS-POLLINATION. In some species of angiosperms stamens and stigmas mature at the same time and pollen may fall directly onto the stigma of the same flower, or the stamens and stigmas may actually touch. *See also* cleistogamy; dichogamy; homogamy; incompatibility; protandry; protogyny.

pollution Any damaging or unpleasant change in the environment that results from the physical, chemical, or biological side-effects of human activities. Pollution can affect the atmosphere, rivers, seas, and the soil.

Air pollution is caused by the domestic and industrial burning of carbonaceous fuels, by industrial processes, and by vehicle exhausts. Industrial emissions of sulfur dioxide cause acid rain, and the release into the atmosphere of chlorofluorocarbons, used in refrigeration, aerosols, etc., leads to the depletion of ozone in the stratosphere (*see* acid rain; ozone layer). Carbon dioxide, produced by burning fuel and by motor vehicle exhausts, is slowly building up in the atmosphere, and is one of the factors leading to an overall increase in the temperature of the atmosphere (*see* greenhouse effect). Vehicle exhausts also contain carbon monoxide, nitrogen oxides, and other hazardous substances, such as fine particulate dusts. Water pollutants include those that are biodegradable, such as sewage effluent and nitrates leached from agricultural land, which if allowed to enter water courses can lead to eutrophication (*see* eutrophic) and algal blooms. Non-biodegradable pollutants, such as certain chlorinated hydrocarbon pesticides (e.g. DDT) and heavy metals, such as lead, copper, and zinc in some industrial effluents, accumulate in the environment. Heavy metals, even when present at low concentrations, are toxic to plants, inhibiting water or nutrient uptake, damaging cell membranes, and inhibiting enzymes.

polyembryony The occurrence of many embryos in one ovule. *See* apomixis; parthenogenesis.

polymer A molecule made up of a series of similar units joined together, e.g. starch, which is made up of glucose units.

polymerase An enzyme that regulates the synthesis of a polymer. Examples include *RNA polymerases* and *DNA polymerases*.

polymerase chain reaction (PCR) A technique for amplifying small samples of DNA rapidly and conveniently. It is used widely in research and forensic science, e.g. to clone genes or to produce a suitable quantity of DNA for genetic fingerprinting from the minute amounts present in traces of blood or other tissue. *See also* genetic fingerprinting.

polymorphism A distinct form of variation in which significant proportions of different types of individuals exist within a population of a species at the same time and in the same place, such that the frequency of the rarest form cannot be explained on the basis of recurring mutation. The term is generally used where the frequency of the least common morph exceeds 1 in 20, less frequent occurrences being usually due to rare mutation or recombination events. If the differences persist over many generations then there is a *balanced polymorphism*. *Compare* pleiomorphism.

polypeptide A compound that contains many amino acids linked together by peptide bonds. The three-dimensional shape (conformation) of the poplypeptide chain is determined by its amino acid sequence, but can be disrupted by extremes of pH or heat. Proteins are made up of one or more polypeptides. *See* peptide; denaturation; conjugated protein.

polyploid The condition in which a cell or organism contains three or more times the haploid number of chromosomes. It occurs when chromosomes fail to separate during anaphase I of meiosis, giving rise to gametes that are diploid instead of haploid. Fertilization results in triploid or tetraploid individuals. Polyploidy is far more common in plants than in animals and very high chromosome numbers may be found; for example in octaploids and decaploids (containing eight and ten times the haploid chromosome number). Some 60% of monocotyledons and 40% of dicotyledons are polyploid, as are 90% of ferns, many bryophytes, and some algae. *See* allopolyploidy; autopolyploidy; heterosis.

Polypodiaceae A family of terrestrial or epiphytic leptosporangiate ferns.

polyribosome *See* ribosome.

polysaccharide A polymer of 10 or more monosaccharides joined by glycosidic links (*see* glycoside). They contain many repeated units in their molecular structures and are of high molecular weight. They can be broken down to smaller polysaccharides (*oligosaccharides*), DISACCHARIDES, and MONOSACCHARIDES by hydrolysis or by the appropriate enzyme. Important polysaccharides are inulin (hydrolyzed to fructose), starch (hydrolyzed to glucose), and cellulose (hydrolyzed to glucose). Polysaccharides may form branching chains, or they may be unbranched. They have many roles in plants, as storage molecules (e.g. starch), structural elements in cell walls (cellulose), and be components of gums and mucilages (glucuronic acid). A polysaccharide containing only one type of sugar residue is called a homopolysaccharide, e.g. starch, which contains only glucose units. Heteropolysaccharides contain two or more different monosaccharides, e.g. hemicelluloses. *See also* carbohydrates; sugar.

polysome *See* ribosome.

polysomy *See* aneuploidy.

polystely Having many steles, as in the stems of some *Selaginella* species. *See also* distely.

pome A fleshy pseudocarp fruit, characteristic of the family Rosaceae, e.g. apple (*Malus*). The fleshy part of the fruit is derived from the greatly enlarged urn-shaped receptacle, which encloses the entire fruit, and which is fused to the tough carpel walls. Thus the apple core is the true fruit.

population dynamics The study of the factors influencing the fluctuations in numbers in a population or of its gene pool: the factors affecting birth and mortality rates, immigration and emigration, and reproductive potential (for example, seed size, dispersal, and dormancy).

population A group of interbreeding organisms of the same species (or other groups within which individuals may exchange genetic information) occupying a particular space. A population is continually modified by increases (birth and immigration) and losses (death and emigration), and is limited by the effects of environmental factors such as disease. Many plants cannot be divided strictly into populations as they may reproduce vegetatively, with the result that one individual can occupy a large area and parts can detach.

porogamy The usual method of fertilization in angiosperms in which the pollen tube enters the ovule by the micropyle. *Compare* chalazogamy.

porphyrins Cyclic organic structures containing tetrapyrrole rings (four joined rings, each consisting of four CH units and one NH unit) that have the important characteristic property of forming complexes with metal ions. Examples of such *metalloporphyrins* are the iron porphyrins (e.g. heme in hemoglobin) and the magnesium porphyrin, chlorophyll, the photosynthetic pigment in plants. In nature, the majority of metalloporphyrins are conjugated to proteins to form a number of very important molecules, such as the cytochromes.

potassium One of the essential elements in plants. It is absorbed by plant roots as the potassium ion, K^+, and in plants is the most abundant cation in the cell sap. Potassium ions are required in high concentrations in the cell for efficient protein synthesis, and for glycolysis in which they are an essential cofactor for the enzyme pyruvate kinase. Potassium ions balance anions and organic acids, and are thus important in osmoregulation, the control of stomatal opening, and leaf movements (through their effects on the turgor of pulvini). A deficiency of potassium leads to a characteristic red or purple coloration of the leaves, poor root growth, and flower and fruit formation.

potometer An apparatus for measuring the rate of water uptake by a cut shoot or whole plant. This is normally closely related to the rate of water loss by transpiration, and the potometer can be used to compare transpiration rates under different conditions. The rate of uptake is measured by the progress of an air bubble in a capillary tube along a scale. More meaningful comparisons can be made by comparing water loss from the potometer with that from an *atmometer* (an apparatus used to measure the rate of evaporation from a porous pot or other nonliving wet surface), which allows uncontrolled evaporation.

PPLO *See* mycoplasmas.

P_R (P660) *See* phytochrome.

Precambrian The time in the earth's geological history that precedes the Cambrian period, i.e. from the origin of the earth, nearly 5 billion years ago, to the start of the Cambrian, around 570 Ma (million years ago). The term 'Precambrian' is now used mainly descriptively, and has been largely discarded as a geological term in the light of greater knowledge of the early evolution of life. Precambrian time is now divided into three eons: Hadean, from the earth's origin to about 3900 Ma; Archaean, 3900–2390 Ma; and Proterozoic, 2390–570 Ma (the Cambrian marks the start of the Phanerozoic eon, which extends to the present day).

The oldest fossils discovered so far are remains of bacterialike organisms, dating from about 3500 Ma. There is abundant evidence of flourishing colonies of cyanobacteria and other bacteria throughout the Archaean and Proterozoic eons. This takes the form of STROMATOLITES, rock structures representing the remains of sediment trapped or precipitated by bacterial communities. Stromatolites are still found in a few special locations today. The ancient rocks also contained spherical carbonaceous aggregates up to 20 mm diameter, that resemble algae or the cysts of flagellates, and delicate carbonaceous filaments up to 150 μm long. The first appearance of multicellular animals is in the so-called Ediacara fauna, in rocks dated to the last 100 million years of Precambrian time. Around this time the stromatolites began to decline, perhaps as they were being eaten by newly evolved multicellular animals.

pressure potential *See* turgor pressure; water potential.

prickle A protective multicellular outgrowth from the surface of a plant. It is a modified trichome and thus completely epidermal in origin. Compare spine, thorn.

primary endosperm nucleus *See* endosperm.

primary growth Growth derived solely from meristems present in the embryo, i.e. apical meristems. Such growth generally increases the length of plant organs. All the tissues of the young plant are the result of primary growth. *Compare* secondary growth.

primary phloem PHLOEM derived from the procambium in the young plant. It consists of the PROTOPHLOEM (formed first) and METAPHLOEM. In nonwoody plants it is the only food-conducting tissue, but in mature plants with secondary growth this function is taken over by the secondary phloem. *See* secondary growth.

primary structure The sequence of amino acids in a polypeptide chain. *See also* conformation; secondary structure, tertiary structure; quaternary structure.

primary tissue Plant tissue that is derived solely from meristems present in the embryo and their derivatives. *See* primary growth.

primary xylem Xylem derived from the procambium in the young plant. It consists of the protoxylem (formed first) and metaxylem. In nonwoody plants the primary xylem is the only water-conducting tissue, but in mature plants with secondary growth this function is mostly taken over by the secondary xylem. *See* secondary growth.

primordium (*pl.* **primordia**) A collection of cells that differentiates into a specialized cell, tissue or organ, e.g. the apical shoot and apical root primordia of the embryo. The term is used particularly in relation to a part of the apical meristem that later differentiates further.

probability The chance that a given event will occur, or that over a series of observations a particular kind of observation will occur regularly as a given proportion of the total number of observations. Statistical probability is usually based on an infinite number of observations. For example, in genetics, if a heterozygous plant is selfed, the probability of finding the double recessive is 1 in 4, or 25%. The greater the number of offspring the better the chance that this actual percentage will be achieved.

procambium (**procambial strand**) The layer of cells that gives rise to the vascular tissue. Part of the apical meristem, it is discernable just below the apex as a strand of flattened cells which, if traced back along the shoot or root, may be seen to give rise to the primary vascular tissues. It is continuous with the intrafascicular cambium. In roots it may also be called plerome. *See* histogen theory; primary growth.

procaryote *See* prokaryote.

producer The first TROPHIC LEVEL in a food chain. Producers are those organisms that can build up foods from inorganic materials, i.e. green plants, algae, and photosynthetic and chemosynthetic bacteria. Producers are eaten by herbivores (primary CONSUMERS).

proembryo A young plant after fertilization but before differentiation into embryo and suspensor tissue.

profundal The deepwater zone of a lake beyond a depth of 10 meters. Little light penetrates this zone and thus the inhabitants are all heterotrophic, depending on the littoral and sublittoral organisms for basic food materials. Commonly found inhabitants include bacteria, fungi, mollusks, and insect larvae. Species found in the profundal zone are adapted to withstand low oxygen concentration, low temperatures, and low pH. *Compare* littoral; sublittoral; photic zone.

prokaryote (**procaryote**) An organism whose genetic material (DNA) is not enclosed by membranes to form a nucleus but lies free in the cytoplasm. Organisms can be divided into prokaryotes and EUKARYOTES, the latter having a true nucleus. In prokaryotes the nuclear material is simply a circular strand of DNA, not complexed with histone proteins, and cell division is amitotic. This is a fundamental division because it is associated with other major differences. Prokaryotes constitute the kingdom Bacteria (formerly the Prokaryotae). Recent molecular evidence suggests the prokaryotes actually comprise two distinct kingdoms, or even domains, the Archaea and the Eubacteria or Bacteria. Eukaryotes comprise all other organisms. Prokaryote cells evolved first and gave rise to eukaryote cells. They are normally much smaller (about 1µm in diameter) than eukaryote cells (about 20µm). Prokaryotes lack organelles such as nucleoli, mitochondria, plastids, Golgi apparatus, endoplasmic reticulum, and Golgi apparatus, and do not exhibit cyclosis. Their ribosomes are smaller (70S) than those of eukaryotes (80S) – similar in size to the ribosomes of chloroplasts and mitochondria. This has led to the theory (endosymbiont theory) that eukaryotes arose as a result of symbiotic associations between prokaryotes. *See* Bacteria; cell; endosymbiont theory.

proline A nonpolar molecule synthesized from glutamic acid and broken down to glutamic acid by the same pathway. The derivative 4-hydroxyproline is an important component of cell walls. Proline has been found in every protein studied to date. *See* amino acids.

promoter A specific DNA sequence within an OPERON that initiates TRANSCRIPTION by binding RNA polymerase. The promoter is situated between the operator and the structural genes or genes. It is the starting point for transcription of the structural genes in the operon, but is not itself transcribed. In *Escherichia coli* the RNA polymerase has a protein 'sigma factor' that recognizes the promoter; in the absence of this factor the enzyme binds to, and begins transcription at, random points on the DNA strand. In eukaryotic cells, binding of RNA polymerase to the promoter involves proteins called *transcription factors*.

prophage *See* lysogeny.

prophase The first stage of cell division in meiosis and mitosis. During prophase the chromosomes become coiled, shortening and thickening and becoming visible, and the nuclear membrane dissolves. Prophase may be divided into successive stages termed *leptotene*, *zygotene*, *pachytene*, *diplotene*, and *diakinesis*. In leptotene the chromosomes, already replicated, start to condense and appear as fine threads, although sister chromatids are not yet distinct. The spindle starts to form around the intact nucleus. Zygotene is characterized by the active and specific pairing (synapsis) of homologous chromosomes leading to the formation of a haploid number of bivalents. In pachytene the paired homologous chromosomes contract. At this point each chromosome consists of a pair of chromatids and the two associated chromosomes are termed a tetrad. In diplotene the pairs of chromatids begin to separate from the tetrad formed by the association of homologous chromosomes. Chiasmata can often be seen at this stage. At diakinesis the sister chromatids of homologous pairs of chromosomes complete their separation, and the chromosomes coil tightly, shortening and thickening. In practice, there is no clear demarcation between these stages: the process is continuous. The events occurring during these stages differ in meiosis and mitosis, notably in that bivalents (pairs of homologous chromosomes) are formed in meiosis, whereas homologous chromosomes remain separate in mitosis. *See also* meiosis; mitosis.

proplastid A self-duplicating undifferentiated plastid, about 0.5–1 μm in diameter and found in the meristematic regions of plants. They grow and develop into plastids of different types.

prop root An adventitious root, arising from one of the lower nodes of the stem close to the soil surface, found at the junction of stem and soil that grows down to the ground and gives additional support to the stem, as seen in maize (*Zea mays*). Some palms, mangroves and other tropical trees form woody prop roots, often called *stilt roots*. *See also* buttress root.

prosthetic group The nonprotein component of a conjugated PROTEIN. Thus the heme group in hemoglobin is an example of a prosthetic group, as are the coenzyme components of a wide range of enzymes. e.g. the heme group is an iron–porphyrin complex that acts as a prosthetic group in the cytochromes, important components of electron-transport chains in both plants and animals. *See* enzyme.

protandry The maturation of the anthers before the stigma, a very common condition seen in many specialized insect-pollinated flowers, especially those pollinated by bees. It helps to promote cross-pollination and is often combined with self-incompatibility. *Compare* protogyny. *See also* dichogamy.

protease (**proteinase**) An enzyme that catalyzes the hydrolysis of peptide bonds in proteins to produce peptide chains and amino acids. Individual proteases are highly specific in the type of peptide bond they hydrolyze.

protein A large, complex molecule made up of one or more polypeptide chains, i.e. it consists of amino acid molecules joined together by peptide links. The molecular weight of proteins may vary from a few thousand to several million. About 20 amino acids are present in proteins. *Simple proteins* contain only amino acids. In *conjugated proteins*, the amino acids are joined to other groups.

The *primary structure* of a protein is the particular sequence of amino acids present; this determines the three-dimensional shape of the molecule. The secondary structure is the way in which this chain is arranged; for example, coiled in an alpha helix or held in beta-pleated sheets. The *secondary structure* is held by hydrogen bonds. The *tertiary structure* of the protein is the way in which the protein chain is folded. This may be held by cystine bonds (disulfide bridges) and by weak polar and hydrophobic attractive forces between atoms. The *quaternary structure* is the combination of more than one polypeptide chain. There are two main groups of proteins: *globular proteins* and *fibrous proteins*. Globular proteins have the more complex tertiary structure, and are mostly enzymes. Fibrous proteins usually have a structural or contractile role. Proteins are vital compounds found in all living organisms. Their functions range from enzymes and hormones to electron-transfer molecules, contractile elements, and important genetic elements such as the histones that coat DNA in eukaryotic chromosomes. By coding for structural proteins and for the enzymes that catalyze the synthesis of other proteins and biological molecules, genes control the functioning and development of all living organisms.

proteinase *See* protease.

protein sequencing The determination of the primary structure of proteins, i.e. the type, number, and sequence of amino acids in the polypeptide chain. The most widely used procedure is the *Edman degradation*. This uses a series of three chemical reactions, each taking place at a specific pH, to remove one amino acid at a time from the amino terminus of a protein. The amino acids are then identified by ion-exchange chromatography. The process is automated in an *amino acid analyzer*. The Edman degradation can sequence peptides with only 5–10 picomoles of protein present.

protein synthesis The process whereby proteins are synthesized on the ribosomes of cells. The sequence of bases in MESSENGER RNA (mRNA), transcribed from DNA, determines the sequence of amino acids in the polypeptide chain: each CODON in the mRNA specifies a particular amino acid. As the ribosomes move along the mRNA in the process of translation, each codon is 'read', and amino acids bound to different transfer RNA molecules are brought to their correct positions along the mRNA molecule. These amino acids are polymerized to form the growing polypeptide chain. *See also* transfer RNA; translation.

proteolysis The hydrolysis of proteins into their amino acids. Enzymes that catalyze this are *proteases* or *proteolytic enzymes*.

proteomics The compilation and characterisation of all the proteins in an organism – its *proteome*.

proteoplast A colorless plastid (leukoplast) that stores protein.

Proterozoic *See* Precambrian.

prothallus (*pl.* **prothalli**) A flattened disk of cells that forms the free-living haploid gametophyte generation of certain ferns, e.g. *Dryopteris*. In homosporous plants, there is only one type of prothallus with both male and female sex organs. In heterosporous plants, the microspores give rise to small male prothalli bearing male sex organs (antheridia), and larger female prothalli bearing female sex organs (archegonia). The prothallus is greatly reduced in spermatophytes. *See* alternation of generations.

Protista In some classifications, a kingdom of simple eukaryotic organisms including the algae and protozoans. It was introduced to overcome the difficulties of assigning such organisms that may show both animal and plantlike characteristics to the kingdoms Animalia or Plantae. Today the grouping is considered to be artificial and many taxonomists support the FIVE KINGDOMS CLASSIFICATION whereby the bacteria and fungi are both assigned to separate kingdoms, while algae and protozoans constitute various phyla of the kingdom PROTOCTISTA.

protocooperation *See* mutualism.

Protoctista A kingdom of simple eukaryotic organisms that includes the algae, slime molds, funguslike oomycetes, and the organisms traditionally classified as protozoa, such as flagellates, ciliates, and sporozoans. The members of this kingdom do not share clear phylogenetic links, but are grouped together simply because they do not not belong in any of the other four kingdoms. Most are aerobic, some are capable of photosynthesis, and most possess undulipodia (flagella or cilia) at some stage of their life cycle. Protoctists are typically microscopic single-celled organisms, such as the amebas, but the group also has large multicellular members, for example the seaweeds and other large algae, some of which are sometimes classified as plants. Protoctists show a wide range of nutritional habits, including photoautotrophs, heterotrophs, phagotrophs and mixotrophs.

protoderm The outermost layer of the apical meristem: the tissue that develops from the tunica initials of the apical meristem and gives rise to the epidermis and, in some plants, the root cap. In the root it may also be called the *dermatogen*. *See* histogen theory.

protogyny The maturation of the stigma before the anthers. It is a mechanism that helps prevent self-pollination and is most common among wind-pollinated and unspecialized insect-pollinated flowers. as in figwort. It is less common than protandry. *Compare* protandry. *See also* dichogamy.

protonema (*pl.* **protonemata**) The young bryophyte gametophyte that develops following spore germination. The mature gametophyte plants develop from buds that form at several points along the protonema.

proton pump An energy-requiring metabolic process that causes protons (hydrogen ions) to move across a membrane against their concentration gradient. This creates a potential difference across the membrane, a source of potential energy that can be used to drive other metabolic processes. For example, the inner membrane of the mitochondrion is impermeable to the movement of protons. During respiratory electron transport, protons are drawn from the internal matrix into the space outside the membrane, creating a potential difference across the membrane. In respiration and photosynthesis, the movement of protons back across a membrane down a diffusion gradient takes place at sites where it can be coupled to ATP synthesis, using the enzyme ATPase. The energy thus trapped can be transported to other parts of the cell. *See* chemiosmotic theory; electron-transport chain.

protophloem The first-formed PRIMARY PHLOEM, differentiated from the procambium in the region just behind the meristem.

protoplasm The living contents of a cell, comprising the cytoplasm plus nucleoplasm. The term usually excludes the large vacuoles. *See* cytoplasm; nucleus.

protoplast The protoplasm and plasma membrane of a cell after removal of the cell wall, where present. This can be achieved by physical means or by enzymic digestion.

protostele A simple form of stele, uninterrupted by leaf gaps, consisting of xylem in the center completely surrounded by phloem. Most roots have protosteles. The stems of some lower plants (e.g. *Lycopodium*) also have protosteles.

protoxylem The PRIMARY XYLEM elements that are formed from the procambium first. In seed plants the protoxylem occurs at the innermost edge of the vascular bundles in stems, but external to the later-formed xylem in roots.

protozoa An old name for animal-like PROTOCTISTA that included some photosynthetic organisms (e.g. *Euglena*, *Volvox*).

proximal Denoting the part of an organ that is nearest the origin or point of attachment. *Compare* distal.

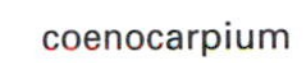

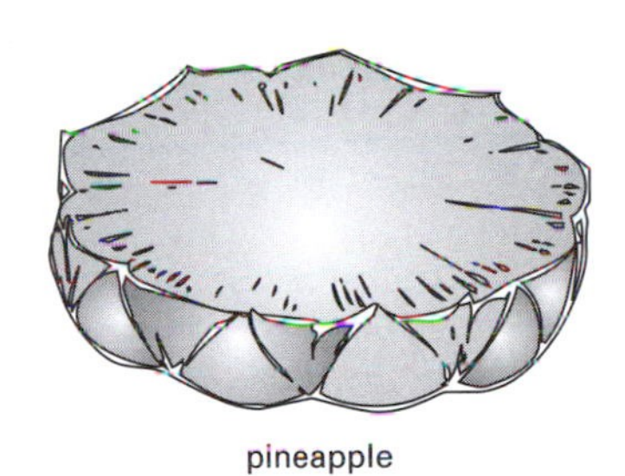

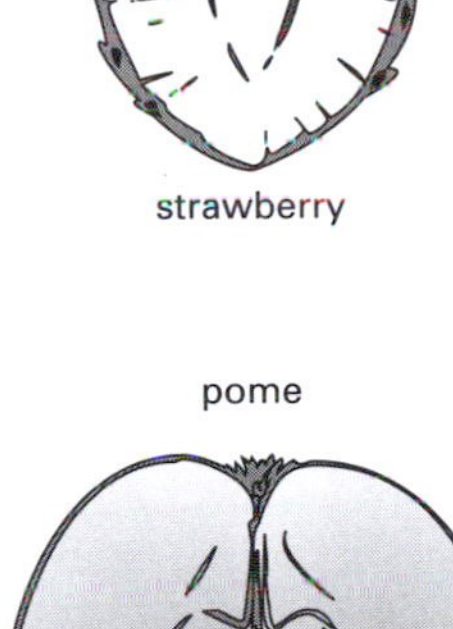

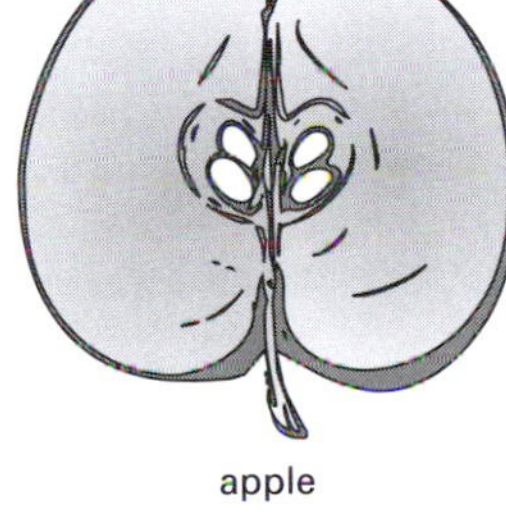

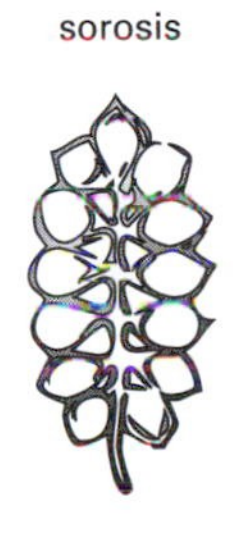

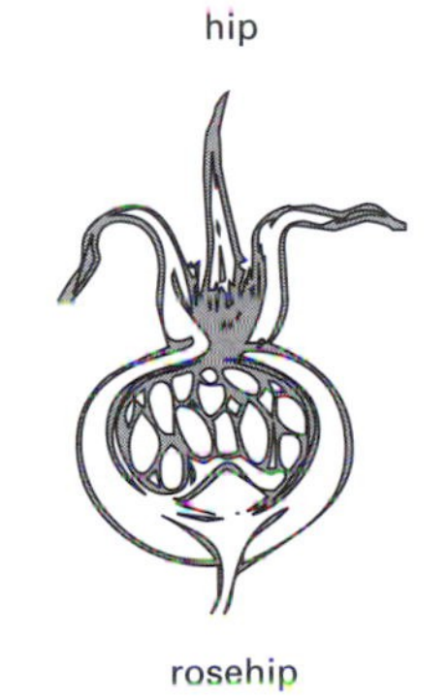

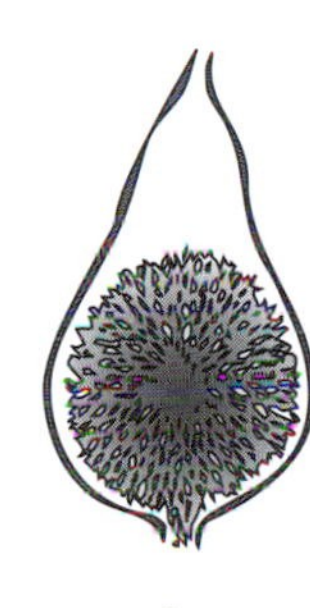

Pseudocarp: various types of pseudocarp

pseudoallele A mutation in a gene that produces an effect similar to another mutation at a different site in the same gene locus. They are closely linked, so recombinations between them are rare. *See cis–trans* effect.

pseudocarp (**false fruit**) A fruit that includes parts of the flower, e.g. the bracts, inflorescence, or receptacle, other than those derived from the ovary. *See also* hip; fruit; pome. See illustration on page 193.

pseudogene A DNA sequence that cannot be transcribed. Although they have no immediate function, pseudogenes have high potential to form new genes by further mutation as they already have useful sequences, such as those signaling transcription. Pseudogenes are genes that have become 'switched off' during the course of evolution.

pseudoparenchyma A fungal or algal tissue resembling parenchyma but made up of interwoven hyphae (fungi) or filaments (algae). The stipe of the mushroom and the thallus of red algae (e.g. *Porphyra*) are pseudoparenchymatous tissues.

pseudoplasmodium *See* Acrasiomycota.

pseudopodium (*pl.* **pseudopodia**) **1.** A leafless stalk that bears the capsule in *Sphagnum* and other mosses that lack a seta.
2. In slime molds, a projection of the body used for locomotion.
3. A temporary fingerlike projection or lobe on the body of an ameboid cell. It is formed by a flowing action of the cytoplasm and functions in locomotion and feeding.

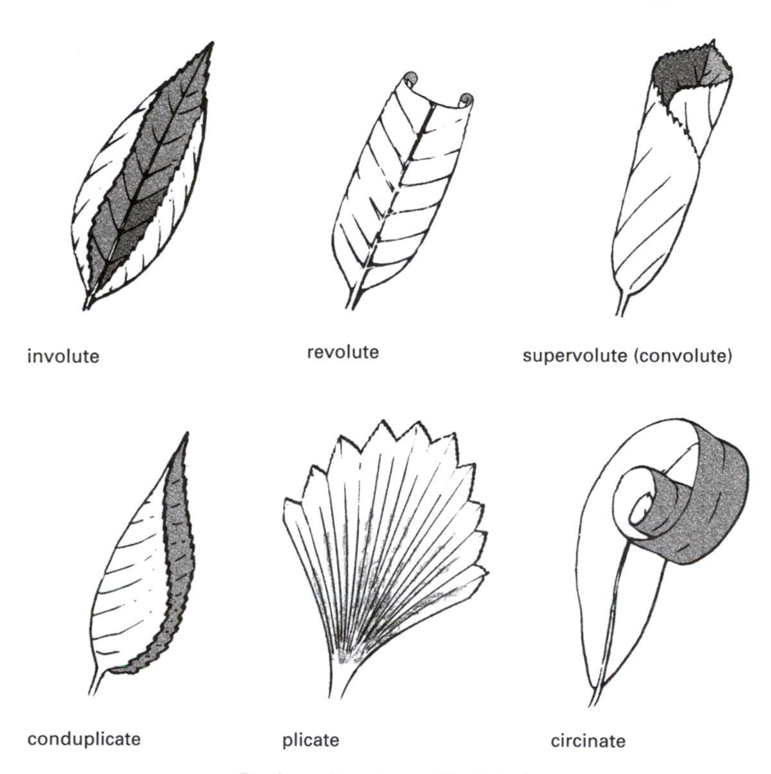

Ptyxis: various types of leaf ptysis

Psilophyta (**whisk ferns**) In the Five Kingdoms classification scheme, a phylum of vascular seedless plants, the whisk ferns. In most modern classifications they are placed in the class Psilopsida of the Filicinophyta. There are only two living subtropical genera (*Psilotum* and *Tmesipteris*).

psychrophilic Describing microorganisms that can live at temperatures below 20° C. *Compare* mesophilic; thermophilic.

pteridophyte (**Pteridophyta**) A general term, now largely obsolete, to include any vascular nonseed- bearing plant. Pteridophytes include the club mosses (phylum Lycophyta), horsetails (Sphenophyta), ferns (Filicinophyta), and whisk ferns (Psilophyta).

ptyxis The way in which young leaves are folded or rolled in the bud. *See also* vernation.

pulvinus (*pl.* **pulvini**) **1.** A specialized group of cells with large intercellular spaces that are located at the bases of leaves or leaflets in certain plants. They are involved in nongrowth nastic movements, bringing these about by rapid changes in turgor through loss of water to the intercellular spaces. *See* nyctinasty.
2. A thickened region at the node of a grass stem, which often contains an INTERCALARY MERISTEM.

punctuated equilibrium A theory of evolution proposing that there have been long periods of geological time, lasting for several million years, when there is little evolutionary change, punctuated by short periods of rapid speciation of less than 100 000 years. This is in contrast to the traditional theory (*see* neo-Darwinism) in which it is postulated that species have evolved gradually throughout geological time.

Punnett square A chequerboard diagram used in genetics to illustrate the frequencies of various genotypes in the offspring of a particular cross.

pure line The succession of descendants of an individual that are identical to each other and continue to breed true, i.e. they produce genetically identical offspring. They are assumed to be homozygous.

purine A simple nitrogenous organic molecule with a double-ring structure – a

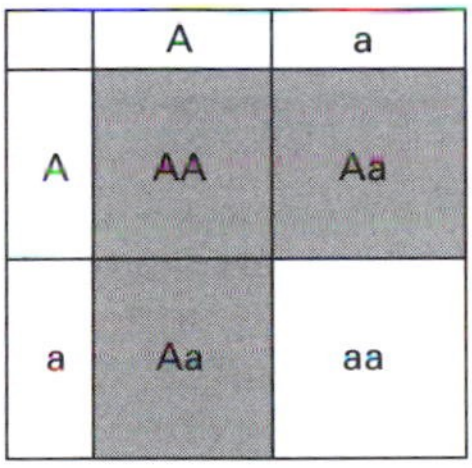

	A	a
A	AA	Aa
a	Aa	aa

monohybrid ratio 3:1

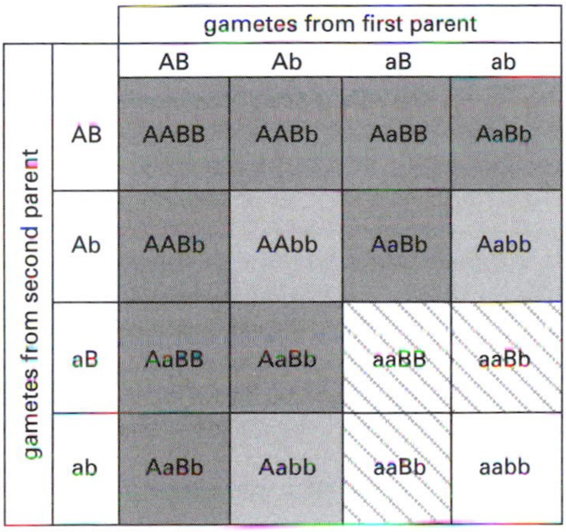

		gametes from first parent			
		AB	Ab	aB	ab
gametes from second parent	AB	AABB	AABb	AaBB	AaBb
	Ab	AABb	AAbb	AaBb	Aabb
	aB	AaBB	AaBb	aaBB	aaBb
	ab	AaBb	Aabb	aaBb	aabb

dihybrid ratio 9:3:3:1

Punnet squares

Purine

6-membered pyrimidine ring fused to a 5-membered imidazole ring. Members of the purine group include adenine and guanine, which are constituents of the nucleic acids, and certain plant alkaloids, e.g. caffeine and theobromine.

pyramid of biomass A type of ecological pyramid based on the total amount of living material at each trophic level in the community, which is normally measured by total dry weight or calorific value per unit area or volume, and shown diagrammatically. The biomass depends on the amount of carbon fixed by green plants and other producers. The pyramid of biomass usually has a more gentle slope than the PYRAMID OF NUMBERS because organisms at successively higher levels in the pyramid tend to be larger than those below.

pyramid of numbers A type of ecological pyramid in which the number of individual organisms at each stage in the food chain of the ecosystem is depicted diagrammatically. The producer level forms the base, and successive levels the tiers. The shape of the pyramid of numbers depends upon the community considered; generally, the organism forming the base of a food chain is numerically very abundant, and each succeeding level is represented by fewer individual organisms, culminating with the final large predator. The pyramid of numbers may be partly inverted (base smaller than one or more of the upper tiers) if the producing organisms are large.

pyrenocarp 1. *See* drupe.
2. *See* ascoma.

pyrenoid A protein structure found in the chloroplasts of green algae and hornworts (Anthocerophyta). Pyrenoids are associated with the storage of starch.

pyridoxine (**vitamin B_6**) One of the water-soluble B-group of VITAMINS. Pyridoxine gives rise to a coenzyme involved in various aspects of amino acid metabolism, especially transamination.

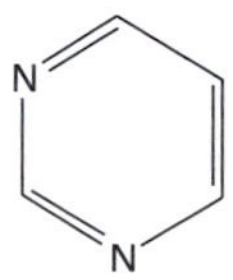

Pyrimidine

pyrimidine A simple nitrogenous organic molecule whose ring structure is contained in the pyrimidine bases cytosine, thymine, and uracil, which are constituents of the nucleic acids, and in thiamine (vitamin B_1).

pyruvic acid A 3-carbon carboxylic acid ($CH_3COCOOH$) that is an intermediate in GLYCOLYSIS. In aerobic respiration it is decarboxylated to form acetyl CoA. In ANAEROBIC RESPIRATION it is irreversibly reduced to form ethanol and carbon dioxide in plant cells, and to other compounds in bacteria. In C_4 PLANTS it is an intermediate of the Hatch–Slack pathway. It is a precursor in the synthesis of various amino acids, such as alanine, and of malic acid and oxaloacetic acid, both intermediates in the KREBS CYCLE.

Q

quadrat A square area of variable size depending on the study (standard sizes are 0.5 or one meter square to 10 m square or more) taken at random, within which the composition of organisms is noted. The quadrat sampling technique is mostly used in plant ecology to study the distribution and abundance of species in plant communities.

qualitative inheritance *See* discontinuous variation.

qualitative variation *See* discontinuous variation.

quantitative inheritance *See* continuous variation.

quantitative variation *See* continuous variation.

quaternary structure The pattern in which two or more polypeptide chains are linked together to form a protein molecule. This involves hydrogen bonds, ionic bonds, disulfide bridges and hydrophobic interactions. See conformation; protein. *Compare* primary structure; secondary structure; tertiary structure.

Quaternary (**Neogene**) The most recent period of the Cenozoic era from about two million years ago to the present day and composed of the Pleistocene and Holocene epochs. It was a period of cyclical climate change, including several ice ages in high and mid-latitudes, cycles of varying aridity and humidity in low latitudes, and significant fluctuations in sea level. This had important implications for plant speciation, as species ranges expanded and contracted, new species radiated out as climate ameliorated, and retreating ice afforded opportunities for colonization of newly exposed land. The Quaternary is also characterized by the emergence of humans and the start of human modification of the environment. Increased rates of evolution and extinction of mammals (due to both climate change and human activity) also affected plant distribution. *See also* Ice Age.

quiescent center A group of cells in the center of the apical meristem in which mitotic divisions are rare or absent. The cells may begin dividing if another part of the meristem is damaged.

quillworts *See* Isoetales.

R

race In classification, a taxon below the rank of SPECIES, sometimes being placed between subspecies and variety. The term is sometimes used instead of ECOTYPE, for groups of individuals that have uniform ecological preferences or physiological requirements.

raceme (racemose inflorescence; indefinite inflorescence) *See* inflorescence.

rachis **1.** The main stalk on which the leaflets are borne in a compound leaf, or on which the pinnae are attached in the fern leaf.
2. The central axis of certain inflorescences, e.g. wheat (*Triticum*), that bears the spikelets.

radial **1.** Aligned along the radius of a circle or cross-section.
2. *See* anticlinal.

radial symmetry The arrangement of parts in an organism in such a way that cutting in any plane across the diameter splits the structure into similar halves (mirror images). Many stems and roots show radial symmetry. The term *actinomorphy* is generally used to describe radial symmetry in flowers. *See also* bilateral symmetry.

radicle The embryonic root. It is the first organ to emerge from the seed on germination. The radicle is joined to the hypocotyl and both tissues are derived from the four octants nearest the suspensor. The root cap over the tip of the radicle is derived from the closest cell of the suspensor. The radicle may persist to form a taproot, or it may be replaced by adventitious roots.

radioactive dating *See* radiometric dating.

radiometric dating (isotopic dating; radioactive dating) Any method of dating that uses the decay rates of naturally occurring radioactive isotopes to assess the age of a specimen or material. Organic matter less than 7000 years old can be dated using radiocarbon dating. This uses the fact that the isotope carbon-14 is found in the atmosphere and taken in by plants when they photosynthesize, and subsequently assimilated by the animals that feed on them. When plants and animals die, no more carbon is taken in and the existing ^{14}C decays to the nonradioactive isotope carbon-12. If the proportion of ^{14}C to ^{12}C in the atmosphere and the decay rate of ^{14}C to ^{12}C are both known, as they are, then the sample may be dated by finding the present proportion of ^{14}C to ^{12}C. Specimens of rocks over 7000 years old can be dated by other radioisotope methods, e.g. potassium–argon dating and rubidium– strontium dating.

rainforest *See* forest.

random amplified polymorphic DNA (RAPD) The generation of copies of segments of DNA by means of the polymerase chain reaction, whereby the segments multiplied are those that lie between two identical binding sites for a particular primer, usually about 10 base pairs long. Several different segments of DNA may fall into this category, and the amplified DNA can be sorted using ELECTROPHORESIS. Different individuals may differ in the number and location of such sites, giving rise to a kind of genetic fingerprint. *See* genetic fingerprinting; molecular systematics; polymerase chain reaction.

Ranunculaceae A highly variable family of about 2450 species of dicotyledonous plants that includes the buttercups, monkshoods, anemones, clematis, delphiniums, and hellebores.

raphe 1. In some seeds, a ridge formed by the portion of the funicle that is fused with the integument.
2. A slit in the valve of motile diatoms, thought to be involved in movement by allowing contact between cytoplasm and substrate.

raphides Bunches of needlelike crystals of calcium oxalate found in certain plant cells.

Raunkiaer's plant classification A classification of growth forms based on the persistence of the shoots and the position of the resting buds. The system simplifies assessment of the percentages of different plant forms in any given type of vegetation, specially in temperate communities. *See also* chamaephyte; cryptophyte; helophyte; hemicryptophyte; phanerophyte; therophyte.

Ray, John (1627–1705) English naturalist and taxonomist. Ray, from a humble background, was appointed a Cambridge fellow in 1648 but lost his fellowship in 1662, having refused to take the oath required by the Act of Uniformity. His friends at Cambridge, especially Francis Willoughby, helped to finance him in his further studies. His life's work was an attempt to describe all the known flora and fauna. With Willoughby, he traveled throughout Europe to study a wide range of plants and animals. In 1667 he published a catalog of British plants and in the same year was elected to the Royal Society. On Willoughby's death in 1672 he inherited enough money to continue their endeavor. His three-volume *Historia plantarum* was published between 1686 and 1704. He is best known for fixing the species as the basic unit in taxonomy and also for realizing the importance of distinguishing between monocotyledons and dicotyledons.

ray floret *See* inflorescence.

ray fungi *See* Actinobacteria.

ray initial *See* initial.

reaction center In PHOTOSYNTHESIS, a hypothetical site at which absorbed light energy is used to transport electrons for photophosphorylation. It contains a molecule of the primary pigment (in plants this is chlorophyll *a*, which transfers the electrons. Plants have two reaction centers in the thylakoids of the chloroplast: Photosystems I and II.

Recent *See* Holocene.

receptacle 1. The tip of the angiosperm

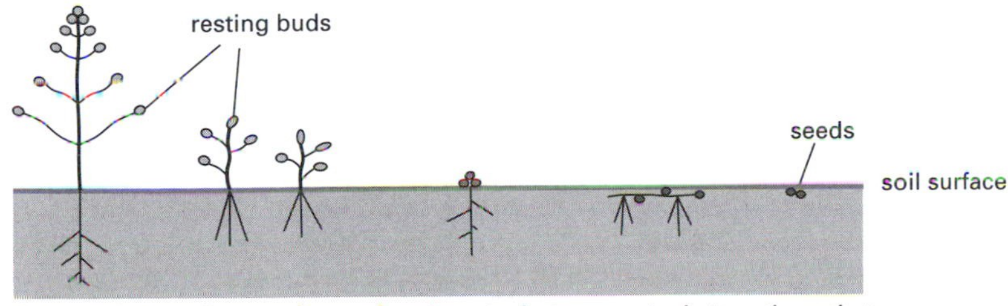

Raunkiaer's plant classification

flower stalk upon which the floral organs are inserted. If the ovary is on top of the receptacle with the other organs it is said to be superior but if it is sunk in the center, e.g. in the daffodil (*Narcissus)*, it is said to be inferior. *See also* epigyny; hypogyny; perigyny.
2. The swollen portion of the lamina (blade) bearing the conceptacles in certain algae (e.g. *Fucus*).

recessive An allele that is only expressed in the phenotype when it is in the homozygous condition. *Compare* dominant. *See also* double recessive.

reciprocal cross A cross that tests whether the inheritance of a particular character is affected by the sex of the parent. The cross is thus made both ways, i.e. the character under consideration is carried by the female in one cross and by the male in the second cross. The procedure can demonstrate which characters are controlled by sex-linked genes.

recombinant DNA **1.** DNA produced by recombinant DNA technology.
2. DNA formed naturally by recombination, e.g. by crossing over in meiosis or by conjugation in bacteria. *See* recombination.

recombinant DNA technology The technique by which foreign DNA, whether from another organism or genetically engineered, is inserted into another DNA fragment or molecule. The product – recombinant DNA – is fundamental to many aspects of GENETIC ENGINEERING, particularly the introduction of foreign genes to cells or organisms. There are now many techniques for creating recombinant DNA, depending on the nature of the host cell or organism receiving the foreign DNA. Particular genes or DNA sequences are cut from the parent molecule using specific type II RESTRICTION ENDONUCLEASES, or are assembled using a messenger RNA template and the enzyme REVERSE TRANSCRIPTASE. In GENE CLONING using cultures of bacterial or eukaryote tissue cells, the foreign gene is inserted into a VECTOR, e.g. a bacterial plasmid or virus particle, which then infects the host cell. Inside the host cell the recombinant vector replicates and the foreign gene is expressed. One of the most common vectors used with plants is the Ti (tumor-inducing) plasmid of the bacterium *Agrobacterium tumefaciens*. This causes crown gall tumors in plants, and its plasmid has been used on a range of crop plants.

recombination The regrouping of genes that regularly occurs during meiosis as a result of the independent assortment of chromosomes into new sets, and the exchange of pieces of chromosomes (crossing over). Recombination results in offspring that differ both phenotypically and genotypically from both parents (*recombinants*) and is the most important means of producing variation in sexually reproducing organisms.

red algae *See* Rhodophyta.

Red Data Book An updatable set of information on rare, threatened, and endangered species worldwide, compiled by the World Conservation Monitoring Center, and published at regular intervals by the Species Survival Commission of the World Conservation Union (WCU).

red desert soil A type of coarse SOIL rich in salts and lime but poor in humus, formed in hot deserts.

red light Electromagnetic radiation of wavelength approximately 630 nm. It is the most effective wavelength for initiating many light-dependent reactions in plants, such as the germination of light-sensitive seeds. *See* photoperiodism; phytochrome.

red tropical soil A type of SOIL formed under rainforests and in savannas in areas of alternating wet and dry seasons. It is heavily leached, but nutrients released by decomposition help to maintain the mineral balance. The red color is due to oxides of iron, magnesium, and aluminum, which in some such soils form a hard crust (LATERITE) at the soil surface.

reducing sugar A sugar that can reduce certain inorganic ions in solution, such as the cupric ions in Benedict's and Fehling's solutions. Examples are glucose, lactose, and meltose.

reduction division The first division of meiosis, including prophase, metaphase I, and anaphase I. It results in a haploid number of chromosomes gathering at each end of the nuclear spindle. *See* meiosis.

reflexed Bent backwards.

regeneration The regrowth by an organism of an organ or tissue that has been lost by injury or other means. It applies to natural regrowth from living roots or coppiced stems as well as artificial regrowth in the laboratory. Many plant propagation techniques make use of the ability of cuttings of shoots, roots, or leaves to regenerate into complete plants.

regulator gene A gene whose product can promote or prevent the TRANSCRIPTION of structural genes, which may or may not be adjacent to other regulator genes, and may even be on another chromosome. The regulator may code for a REPRESSOR protein that switches off an operator gene (in bacteria). In eukaryotes it may bind to proteins in a transcription complex to inhibit its action, or it may bind to a TRANSCRIPTION factor. *See also* operon; promoter.

relative humidity The ratio of the amount of water vapor present in a body of air at a given temperature compared to the maximum quantity of water vapor that body can contain at the same temperature, i.e. when saturated at that temperature. Relative humidity affects the diffusion gradient for TRANSPIRATION and evaporation, which proceed more slowly with increase in relative humidity.

release factor *See* translation.

rendzina A type of brown earth soil typical of humid to semi-arid grasslands over calcareous rocks. The rendzina is rich in lime and has an upper humus-rich horizon.

repetitive DNA DNA that consists of multiple repeats of the same nucleotide sequences (*repeated sequences*). In some plants, such as lilies, repeats may constitute over 50% of the genome, leading to unexpectedly large genomes. The repeats may occur together, or they may be widely scattered. Unlike prokaryotic cells, eukaryotic cells contain appreciable amounts of repetitive DNA. For example, *satellite DNA* (*tandem DNA*) is made up of up to one million adjacent repeats of the same short sequence, usually not more than 10 base pairs long. It occurs in the region of centromeres and telomeres, and is described as *highly repetitive DNA*. This DNA often has a nucleotide composition significantly different from the rest of the cell's DNA, giving it a different density, so it shows up as a separate band (the 'satellite') when centrifuged. *Microsatellite DNA*, with short sequences repeated only 10 to 100 times, is used in DNA fingerprinting. Some repetitive DNA is accounted for by multiple copies of particular genes, e.g. genes encoding histones and ribosomal RNA. *See also* junk DNA; transposon.

replica In a general sense an exact copy. In biology specifically used for a thin detailed copy of a biological specimen, obtained by spraying the surface with a layer of plastic and carbon. Replicas are used in electron-microscope work.

replication The mechanism by which exact copies of the genetic material are formed. Replicas of DNA are made when the double helix unwinds as a result of helicase enzyme action and the separated strands serve as templates along which complementary nucleotides are assembled through the action of the enzyme DNA polymerase. The result is two new molecules of DNA each containing one strand of the original molecule, and the process is termed *semiconservative replication*. Replication of DNA occurs in interphase; by the end of interphase the replicated DNA and associated histones are in the form of chromosomes each consisting of two chromatids joined at the centromere.

In RNA viruses an RNA polymerase is involved in the replication of the viral RNA.

repressor A protein molecule that prevents protein synthesis by binding to the operator sequence of the gene and preventing transcription. The molecule is produced by a regulatory gene and may act either on its own or in conjunction with a second protein called a *corepressor*. In some cases another molecule, an *inducer*, may bind to the repressor, weakening its bonds with the operator and derepressing the gene, allowing transcription to proceed. Such a molecule is termed an *effector*. *See* operon.

reproduction *See* asexual reproduction; sexual reproduction.

resin One of a group of substances occurring in many trees and shrubs, including conifers and certain families of flowering plants. They are usually secreted by special cells into long *resin ducts* or *canals*. Sometimes they are produced in response to injury or infection. They sometimes form a sticky covering to buds (e.g. horse chestnut), reducing transpiration and giving protection.

resolving power (**resolution**) The ability of an optical system to form separate images of closely spaced objects. It depends upon the perfection of the lens, and is theoretically limited by the wavelength of light, being greater with decreasing wavelength. *See* microscope.

respiration The oxidation of organic molecules to provide energy in plants and animals. Autotrophic plants respire molecules that they have synthesized by photosynthesis. Respiration occurs in all cells, whether or not they are capable of photosynthesis. The energy from respiration is used to attach a high-energy phosphate group to ADP to form the short-term energy carrier ATP, which can then be used to power energy-requiring processes within the cell. ATP is not transported between cells, but is made in the cell where it is required.

The chemical reactions of respiration normally require oxygen from the environment (AEROBIC RESPIRATION). Some organisms are able to respire, at least for a short period, without the use of oxygen (ANAEROBIC RESPIRATION), although this process produces far less energy than aerobic respiration. A few bacteria can survive indefinitely in anaerobic conditions. The complex reactions of cell respiration fall into two stages, GLYCOLYSIS and the KREBS CYCLE. The first stage, glycolysis, occurs within the cytoplasm, but the Krebs cycle enzymes are localized within the mitochondria of eukaryotes; cells with high rates of respiration have many mitochondria. Glycolysis results in partial oxidation of the respiratory substrate to the 3-carbon compound pyruvate. It does not require oxygen, so can occur in anaerobic conditions, when the pyruvate is converted to ethanol in plants and most bacteria, or to lactic acid in certain bacteria. The yield of glycolysis is 2 molecules of ATP plus 2 molecules of the reduced coenzyme $NADH_2$ for each molecule of glucose respired. In the presence of oxygen, the pyruvate is further oxidized in the Krebs cycle. In the Krebs cycle, which requires free oxygen, pyruvate is converted into the 2-carbon acetyl group, which becomes attached to a coenzyme forming acetyl coenzyme A. This then enters a cyclic series of reactions during which carbon dioxide is evolved and hydrogen atoms are transferred to the coenzymes NAD and FAD. The energy released by the Krebs cycle is transferred via the reduced coenzymes $NADH_2$ and $FADH_2$ to an ELECTRON-TRANSPORT CHAIN embedded in the inner mitochondrial membrane. According to the chemiosmotic theory, electrons flow through the various components of the chain, while a gradient of hydrogen ions is maintained by an energy-requiring (active) proton pump that drives protons into the intermembrane space. This gradient drives the formation of ATP. Overall, 38 molecules of ATP are generated for each molecule of glucose oxidized during aerobic respiration, compared with only 2 molecules of ATP during anaerobic respiration. At the end of the electron-transport chain the electrons react with protons and oxygen to give water. In aerobic respiration, therefore, pyruvate is completely oxidized to carbon dioxide and water.

respiratory chain *See* electron-transport chain.

respiratory quotient (RQ) The ratio of the volume of carbon dioxide evolved during RESPIRATION to the volume of oxygen absorbed:

$$RQ = \text{volume } CO_2/\text{volume } O_2$$

The complete breakdown of sugars by aerobic respiration gives an RQ of 1. This rises if anaerobic respiration is involved. The respiration of proteins or fats produces less carbon dioxide, so the RQ falls below one.

response A change in an organism or in part of an organism that is produced as a reaction to a stimulus.

restriction endonuclease (restriction enzyme) A type of enzyme, found mainly in bacteria, that can cleave and fragment DNA internally (*see* endonuclease). Their function is to protect the cell against foreign DNA, such as the DNA of an invading virus. Some restriction endonucleases cleave DNA at random, but a particular group of enzymes, known as *class II restriction endonucleases*, cleave DNA at specific sites. The discovery of these enzymes formed the basis for the development of genetic engineering, since they enable the isolation of particular gene sequences and the DNA fragments are replicated by means of base pairing and DNA ligases. *See also* recombinant DNA technology.

restriction fragment length polymorphism (RFLP) Variation among the members of a population in the sites at which restriction enzymes cleave the DNA, and hence in the size of the resulting DNA fragments. It results from differences between individuals in nucleotide sequences at the cleavage sites (restriction sites). The presence or absence of particular restriction sites can be ascertained using DNA probes in the technique called SOUTHERN BLOTTING. Restriction sites vary enormously, and this variation is exploited in analyzing and comparing the genomes of different individuals, e.g. to establish how closely related they may be. Restriction sites are also invaluable as genetic markers in chromosome mapping, and can be used to track particular genes. RFLPs are used in GENETIC FINGERPRINTING. *See* chromosome map.

restriction map A map of a segment of DNA showing the cleavage sites of restriction endonucleases and their physical distance apart, usually measured in base pairs. It can be used to reveal variations in restriction sites between individuals of the same species or between different species (*see* restriction fragment length polymorphism). This variation serves as a key to the organism's genes, since the restriction sites can be used as markers to identify closely linked genetic loci and allow investigation of deletions, insertions, or other mutations. RFLPs are an essential tool in chromosome mapping.

reticulate thickening The type of secondary-wall formation in which an irregularly branching mass of lignified deposits covers the inner wall of the cell. It is usually found in metaxylem tracheids and vessels. *Compare* annular thickening; scalariform thickening.

reverse transcriptase An enzyme that catalyzes the synthesis of DNA from RNA (i.e. the reverse of transcription, in which mRNA is synthesized from a DNA template). The enzyme occurs in certain RNA viruses and enables the viral RNA to be 'transcribed' into DNA, which is then integrated into the host DNA and replicates with it. It is also used in genetic engineering to make complementary DNA (cDNA) from an RNA template.

RFLP *See* restriction fragment length polymorphism.

Rf value In chromatography, a ratio that expresses the degree to which a solute moves in a solvent. It is calculated by dividing the distance moved by the spot of solute on the chromatogram column by the distance traveled by the solvent front. The Rf value is constant for a particular molecule and serves to identify it.

Rhizobium A spherical or rod-shaped nitrogen-fixing bacterium that can live either freely in the soil or symbiotically in the root nodules of leguminous plants and a few other species, such as alder (*Alnus*). *See also* nitrogen cycle; nitrogen fixation.

rhizoid A unicellular rootlike structure found in certain algae and in the gametophyte generation of the bryophytes and some ferns. It serves to anchor the plant and absorb water and nutrients.

rhizome A stem that grows horizontally below ground. Rhizomes may be fleshy, e.g. *Iris*, or wiry, e.g. couch grass (*Elytrigia repens*) and may serve as an organ of perennation or vegetative propagation, or occasionally both. *Compare* corm; tuber; stolon. *See also* rootstock.

Rhodophyta (**red algae**) A phylum of protoctists comprising aquatic, mainly marine, algae characterized by their red color. The color results from photosynthetic pigments, phycoerythrin and phycocyanin (*see* phycobilins). Most species are marine, a few freshwater or terrestrial; their distribution ranges from the shore to quite deep water.

rhytidome *See* bark.

riboflavin (**vitamin B_2**) A flavin pigment that is thought to be involved in the reception of the light stimulus in phototropism, and in the photooxidation of endogenous auxins. It is one of the water-soluble B-group of VITAMINS. It is found in cereal grains, peas, and beans. Riboflavin is a constituent of several enzyme systems (flavoproteins), acting as a coenzyme for hydrogen transfer in the reactions catalyzed by these enzymes.

ribonuclease (**RNase**) Any nuclease enzyme that cleaves the phosphodiester bonds between adjacent nucleotides in RNA. *Exoribonucleases* cleave nucleotides from one or both ends of the RNA molecule, while *endoribonucleases* cleave bonds within the molecule. RNase is thought to be involved in the incompatibility system that prevents pollen of certain genetic composition from germinating on the style of a flower.

ribonucleic acid *See* RNA.

ribose A monosaccharide, $C_5H_{10}O_5$; a component of RNA.

ribosomal RNA (**rRNA**) *See* ribosome.

ribosome A small organelle that is the site of protein synthesis in the cell. Ribosomes are found in large numbers in all cells, free in the cytoplasm or bound to ENDOPLASMIC RETICULUM. In most species they are composed of roughly equal amounts of protein and ribosomal RNA (rRNA). The ribosome consists of two unequally sized rounded subunits arranged on top of each other like a cottage loaf. Eukaryotic cells have larger ribosomes than prokaryotic cells but the ribosomes in mitochondria and chloroplasts are about the same size as prokaryotic ribosomes.

The ribosome comprises different types of rRNA and associated proteins. The rRNA interacts directly with the mRNA during initiation of translation, and there are specific sites on the rRNA that interact with anticodons on the transfer RNAs (tRNA). The proteins have various roles, including keeping the mRNA in a single-stranded state so that the bases are exposed for TRANSLATION. During translation the ribosome moves along the messenger RNA (mRNA), enabling the peptide linkage of amino acids delivered to the site by transfer RNA molecules according to the code in mRNA. Several ribosomes may be actively engaged in protein synthesis along the same mRNA molecule, forming a *polyribosome*, or *polysome*.

ribozyme Any RNA molecule that acts as an enzyme.

ribulose A 5-carbon sugar that is phosphorylated to form RIBULOSE BISPHOSPHATE.

ribulose bisphosphate (RUBP) A 5-carbon compound that accepts carbon dioxide during PHOTOSYNTHESIS. Each molecule is then converted into two molecules of 3-carbon phosphoroglyceric acid. The ribulose bisphosphate is regenerated in the Calvin cycle when the carbon dioxide is converted into carbohydrate.

ribulose bisphosphate carboxylase (RuBP carboxylase; rubisco) An enzyme that catalyzes the carboxylation of ribulose bisphosphate to two molecules of glycerate 3-phosphate, the carbon dioxide-fixing first step in the Calvin cycle of PHOTOSYNTHESIS. It may be as much as 15% of the total protein in the chloroplast. Rubisco can also act as an oxygenase, breaking down ribulose bisphosphate in the presence of oxygen to glycerate 3-phosphate and phosphoglycolic aid. This reaction forms the basis of PHOTORESPIRATION. The relative concentrations of oxygen and carbon dioxide in the surrounding air determine the balance of photosynthesis and photorespiration and thus the efficiency of carbon fixation. *See also* C_3 plant; C_4 plant.

ring-porous Describing wood in which the largest vessels are in the early wood, giving distinct growth rings, as seen in elm. *Compare* diffuse-porous. *See also* annual ring.

RNA (ribonucleic acid) A nucleic acid comprising a single polynucleotide chain similar in composition to a single strand of DNA except that the sugar ribose replaces deoxyribose and the pyrimidine base uracil replaces thymine. RNA is synthesized in the nucleus, using DNA as a template, and exists in three main forms (messenger RNA; transfer RNA; ribosomal RNA). In certain viruses RNA is the genetic material.

RNA polymerase *See* polymerase.

RNase (ribonuclease) An enzyme that catalyzes the hydrolysis of the sugar–phosphate bonds of RNA. There are several types, each having a specific action. For example ribonuclease T1 degrades RNA to mono- and oligonucleotides terminating in a 3′-guanine nucleotide, while those produced by ribonuclease T2 terminate in a 3′-adenine nucleotide.

RNA splicing In eukaryote cells, the process whereby noncoding segments (INTRONS) of the primary RNA transcript are removed and the remaining coding segments (EXONS) joined together to form the functional mRNA molecule. It is catalyzed by a complex of proteins and RNA molecules (a *spliceosome*), or sometimes by mRNA itself (*see* ribozyme). *See* transcription.

root The organ that anchors a plant to the ground and that is responsible for the uptake of water and mineral nutrients from the soil. Roots develop from the radicle of the embryo and, according to the nature of branching from the seedling root, a fibrous or tap root system develops. Roots differ from shoots in lacking chlorophyll and not producing buds or leaves. Roots also differ from shoots in the arrangement of xylem and phloem, having a solid central strand of vascular tissue rather than a hollow cylinder of conducting tissue.

The growing point of the root is protected by a *root cap* to withstand the abrasion that occurs as the root grows through the soil. The direction of root growth is controlled by both gravity and water supply. The main absorptive region of the root is just beyond the zone of elongation behind the root tip, where the ROOT HAIRS are formed. Further back along the root, lateral roots are formed, which develop from within the vascular tissue and grow out through the cortex (endogenous root formation). All roots whose derivation can eventually be traced back to the radicle are called *primary roots*. Roots that arise in any other way are termed *adventitious roots*, for example those that develop from the stem in bulbs and corms. In many plants, particularly biennials, the root may become swollen with carbohydrates and act as an underground food store during the winter. Many other root modifications are seen (*see* buttress root; contractile root; prop root). The roots of many plant species exist in association with fungi as mycorrhizae or, as in the Fabaceae, with nitrogen-fixing bacteria to form *root nodules* (*see* nitrogen fixation).

root hair An outgrowth, or trichome, from the piliferous layer of the root. Root hairs are projections of single epidermal cells that make contact with the soil for uptake of water and solutes, increasing the surface area of the root for absorption. They may also have an anchoring role, helping to retain contact between the root tip and the soil under dry conditions when the soil tends to shrink away from the roots.

root cap *See* calyptra.

root nodule *See* nitrogen fixation.

root pressure The pressure that may build up in a plant root system due to the osmotic potential of the root cells, which is thought to help force water upward in the xylem vessels. It is demonstrated by the oozing of water from the surface of a cut stem. Root pressure tends to build up at night, when the rate of transpiration is low, and is a cause of GUTTATION in some species, especially grasses. *See* osmosis; water potential.

rootstock A vertical, usually short, underground stem or rhizome. It is found in many angiosperms, e.g. rhubarb (*Rheum* spp) and strawberry (*Fragaria ananassa*), and in certain ferns, e.g. *Osmunda*. It can reproduce vegetatively.

Rosaceae A large family of dicotyledonous trees, shrubs, and herbs with alternately arranged (opposite in *Rhodotypos*), simple, compound, or dissected leaves with stipules, and actinomorphic flowers with parts in fives and stamens in one to several whorls of five or more. The ovary is inferior or partly inferior or the carpels lie free in a hollow receptacle. The fruit may be a head of follicles or achenes, as in avens (*Geum*), achenes on a fleshy receptacle, as in strawberry (*Fragaria*); a drupe, as in cherries, plums, and peaches (*Prunus*); a head of drupelets, e.g. blackberry and raspberry (*Rubus*); or a pome, as in apple (*Malus*). There are about 2825 species. Many species and cultivated hybrids, especially roses and cherry trees, are grown as ornamentals for their showy flowers. *Cotoneaster* and *Spiraea* are also popular garden shrubs. Other edible fruits include loquats (*Eriobotrya japonica*), pears (*Pyrus communis*), quinces (*Cydonia oblonga*), and strawberries (*Fragaria*).

rosette plant Any plants whose leaves radiate outward from a short stem at soil level, e.g. daisy (*Bellis perennis*). This growth form helps plants to withstand trampling or grazing and in exposed habitats avoids exposure to strong winds.

royal ferns *See* Osmundales.

Rubiaceae A large family of dicotyledonous trees, shrubs, and herbs, including some lianas, epiphytes, and ant plants (e.g. *Myrmecodia*). They have decussately arranged simple leaves often bearing elaborate stipules, and flower parts fused, in fours or fives (sometimes including one extra-large, brightly colored sepal), and 4–5 stamens that alternate with the corolla lobes. The ovary is usually inferior, with two fused carpels. The fruit is a capsule, berry, or schizocarp. There are about 10 200 species, mostly tropical, but with some in temperate and arctic regions. Many species have long corolla tubes, an adaptation to pollination by butterflies, moths, or birds. Most temperate species are herbs, e.g. bedstraws (*Galium* spp.), typically with square stems and whorls of leaves. The family includes a few light timber trees, coffee (*Coffea arabica*), and *Cinchona*, whose bark is a source of quinine. There are some cultivated ornamentals, such as *Gardenia jasminoides* and *Ixora*.

rubisco *See* ribulose bisphosphate carboxylase.

RuBP carboxylase *See* ribulose bisphosphate carboxylase.

runner A branch, formed from an axillary bud, that grows horizontally along the ground. Runners often have greatly elongated inernodes. The axillary buds of the runner may develop into daughter plants with adventitious roots growing from the node, as seen in the creeping buttercup. Conversely, only the terminal bud may form a new plant, as in the strawberry. A new runner then develops from a branch of this daughter plant. If the intervening runner breaks or rots away, the daughter plants become independent, so this is also a means of vegetative reproduction. *See also* offset; stolon.

rusts Parasitic basidiomycete fungi of the order Uredinales. The name derives from the characteristic yellow-brown streaks that appear on the host plant following the eruption of masses of spores (*uredospores*) through the host epidermis. An economically important rust is *Puccinia*, which infects many crops, notably cereals.

Rutaceae A family of dicotyledonous trees and shrubs. They include the citrus fruits (*Citrus* spp.), kumquat (*Fortunella*), and a number of ornamental shrubs.

S

saccharide *See* sugar.

Saccharomycetales An order of the Ascomycota, containing the yeasts, in which the mycelium is poorly developed or absent, and the vegetative stage comprises mainly single cells that reproduce asexually by budding or fission. The commercially important yeasts of the genus *Saccharomyces* are used in brewing (*S. cerevisiae*), wine-making (*S. ellipsoideus*), and bread-making (*s.cervisiae*): they feed on sugars, and in anaerobic conditions produce ethanol (alcohol) and carbon dioxide.

Sachs, Julius von (1832–97) German botanist. Having obtained his PhD in botany at the University of Prague in 1856, Sachs occupied a number of posts, culminating in his appointment as professor of botany at Würzburg University. He established that chorophyll is not univerally distributed in plants, but is confined inside definite structures that he named 'chloroplasts', and showed that starch is the first product of photosynthesis. He also discovered that plants respire in the same way as animals, taking up oxygen and giving out carbon dioxide. His work on plant growth movements led him to invent the KLINOSTAT.

safranin In microscopy, a permanent red stain used to stain nuclei in plant cells. It also stains lignin and cutin red and chloroplasts pink. It is often used with a green counterstain such a fast green or a blue counterstain, e.g. hematoxylin. *See* staining.

Salicaceae A family of dicotyledonous trees, shrubs, and creeping forms including the willows (*Salix*) and poplars (*Populus*).

S alleles *See* multiple allelism.

salt gland *See* hydathode.

saltmarsh Vegetation growing in saline or brackish marshy places, such as river estuaries and sheltered, muddy coasts, and consisting mainly of HALOPHYTES. Salt marshes can be productive grazing land, and serve as important barriers against inundation of the hinterland by the sea.

samara A nut or achene whose pericarp is extended to form a wing for wind dispersal, e.g. ash (*Fraxinus excelsior*).

sand Mineral particles consisting mainly of quartz, felspar, and mica, and measuring between 2 and 0.05 mm in diameter. Sandy soils contain at least 85% sand and not more than 10% clay. They are light and drain well. However, retention of nutrients and water is poor. *Compare* silt.

sap **1.** The solution of mineral salts and sugars that is found in xylem and phloem vessels and oozes out of cut stems.
2. The liquid contents of a plant cell vacuole.

saponins Any of a class of bitter-tasting GLYCOSIDES that form colloidal solutions in water and foam when shaken. Saponins occur in many plants, such as soapwort (*Saponaria offinicalis*).

saprobe (**saprophyte**) An organism that derives its nourishment by absorbing the products or remains of other organisms. Saprobes usually secrete enzymes on their food and absorb the soluble breakdown products. Many fungi and bacteria are saprobes, and are important decomposers in food chains, returning nutrients to the soil by putrefaction and decay.

Saprolegniales (**water molds**) An order of the OOMYCOTA containing aquatic or soil-dwelling funguslike microorganisms, e.g. the water mold, *Saprolegnia*, which may be saprobes or parasites. Some form mycelia; others are unicellular.

saprophyte *See* saprobe.

saprotrophic (**saprophytic**) The mode of nutrition of a SAPROBE. *See* decomposer.

sapwood (**alburnum**) The outer living xylem cells in a tree trunk, consisting of secondary xylem elements, parenchyma, and medullary rays, that are actively involved in water transport and food storage. *Compare* heartwood.

satellite DNA *See* repetitive DNA.

savanna Tropical GRASSLAND with scattered trees and/or tall bushes, found in all parts of the tropics where there is a pronounced dry season. Some savannas have many grazing animals that affect the vegetation structure; these and other savannas have periodic or irregular fires in the dry season. *See* climax.

scab Any plant disease that produces dry, raised, scablike lesions due to the formation of cork layers. The disease is usually caused by a bacterium, e.g. common scab of potatoes (*Streptomyces scabies*), or a fungus, e.g. apple scab (*Venturia inaequalis*).

scalariform thickening A type of secondary wall formation consisting of interlaced helical bands of thickening giving a ladderlike formation. It allows for very little further extension, and is found in tissues such as metaxylem tracheids and vessels, which do not elongate after maturation. *See* xylem. *Compare* pitted thickening; spiral thickening.

scanning electron microscope *See* microscope.

Schiff's reagent A reagent used to detect the presence of aldehyde and ketone groups in certain compounds. It is produced by the reduction of fuchsin by sulfurous acid, and is oxidized by aldehydes and ketones, restoring the magenta color of the fuchsin.

schizocarp A dry fruit, formed from two or more carpels, that divides at maturity into one-seeded achenelike segments termed mericarps. Such fruits are seen in hollyhock (*Alcea*).

Schleiden, Matthias Jakob (1804–81) German botanist. Having abandoned a career in law, Schleiden eventually became professor of botany at the University of Jena in 1839. After years studying the microscropic structure of plants, he proposed his cell theory *Contributions to Phytogenesis* (1838), in which he stated that the various plant structures are all composed of individual cells. He also recognized the significance of the nucleus. Mistakenly, he believed that new cells formed by being budded off the nuclei. His cell theory was also adopted and expanded by SCHWANN, who applied it to animals also.

Schultze, Max Johann Sigismund (1825–74) German zoologist. During his time as professor of anatomy at Bonn University, Schultze studied the cells of many different species of animals. This led to the publication of the famous paper (1861) in which he concluded that protoplasm is the physical basis of life. He described cells as being composed of 'nucleated protoplasm' and maintained that this constituent of cells was more important than cell walls (of plants) or membranes. He supported his conclusion by pointing out that certain embryonic cells do not possess membranes.

Schultze's solution A solution of zinc chloride, potassium iodide, and iodine used mainly for testing for cellulose and hemicellulose. Both materials stain a blue color with the reagent, that of hemicellulose being weaker. *See also* staining.

Schwann, Theodor (1810–82) German physiologist. Schwann was educated at a number of German universities before eventually moving to Belgium. He is best known for his work on different animal tissues, which he suggested were composed of individual cells, rather than forming directly from molecules. He thus confirmed the cell theory of Schleiden and extended it to animals. His earlier work in Germany, attempting to disprove the theory of spontaneous generation, led him to discover that yeast is involved in fermentation. He was so discredited for this work that he left Germany. When Pasteur's work on fermentation was published in the 1850s, his good name was restored.

scion A shoot or bud from one plant that is joined to another plants with roots (the STOCK) by grafting or budding. The scion supplies only aerial parts to the GRAFT.

sclereid Any SCLERENCHYMA cell except the long fibers. The cell wall has undergone secondary thickening, and is often lignified. The various forms of sclereid include the star-shaped *astrosclereid*, the rod-shaped *macrosclereid*, and the isodiametric *stone cell.*

sclerenchyma The main supporting tissue in plants, made up of cells with heavily thickened, often lignified, walls and empty lumina. The cells usually have simple unbordered pits. Unlike collenchyma, it is not very extensible and is thus not formed in quantity until after the young tissues have fully differentiated. Sclerenchyma is often found associated with vascular tissue and exists as two distinct types of cell: the long FIBER and the shorter SCLEREID. It may develop by thickening of the secondary walls of parenchyma cells, or it may arise directly from meristematic cells.

sclerotium (*pl.* **sclerotia**) The resting body of certain fungi, e.g. ergot (*Claviceps purpurea*) and root-rot (*Sclerotinia* spp), formed from a mass of hyphae. Sclerotia are often rounded or club-shaped. They do not contain spores, and eventually produce either a mycelium or fruiting bodies. *See also* stroma.

Scrophulariaceae A family of dicotyledonous plants, mainly herbs, but including some trees, shrubs, climbers, and aquatics with a wide range of leaf form and flower structure. A few are semiparasitic on the roots of angiosperms, especially grasses, such as eyebright (*Euphrasia*) and lousewort (*Pedicularis*). There are about about 5100 species. Leaves of the foxglove (*Digitalis purpurea*) are the source of drugs such as digitalin.

scutellum The part of the embryo of Poaceae (grasses) that lies next to the endosperm. It is thought to be the modified cotyledon.

secondary cell wall *See* apposition.

secondary growth Plant growth derived from secondary or lateral meristems, i.e. the vascular and cork cambia. It is usually absent in monocotyledons, but occurs in most dicotyledons and conifers, and in a few lower plants. In dicotyledons, the result of secondary growth is termed *secondary thickening* since there is usually an increase in width rather than length. The activity of the vascular CAMBIUM gives rise to the *secondary xylem* and *secondary phloem*. Wood is mostly secondary xylem. The cork cambium (PHELLOGEN) gives rise to the PERIDERM, a protective layer of tissue on the outside of the stem or root, which consists of the cork (PHELLEM), the phellogen, and the PHELLODERM (secondary cortex). The parts of the plant formed by secondary growth are called the *secondary plant body. Compare* primary growth. *See* annual ring.

secondary phloem *See* secondary growth.

secondary plant body *See* secondary growth.

secondary structure In a PROTEIN, the basic shape of the polypeptide, i.e. the way it folds. There are two main forms: the α-helix and the β-pleated sheet. The α-helix is an extended spiral with one turn for every 3.6 amino acids; the turns are held in position by HYDROGEN BONDS between adjacent C=O and NH groups. Such a secondary structure allows for flexibility, the extent of which is determined by the amount of further linking of the structure by DISULFIDE BRIDGES. In the β-pleated sheet the polypeptides are more extended, and adjacent chains of amino acids are arranged antiparallel to each other, linked by hydrogen bonds between the C=O and NH groups of adjacent chains. β-pleated sheets have high tensile strength, but are inflexible. Many proteins contain both α-helices and β-pleated sheets in different parts of the molecule. *See* conformation; primary structure; quaternary structure; tertiary structure.

secondary thickening *See* secondary growth.

secondary xylem *See* secondary growth.

secretion The process of discharging materials from cells. The term is usually applied to materials that have been synthesized by the cell.

seed The structure that develops from the OVULE following fertilization in angiosperms or gymnosperms. In flowering plants one or more seeds are contained within a fruit developed from the ovary wall. The individual seeds are composed of an embryo and, in those seeds in which food is not stored in the embryo cotyledons, a nutritive endosperm tissue. This difference enables seeds to be classified as nonendospermic or endospermic. The whole is surrounded by a testa developed from the integuments of the ovule. In gymnosperms the seeds do not develop within a fruit but are shed 'naked' from the plant. Following dispersal from the parent plant, seeds may germinate immediately to form a seedling or may remain in a relatively inactive dormant state until conditions are favorable for germination. In annual plants, seeds provide the only mechanism for surviving the cold or dry seasons. Seeds may be formed asexually in certain plants by APOMIXIS, e.g. in dandelion.

The development of the seed habit, which makes water unnecessary for fertilization, is one of the most significant advances in plant evolution. It has enabled gymnosperms and angiosperms to colonize dry terrestrial habitats where lower plants are unable to establish themselves.

seed ferns Seed-bearing gymnosperms of the extinct order Pteridospermales (Cycadofilicales), represented only by fossil forms that flourished in the Carboniferous but became extinct in the Cretaceous. The plant body resembled a fern and did not produce flowers, the seeds developing from megasporangia borne on the margins of the fronds. Their stems showed some secondary thickening.

seed plants *See* spermatophyte.

segregation The separation of the two alleles of a gene into different gametes, brought about by the separation of homologous chromosomes at anaphase 1 of meiosis.

seismonasty **(seismonastic movements)** A NASTIC MOVEMENT in response to shock. An example is the collapse of leaves of the sensitive plant (*Mimosa pudica*) when touched or shaken.

Selaginella *See* Selaginellales; Lycophyta.

Selaginellales An order of the Lycophyta (clubmosses) containing the single living genus *Selaginella*, sometimes called spike-mosses, with about 700 species, found worldwide, but especially in the tropics. They range from erect, sometimes tufted, forms to prostrate creeping plants and climbers.

self-compatibility The capacity to self-fertilize. Many flowers are capable of self-fertilization if pollination fails, and a few species do it regularly.

self-incompatibility (**self-sterility**) The condition in whereby male gametes cannot fertilize female gametes from the same individual and give rise to a viable embryo. The pollen tube may be blocked by callose as it tries to grow down the style, be prevented from penetrating the stigma surface, abort on reaching the ovule or the pollen grain may fail to germinate. *See* incompatibility; heterostyly.

selfish DNA DNA that can move around within the genome of an organism or insert copies of itself at various sites without serving any apparent useful function. The prime examples of selfish DNA are the mobile genetic elements called *transposons*. Some biologists also regard INTRONS as selfish DNA. Its name derives from the hypothesis that selection acts within the genome, favoring anything that results in increased replication of DNA. The apparently surplus DNA does not appear to confer any advantage on the organism, so the DNA is considered to be acting 'selfishly'. *Compare* junk DNA; repetitive DNA. *See* transposon.

self-pollination *See* pollination.

self-sterility *See* self-incompatibility.

Seliwanoff's test A standard test for the presence of fructose or other ketose sugars in solution. A few drops of Seliwanoff's reagent, resorcinol in 50% hydrochloric acid, are heated with the test solution. A red color or red precipitate indicates fructose.

semiconservative replication *See* replication.

semipermeable membrane *See* osmosis.

senescence The phase of the aging process of an organism or part of an organism between maturity and natural death. It is usually characterized by a reduction in capacity for self-maintenance and repair of cells, and hence deterioration. There is often accumulation of waste metabolic products, a decrease in dry weight (as substances are withdrawn from the affected part), and a rise in the rate of respiration (*see* climacteric).

sensitive plant A plant (*Mimosa pudica*) of the family Fabaceae (pea family) whose leaves are divided into many small leaflets or pinnae, which progressively collapse and fold when the leaf is touched. This is the result of turgor changes in cells in swellings (*see* pulvinus) in the petiole at the bases of the leaflets. *See also* nastic movements.

sepal One of the structures situated immediately below the petals of a flower. Their collective name is the *calyx*. They are often green and hairy and enclose and protect the flower bud. Sometimes they are brightly colored and attract insects for pollination. Sepals contain several vascular bundles, and are thought to be derived from leaves.

septum (*pl.* **septa**) A wall, partition, or membrane separating two cavities. For example, the capsule of a poppy (*Papaver*) is divided by septa, as are the individual compartments of the mycelia of many fungi.

sere Any plant community in a succession in which each community itself effects changes in the habitat that determine the nature of the following stage. The successive stages are known as *seral stages*. Seres result eventually in a climax community. Stages in a secondary succession, which appear when the biotic components of a primary sere are destroyed, e.g. by fire, are called *subseres*; seres in microhabitats are called *microseres*. The initial (pioneer) community in a succession is termed a *prisere*. *See also* climax; halosere; hydrosere; succession.

serine An AMINO ACID synthesized from glycerate 3-phosphate. Serine is also a product of photorespiration and other metabolic reactions. It is a component of several phosphoglycerides. It is broken down by removal of the amino group to form pyruvic acid.

sessile Unstalked, for example an acorn, leaf, flower, or other organ that is attached to the main body of the plant.

seta (*pl.* **setae**) The part of the SPOROGONIUM that forms the stalk between the foot and the capsule in bryophytes, e.g. *Funaria* and *Pellia*. It may contain vascular tissue, enabling the sporophyte capsule to draw nutrients from the parent gametophyte.

Sewall Wright effect *See* genetic drift.

sex chromosomes Chromosomes that carry sex-determining genes. They are rare in the plant kingdom, and many plants are hermaphrodite. Sex chromosomes occur in some dioecious plants, e.g. *Silene*. In other dioecious plants a single pair of alleles determines sex.

sexual reproduction The formation of new individuals by fusion of two nuclei or sex cells (GAMETES) to form a diploid ZYGOTE. In unicellular organisms whole individuals may unite but in most multicellular organisms only the gametes combine. In organisms showing sexuality, the gametes are of two types: male and female. They may be produced in special organs (e.g. carpel and anther in angiosperms, archegonia and antheridia in lower plants and some algae). The gametes may be derived from the same parent (AUTOGAMY) or from two different parents (ALLOGAMY). Individuals producing both male and female gametes are termed hermaphrodite if in the same flower or MONOECIOUS if in separate flowers; those in which male and female gametes are borne on different individuals are termed DIOECIOUS.

Generally meiosis occurs before gamete formation, resulting in the gametes being haploid (having half the normal number of chromosomes). At fertilization, when the haploid gametes fuse, the diploid number of chromosomes is restored. In this way sexual reproduction permits genetic recombination, which results in greater variety in offspring and so provides a mechanism for evolution by natural selection. In organisms showing ALTERNATION OF GENERATIONS, the gametes may be produced by mitosis in the haploid generation, or by meiosis in the diploid generation and give rise directly to gametes or gamete-producing cells.

shade plant A plant that can tolerate and thrive in low light intensity. Some shade plants are sensitive to very bright light and cannot live in open habitats.

shadowing A method of preparation of material for electron microscopy enabling surface features to be studied. It can be used for small entire structures, subcellular organelles, or even large molecules (e.g. DNA). The specimen is supported on a plastic or carbon film on a small grid and sprayed with vaporized metal atoms from one side while under vacuum. The coated specimen appears blacker (more electron-opaque) where metal accumulates, and the lengths and shapes of 'shadows' cast (regions behind the objects not coated with metal) give structural information. It is often used in association with freeze fracturing. *See* freeze fracturing; microscope.

shoot The aerial photosynthetic portion of a plant that generally consists of a stem upon which leaves, buds, and flowers are borne.

short-day plant (**SDP**) A plant that appears to require short days (days with less than a critical length of daylight) in order to flower. In fact, the plants require a minimum period of darkness (night length). Short-day plants are typically found in temperate latitudes. Examples include spring-flowering plants such as strawberry (*Fragaria*) and autumn-flowering plants such as *Chrysanthemum*. *See* critical day length; photoperiodism. *Compare* long-day plant; day-neutral plant.

siblings (**sibs**) Two or more offspring from the same cross. The term *sibs* is usually used in a narrower sense to mean plants derived by selfing or by crossing between genetically similar parents. The term *sibling species* is sometimes used to describe species that are extremely similar because of recent common ancestry. Such species may be almost impossible to distinguish in the field.

sieve cell *See* sieve element.

sieve element A vascular cell in the PHLOEM, whose function is to transport sugars and other nutrients from the site of production to the site of utilization or storage. The term includes both sieve cells and sieve tubes. *Sieve cells* are long, narrow cells with specialized *sieve areas* in the primary wall, perforated by several pores, derived from pits. The pores are lined with callose, and allow cytoplasmic strands to pass from one sieve cell to the next. They are found in gymnosperms and lower vascular plants. The *sieve tubes* of angiosperms are shorter and wider, and have highly specialized sieve areas called *sieve plates*, in some species (e.g. *Nicotiana*) with several sieve areas, usually on the end walls of the elements. They are usually joined end to end, allowing continuous passage of materials up and down the plants. In contrast to xylem components, sieve tube elements have nonlignified cell walls, and a living enucleate (without a nucleus) protoplast. They are usually associated with specialized parenchyma cells called COMPANION CELLS, whose protoplasts are connected with the protoplasm of the sieve tube by plasmodesmata.

sieve plate *See* sieve element.

sieve tube *See* sieve element.

silicon A MICRONUTRIENT found in many animals and plants, although not essential for growth in most plants. It is found in large quantities in the cell walls of certain algae (e.g. desmids, diatoms) and horsetails, and in smaller amounts in the cell walls of many higher plants, especially grasses, where it can be a useful taxonomic feature.

silicula A capsular fruit that is formed from a bicarpellary ovary. It is flattened, short, and broad and is divided into two or four loculi (valves) by a false septum, which bears the seeds. Siliculas are typical of the Brassicaceae, e.g. honesty (*Lunaria annua*). *Compare* siliqua.

siliqua (**silique**) A fruit of some Brassicaceae, similar to the SILICULA but longer and thinner, for example the wallflower fruit (*Erysimum cheiri*).

silt Mineral particles between 0.05 and 0.002 mm in diameter. A silt soil is one with more than 80% silt and less than 12% clay, and has a smooth, soapy texture. *See* soil. *Compare* sand.

Silurian The period, some 440–405 million years ago, between the Ordovician and the Devonian periods of the Paleozoic. It was a relatively warm period in the Earth's history, during which the first land plants appeared. It is characterized by simple plants such as liverworts; fossils of plants similar to vascular plants, e.g. *Cooksonia*, are known from the later part of the period, some of which had simple strands of tracheids. *Cooksonia* had a slender dichotomously branching stem with no leaves, terminal sporangia and spores impregnated with cutin. It resembles the psilophyte *Rhynia* from the Rhynie chert of the Devonian period. *See also* geological time scale.

simple pit *See* pit.

siphonostele A stele with a central core of pith internal to the xylem. The phloem may lie external to the xylem (ectophloic), or both internal and external to the xylem (amphiphloic). *See* stele.

SI units (**Système International d'Unités**) The internationally adopted system of units used for scientific purposes. It has seven base units (the meter, kilogram, second, kelvin, ampere, mole, and candela) and two supplementary units (the radian and steradian). Derived units are formed by multiplication and/or division of base units; a number have special names. Standard prefixes are used for multiples and submultiples of SI units. *See Appendix.*

sliding growth A pattern of plant growth seen, for example, in many epidermal cells where, in order to accommodate growth by adjoining cells, expanding cell walls of adjacent cells slide over each other. Thus growth is achieved without disruption of neighboring cells, although it does result in breakage of plasmodesmata. *Compare* symplastic growth.

slime bacteria *See* myxobacteria.

slime molds *See* Acrasiomycota; Myxomycota.

smuts Plant diseases caused by basidiomycete fungi of the order Ustilaginales. Many are important parasites of cereals and form a mass of sooty black spores in place of the grain. In other smuts the spores form an uncovered mass of black powder. Examples of smuts are *Tilletia* and *Ustilago*.

snake's tongue ferns *See* Ophioglossales.

snurp *See* spliceosome.

society A minor plant community within a larger community, characterized by a specific dominant species, for example, a *Trillium*-dominated society in a community such as an oak woodland.

sodium An element found in all terrestrial plants, although it is believed not to be essential in most, with the exception of some salt-tolerant C_4 plants.

sodium bicarbonate indicator A mixture of the dyes cresol red and thymol blue in a sodium bicarbonate solution, which changes from red to orange and yellow with a slight increase in acidity, and thus indicates change in pH.

soil The accumulation of mineral particles and organic matter that forms a superficial layer over large parts of the Earth's surface. It provides support and nutrients for plants and is inhabited by numerous and various microorganisms and animals. A section down through the soil is termed a *soil profile* and this can characteristically be divided into three main layers or *horizons*. Horizon A, the topsoil, is darker than the lower layers due to the accumulation of organic matter as humus. It is the most fertile layer and contains most of the soil population and a high proportion of plant roots. Horizon B, the subsoil, contains materials washed down from above and may be mottled with various colors depending on the iron compounds present. Horizon C is relatively unweathered parent material from which the mineral components of the above layers are derived. The depth and content of the horizons are used to classify soils into various types, e.g. podsols and brown earths. The texture, structure, and porosity of soil depends largely on the sizes of the mineral particles it contains and on the amount of organic material present. Soils also vary depending on environmental conditions, notably rainfall. *See* soil structure.

soil profile *See* soil.

soil structure The arrangement of the mineral particles in the soil – whether they are free or bound into aggregates by decomposing organic matter. These units may be further grouped into crumbs, blocks, plates, and so on. The structure affects fertility, drainage, aeration, and ease of cultivation. A blocky soil will tend to dry out and suffer from leaching of minerals from the upper horizons. The soil structure is affected by *soil texture* – the sizes of the different mineral particles, which also influences drainage and aeration, and the retention of water by capillarity. The size of the particles depends in part on the parent rock and how it weathers. The main size classes of mineral particles are gravel and stones (greater than 2 mm in diameter), sand (0.05 to 2 mm in diameter), silt (0.002 to 0.05 mm in diameter), and clay (smaller than 0.002 mm in diameter). Clay soils, which have small particles, tend to be wet and sticky, with poor drainage and aeration. However, they tend to be electrostatically charged and attract humus, forming a clay–humus complex that can attract and retain minerals. The humus sticks the clay particles together, forming larger aggregates and improving drainage and aeration. Clay soils can be further improved by adding lime, which promotes clumping of particles. Sandy soils have large particles, and therefore larger spaces between them for drainage and aeration. They may need addition of humus or mulches to improve water and mineral retention and prevent leaching.

soil texture *See* soil structure.

Solanaceae A family of dicotyledonous herbs, shrubs, lianas, and small trees. They include commercial food plants such as aubergine (*Solanum melongena*), potato (*Solanum tuberosum*), sweet peppers (*Capsicum*), tomato (*Lycopersicon esculentum),* and *Nicotiana* (the source of tobacco).

solenostele *See* stele.

solute potential *See* water potential.

somatic Describing the cells of an organism other than germ cells. Somatic cells divide by mitosis producing daughter cells identical to the parent cell.

soredium (*pl.* **soredia**) *See* lichens.

sorus (*pl.* **sori**) **1.** A reproductive structure found in ferns comprising a collection of sporangia borne on a cushion of tissue termed the *placenta*. The placenta develops over a vein ending on the underside of a foliage leaf or on a specialized fertile leaf. The sorus is covered by a flap of tissue, the indusium.
2. The reproductive area of certain algal thalli, e.g. *Laminaria*.
3. The spore-bearing body of a rust fungus.

Southern blotting A technique for transferring DNA fragments from an electrophoretic gel to a nitrocellulose filter or nylon membrane, where they can be fixed in position and probed using DNA probes. Named after its inventor, E. M. Southern (1938–), it is widely used in genetic analysis. The DNA is first digested with restriction enzymes and the resulting mixture of fragments separated according to size by electrophoresis on an agarose gel. The double-stranded DNA is then denatured to single-stranded DNA using sodium hydroxide, and a nitrocellulose filter pressed against the gel. This transfers, or blots, the single-stranded DNA fragments onto the nitrocellulose, where they are permanently bound by heating. The DNA probe can then be applied to locate the specific DNA fragment of interest, while preserving the electrophoretic separation pattern. *See* DNA probe. *Compare* Western blotting.

SP (**suction pressure**) *See* osmosis.

spadix (*pl.* **spadices**) *See* inflorescence.

spathe *See* inflorescence.

speciation The formation of one or more new SPECIES from an existing species. Speciation occurs when a population separates into isolated subpopulations that develop distinctive characteristics as a result of NATURAL SELECTION or random GENETIC DRIFT, and cannot then reproduce with the rest of the population, even if there are no geographical or other physical reasons to prevent them from doing so. Speciation occurs when barriers to interbreeding arise, such as geographical changes (e.g. in river courses), habitats becoming uninhabitable (e.g. following urbanization), changes in timing of flowering, or when populations at the margins of the main population experience different selection pressures as they colonize different environments. Another cause of rapid speciation in plants is polyploidy. *See also* adaptive radiation.

species A taxonomic unit comprising one or more populations, all the members of which are able to breed amongst themselves and produce fertile offspring. They are normally isolated reproductively from all other organisms, i.e. they cannot breed with any other organisms. Two or more related species unable to breed because of geographical separation are called *allopatric species*. Related species that are not geographically isolated, and which could interbreed but in practice do not because of differences in behavior, breeding season, etc., are called *sympatric species*. Some species can interbreed with other related species, but the hybrids are often wholly or partially sterile. For some species there may be an almost complete continuum between closely related species, as in certain orchids. This may also happen in disturbed areas or in regions where the range of two or more species or even genera overlap, producing *hybrid swarms*. Where species form apomictic clones that are very difficult to distinguish from each other, they are termed *aggregate species*, e.g. bramble (*Rubus fruticosus* agg.). In general, species are regarded as distinct if they remain reproductively isolated for most of their geographical range and have recognizably distinct morphological characteristics. Within a species, there may be subgroups with distinct morphological or (especially in microorganisms) physiological characteristics; these groups are termed SUBSPECIES or RACES. Groups of similar species are classified together in genera.

spectrophotometer An instrument for measuring the amount of light of different wavelengths absorbed by a substance. It gives information about the identity or concentration of the specimen and can be used to plot absorption spectra. It can also be used to investigate how electromagnetic radiation interacts with matter, e.g. the absorption of light in photosynthesis.

spermatophyte (seed plant) Any seed-bearing plant. In many older classifications they constituted the division (phylum) Spermatophyta, subdivided into the classes Angiospermae and Gymnospermae. *Compare* tracheophyte.

spermatozoid *See* antherozoid.

S phase *See* cell cycle.

Sphenophyta A phylum of vascular nonseed-bearing plants that contains one living order, the Equisetales, comprising one genus, *Equisetum* (horsetails or scouring rushes), and three extinct orders, the Calamitales, Sphenophyllales, and Pseudoborniales. Sphenophytes were particularly abundant in the Carboniferous period when the genus *Calamites* formed a large proportion of the forest vegetation, some treelike species reaching heights of 30 m.

spherosome A small spherical organelle of plant cells, about 0.5–1.0 μm in diameter, bounded by a single membrane and storing lipid.

spike *See* inflorescence.

spindle The spindle-shaped structure formed in the cytoplasm during MITOSIS and MEIOSIS that is responsible for moving the chromatids and chromosomes to opposite poles of the cell. The spindle consists of a longitudinally orientated system of protein MICROTUBULES whose synthesis starts late in interphase. A special region of the centromeres of each pair of sister chromatids, the *kinetochore*, becomes attached to one or a bundle of spindle microtubules. During anaphase, the kinetochore itself acts as the motor, disassembling the attached microtubules and hauling the chromatid toward the spindle pole. Later in anaphase, the unattached interpolar microtubules actively slide past each other, elongating the entire spindle.

spine A modified leaf reduced to a sharply pointed structure. In some xerophytes, such as cacti, most of the leaves are modified into spines, greatly reducing the leaf area for transpiration; the photosynthetic function of the leaves is taken over by the green stems. The dense mass of whitish spines also helps to reflect incoming radiation, further reducing transpiration.

spiral thickening (helical thickening) The type of secondary wall formation in which a spiral band of lignified deposits is formed on the inner wall of the cell. It is found in protoxylem and first metaxylem tracheids and vessels. Like annular thickening, it allows for continued elongation of the xylem. *See* annular thickening; xylem.

Compare reticulate thickening; scalariform thickening.

spirillum A helically shaped bacterium.

spirochetes Long spirally twisted bacteria surrounded by a flexible wall. An axial filament, with a similar structure to a bacterial flagellum, is spirally wound about the protoplast inside the cell wall. They swim actively by flexing the cell. They are found in mud and water and can withstand low oxygen concentrations. Many spirochetes are pathogens causing yaws, syphilis, and relapsing fever.

Spirogyra A genus of filamentous green algae found in freshwater and having a characteristic spiral chloroplast. Reproduction is by conjugation: two filaments become aligned and pairs of cells, one from each filament, become joined by a conjugation tube. The contents of one cell pass through the tube and fuse with the contents of the second cell to form a zygote. *See* Gamophyta.

spliceosome A unit in the nucleus that removes the noncoding RNA (INTRONS) from the initial RNA transcript and splices the RNA molecule back together again. It consists of several small ribonucleoproteins called SNURPS, comprising RNA combined with proteins. The RNA molecules act as RIBOZYMES, catalyzing both the splicing process and the assembly of the spliceosomes. *See* transcription.

spongy mesophyll *See* mesophyll.

spontaneous generation The erroneous belief that modern living organisms can be formed directly and spontaneously from inorganic material, given the right conditions (*abiogenesis*). This belief was disproved by Redi and Pasteur in the 17th and 19th centuries.

sporangiophore A structure bearing sporangia (*see* sporangium). In fungi it may be a simple hypha or erect branch of a hyphae; in lower vascular plants it may be a leaf (SPOROPHYLL) or a STROBILUS.

sporangium (*pl.* **sporangia**) A reproductive body in which asexual spores are formed. It may be unicellular, as in simple algae and fungi, or multicellular, as in plants. The spores are liberated by rupture of the sporangium wall.

spore A unicellular or multicellular plant reproductive body. Generally the term is applied to reproductive units produced asexually, such as the spores of bryophytes and ferns. A prefix is often added, providing information as to the nature of the spore, for example conidiospores arise on a conidium, megaspores and microspores are the larger and smaller spores produced by heterosporous plants, etc. In plants that undergo ALTERNATION OF GENERATIONS, spores are produced by meiosis by the sporophyte generation, so are usually haploid; they germinate to produce the haploid gametophyte generation.

spore mother cell A cell that gives rise to four haploid spores by meiosis. In heterosporous species, a distinction is usually made between megaspore mother cells and microspore mother cells. Of the four products of a megaspore mother cell, three usually abort.

sporocarp A hard multicellular spore-containing structure found in water ferns (e.g. *Marsilea* and *Pilularia*). The sori become enclosed by the growing together of fertile fronds; and the spores are not liberated until the sporocarp decays and ruptures.

sporogonium The sporophyte generation in mosses and liverworts. It develops from the zygote and comprises the foot, seta, and capsule. The sporogonium is parasitic on the gametophyte generation.

sporophore The aerial spore-producing body of certain fungi, e.g. the mushroom of *Agaricus*.

sporophyll The sporangium-bearing structure of vascular plants. In some ferns sporangia develop on the normal foliage leaves but in higher plants, the sporophylls are highly modified leaves. They may be grouped together in a strobilus or, in the angiosperms, located in the flowers. *See* megasporophyll; microsporophyll.

sporophyte The diploid generation in the life cycle of a plant, which gives rise asexually to haploid spores. It arises by fusion of haploid gametes produced by the haploid generation (GAMETOPHYTE). In vascular plants, the sporophyte is the dominant generation, while in bryophytes it is parasitic on the gametophyte. *See* alternation of generations.

staining A procedure that is designed to heighten contrast between different structures. Normally biological material is lacking in contrast, protoplasm being transparent, and therefore staining is essential for an understanding of structure at the microscopic level. VITAL STAINS are used to stain and examine living material. Most stains require dead or nonliving material. Staining is done after fixation and either during or after dehydration. *Double staining* involves the use of two stains; the second is called the *counterstain. Acidic stains* have a colored anion, *basic stains* have a colored cation. Some stains are neutral. Materials can be described as *acidophilic* or *basophilic* depending on whether they are stained by acidic or basic dyes respectively. Basic stains are suitable for nuclei, staining DNA. Stains for light microscopy are colored dyes; those for electron microscopy contain heavy metals, e.g. osmium tetroxide.

For immediate observation, *temporary stains* may be used. These stains may fade in time, e.g. ruthenium red. Or the stain may damage the specimen if left, for example SCHULTZE'S SOLUTION, which dissolves cell walls and causes swelling. *Permanent stains* are used for specimens that are to be kept. They do not fade or damage specimens. Examples include fast green and the counterstain safranin. *See also* negative staining.

stamen The male reproductive organ in flowering plants consisting of a fine stalk, the *filament*, bearing the pollen producing *anther*. It is equivalent to the microsporophyll present in the gymnosperms and heterosporous members of the Filicinophyta and Lycophyta. The collective term for the stamens is the *androecium. See also* microsporophyll.

staminode A sterile stamen. It may be rudimentary, consisting of only the filament, as in figwort (*Scrophularia*), or it may form a conspicuous part of the flower, as in *Iris*.

standard deviation In statistics, a measure of the dispersion of a frequency distribution: it is the average magnitude of deviations from the center of a normal curve, calculated by squaring all the deviations, calculating their mean, then finding the square root of the mean. This gives a value σ, which is the point of maximum slope either side of the center of the curve. It is more suitable for use than the mean deviation, because it avoids problems of positive and negatives values on either side of the center.

standing crop The total weight of all the living organisms present in an ecosystem at a given moment, usually expressed as dry weight per unit area.

starch A polysaccharide that occurs exclusively in plants. It is an early end product of photosynthesis, and is stored temporarily in the chloroplasts. In the dark this starch is broken down by enzymes to sucrose and transported to other parts of the plant. Starch is a storage reservoir for the plant: it is broken down by enzymes to simple sugars and then metabolized to supply energy needs. It is present in plant cells as granules dispersed in the cytoplasm. The shape of these granules is used in taxonomy. In most plants it is found mainly in the root cortex, and in storage organs such as swollen taproots, corms, etc. Starch is extracted commercially from maize, wheat, barley, rice, potatoes, sorghum, cassava, and other crops. It is also the main storage material in many seeds. Starch is a polymer of glucose. It is not a single molecule, but a mixture of amylose (straight chains of glucose subunits) and amylopectin (branching chains of subunits).

starch sheath The innermost layer of cells of the cortex replacing the endodermis in some stems, especially young herbaceous dicotyledonous stems. Its cells contain prominent starch grains. *See also* endodermis.

starch–statolith hypothesis A hypothesis concerning the mechanism of gravity perception in plants. *See* gravitropism.

statocyte A gravity-sensitive plant cell containing STATOLITHS. *See* gravitropism.

statolith One of a number of large starch grains found in the statocytes, plant cells that are thought to be gravity sensitive. They move through the cytoplasm to the lowermost cell surface. *See* gravitropism.

stearic acid (**octadecanoic acid**) A saturated carboxylic acid, which is widely distributed in nature as the glyceride ester. It is present in most fats and oils of plant and animal origin, particularly the so-called hard fats, i.e. those of higher melting point. Stearic acid is formed from palmitic acid. It is an intermediate in the synthesis of oleic, linoleic, and linolenic acids.

stele The vascular tissue and associated parenchyma tissue and (if present) the surrounding pericycle and endodermis of a stem or root. The arrangement of tissues in the stele changes from stem to root. In dicotyledons, for example, the stem has a ring of discrete vascular bundles, which allows it to resist bending, while the root has a central core of vascular tissue, adapting it to resist pulling stresses. Stelar arrangements vary considerably in plants from the simple protostele, in which leaf gaps are absent, to the complex dictyostele, in which there are many closely spaced LEAF GAPS. Intermediate between these is the *solenostele* in which the leaf gaps are more widely spaced vertically so that only one is seen per cross section of the stem. *See* polystely.

stem A longitudinal axis upon which are borne the leaves, buds, and reproductive organs of the plant. The stem is generally aerial and erect but various modifications are found, for example underground stems like RHIZOMES, BULBS, and CORMS, and horizontal structures, such as RUNNERS. The stem serves to conduct water and food materials up and down the plant and, particu-

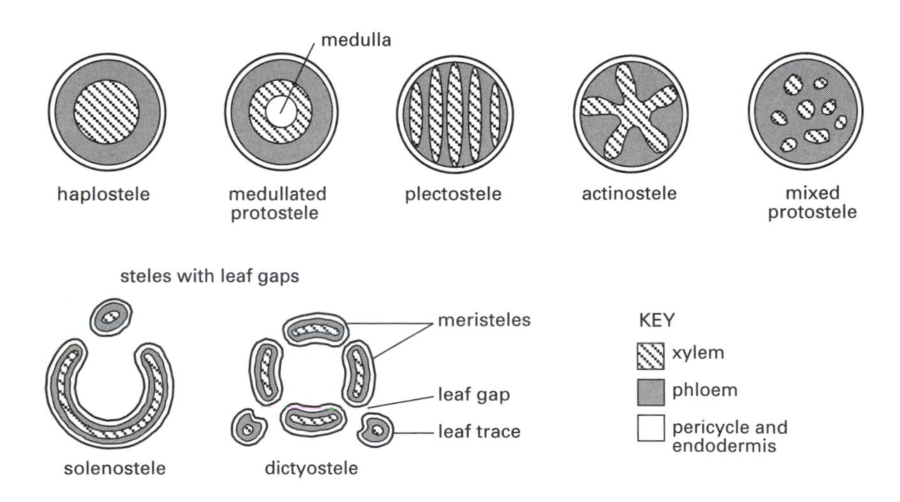

Stele: types of stele

larly in young plants and some succulents, it may serve as a photosynthetic organ. The stem is generally cylindrical and consists of regularly arranged conducting (vascular), strengthening, and packing cells, the whole being surrounded by a protective epidermis.

steppe Grassland dominated by drought-resistant species of perennial grasses, found in regions of LOESS soil and extreme temperature range, often on CHERNOZEM soils, in a zone from Hungary eastwards through the Ukraine and southern Russia to Central Asia and China. There are pronounced seasons, with hot summers and cold winters, with greater temperature extremes and lower rainfall in the east. The species composition varies with the climate. Large areas of the steppe are under cultivation for grain production.

sterigma (*pl.* **sterigmata**) A fingerlike projection upon which spores are formed in most basidiomycete fungi. Usually four sterigmata, each bearing one basidiospore, are borne on each basidium. The cultivated mushroom is an exception, with only two sterigmata on each basidium.

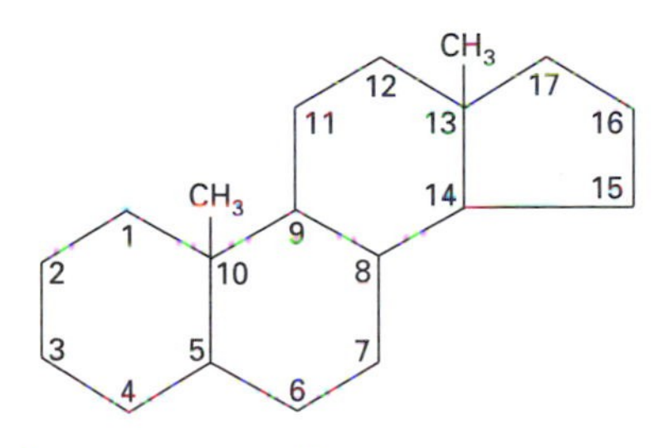

Steroid: skeleton of the molecular structure

steroid Any member of a group of compounds having a complex basic ring structure. Many plant steroids are alcohols (sterols); some are alkaloid-like compounds. *See also* sterol.

sterol A steroid with long aliphatic side chains (8–10 carbons) and at least one hydroxyl group. They are lipid-soluble and often occur in membranes (e.g. cholesterol and ergosterol).

stigma **1.** The receptive tip of the carpel of plants. It may be elevated on a stalklike protrusion called a style.
2. *See* eyespot.

stilt root An enlarged form of PROP ROOT, seen in some mangroves and a few palms and other trees, that helps support plants in unstable soils.

stimulus A change in the external or internal environment of an organism that elicits a response in the organism. The stimulus does not provide the energy for the response.

stipe **1.** In the more highly differentiated Phaeophyta (brown algae), such as wracks and kelps, the stalk between the holdfast and the blade.
2. The stalk of a mushroom or toadstool, which bears the cap or pileus.

stipule A modified leaf found as an outgrowth from the petiole or leaf base.

stock A plant onto which SCIONS (shoots or buds) of another plant are grafted. The stock provides the underground parts of the plant, and affects its size, vigor, and the timing of flowering and fruiting. *See* graft.

stolon An initially erect branch that, due to its great length, eventually bends over toward the ground. Where a node touches the soil adventitious roots develop and the axillary bud at that node grows out to form a new plant. *See also* layering; offset; runner.

stoma (*pl.* **stomata**) One of a large number of pores in the epidermis of plants through which gaseous exchange occurs. Each stoma is surrounded by two crescent-shaped GUARD CELLS, attached to each other at their ends, which regulate the opening and closing of the pore by changes in their turgidity. The arrangement of cellulose microfibrils in the guard-cell walls causes the guard cell to increase in length rather than width as it takes in water. This extra length is accommodated by the cell bulging away from the aperture, so widening the pore. Adjacent to the guard cells there are usually modified epidermal cells called subsidiary or accessory cells. The stoma and guard cells are collectively called the *stomatal apparatus*. In most plants stomata are located mainly in the lower epidermis of the leaf, where evaporation rates are lower. In many xerophytes and in some conifers of cooler climates, where soil water may be difficult to obtain in winter, the guard cells – and hence the stoma – are sunk in pits in the epidermis, often protected by hairs from passing air currents. TRANSPIRATION rates are thus greatly reduced, while allowing gaseous exchange for photosynthesis. In vertical leaves, such as those of grasses, they may be equally distributed on both surfaces.

stomatal apparatus *See* stoma.

stomium (*pl.* **stomia**) The site at which a sporangium or pollen sac ruptures to release the spores or pollen.

stone cell (**brachysclereid**) *See* sclereid.

strain Any group of similar or identical individuals, such as a CLONE, mating strain, physiological RACE, or PURE LINE.

stratification **1.** A seed treatment that enables seeds that require vernalization to germinate the following spring: the seeds are placed between layers of moist sand or peat and exposed to low temperatures, usually by leaving them outside through the winter.
2. The existence of layers of water of different density in a lake or other body of water. The differences in density may be due to temperature or to salinity. In many lakes in summer that is a warm low-density layer (EPILIMNION) lying above a colder denser layer (the HYPOLIMNION), the zone of rapid temperature change between the two layers being called the *thermocline*.

strobilus (*pl.* **strobili**) A group of sporophylls (spore-bearing leaves) arranged around a central axis, found in gymnosperms, Lycophyta (clubmosses), and Sphenophyta (horsetails). In the clubmosses and horsetails only one type of strobilus is formed, while in the gymnosperms female megastrobili and male microstrobili both develop. The strobili of gymnosperms and cycads are also termed *cones*.

stroma (*pl.* **stromata**) **1.** The colorless ground matter between the grana lamellae in a CHLOROPLAST. It is the site of the dark (light-independent) reactions of photosynthesis.
2. A solid mass of fungal hyphae, sometimes including host tissue, in which fruiting bodies may be produced. An example is the compact black fruiting body of the ergot fungus (*Claviceps purpurea*).

stromatolite A layered cushionlike mass of carbonate-rich rock formed by the actions of CYANOBACTERIA. Communities of these organisms secrete calcium carbonate, often building up concentric rings of deposits forming cushions up to a meter in diameter. Modern stromatolite-building communities are confined to salt flats or shallow salty lagoons where invertebrate predators cannot survive, but the fossil record demonstrates a much more widespread distribution in the Precambrian period. Some stromatolites date back nearly 4000 million years, making them the oldest known fossils.

style The stalklike portion of a carpel, joining the ovary and the stigma. The style may be elongated in plants relying on wind, insect, or animal pollination so that the stigma has a greater chance of coming into contact with the pollinating agent.

subclimax *See* climax.

suberin A fatty acid polyester produced in the walls of the endodermis and in bark, where it renders the tissue impermeable to water and resistant to decay. The Casparian strip in roots and some stems contains suberin, lignin, or similar substances. By preventing the passage of water through the cell walls of the endodermis, the Casparian strip forces it to pass through the cytoplasm, thus allowing selective uptake or exclusion of dissolved substances. *See* cutin.

sublittoral 1. (neritic) The marine zone extending from low tide to a depth of about 200 m, usually to the edge of the continental shelf. Light penetrates to the seabed, and the water is well oxygenated. Large algae (e.g. kelps) are found in shallower waters while certain Rhodophyta (red algae) may be found in deeper water. *Compare* littoral; benthic.
2. The zone in a lake or pond between the LITTORAL and PROFUNDAL zones, extending from the edge of the area occupied by rooted plants to a depth of about six to ten meters, where the water temperature declines. Its depth is limited by the *compensation level* – the depth at which the rate of photosynthesis is equalled by the rate of respiration, and below which plants cannot live (but some phytoplankton can).

subsere *See* sere.

subsidiary cell (**accessory cell**) One of a number of specialized epidermal cells of a plant that are found adjacent to the guard cells, and may help in opening and closing the stomata. The number, arrangement and character of subsidiary cells is often of value in taxonomy.

subspecies The taxonomic group below the SPECIES level. Crosses can generally be made between subspecies of a given species but this may be prevented in the wild by various isolating mechanisms, e.g. geographical isolation or different flowering times. Subspecies usually differ from each other in morphology, and occupy distinct habitats or regional locations.

substrate 1. The molecule or molecules upon which an enzyme acts.
2. The nonliving material upon which an organism lives or grows.

succession A progressive series of changes in vegetation and animal life of an area over time from initial colonization to the final stage, or CLIMAX. The climax is a dynamic equilibrium because, although the succession can progress no further under the environmental factors present at the time, the populations present change, e.g. trees die, creating gaps for other species to colonize. In addition, the climate is seldom completely stable – at best it is cyclical, with variations from year to year. *See* sere.

succinic acid (**butanedioic acid**) A dicarboxylic acid formed by fermentation of sugars. It is an intermediate in the KREBS CYCLE, and is formed from succinyl CoA in a reaction linked to the formation of GTP.

succulent A fleshy plant. Succulents, such as cacti, store water in large parenchyma cells in swollen stems and leaves.

sucker An adventitious underground shoot that at some stage emerges above the soil surface and gives rise to a new plant, which initially is nourished by the parent plant until it becomes established.

sucrase An enzyme that breaks down sucrose into glucose and fructose.

sucrose A SUGAR that occurs in many plants. Sucrose is a disaccharide formed from a glucose unit and a fructose unit. Sucrose is the main transport sugar in higher plants. It is formed in the chloroplasts from fructose 6-phosphate and glucose, and transported in the phloem to the rest of the plant. It is hydrolyzed to a mixture of fructose and glucose by the enzyme invertase. Since this mixture has a different optical rotation (levorotatory) from the original sucrose, the mixture is called *invert sugar*. Invert sugar is found in many fruits.

suction pressure (**SP**) *See* osmosis.

Sudan stains Any of various stains based on aniline dyes, and used to color fats and waxes, and hence cutinized (e.g. cuticle) and suberized (e.g. Casparian strip) tissues. *See* staining.

sugar (**saccharide**) One of a class of carbohydrates that are soluble in water. Sugar molecules consist of linked carbon atoms with –OH groups attached, and either an aldehyde or ketone group. The simplest sugars are the *monosaccharides*, such as glucose and fructose, which cannot be hydrolyzed to sugars with fewer carbon atoms. They can exist in a chain form or in a ring formed by reaction of the ketone or aldehyde group with an –OH group on one of the carbons at the other end of the chain. It is possible to have a six-membered (*pyranose*) ring or a five-membered (*furanose*) ring. Monosaccharides are classified according to the number of carbon atoms: a *pentose* has five carbon atoms and a *hexose* six. Monosaccharides with aldehyde groups are *aldoses*; those with ketone groups are *ketoses*. Thus, an *aldohexose* is a hexose that has an aldehyde group; a *ketopentose* is a pentose with a ketone group, etc.

Two or more monosaccharide units can be linked in *disaccharides* (e.g. sucrose), *trisaccharides*, etc. *See also* fructose; glucose; polysaccharide; sucrose.

sugar acid An acid formed from a monosaccharide by oxidation. Oxidation of the aldehyde group (CHO) of the aldose monosaccharides to a carboxyl group (COOH) gives an *aldonic acid*; oxidation of the primary alcohol group (CH_2OH) to COOH yields *uronic acid*; oxidation of both the primary alcohol and carboxyl groups gives an *aldaric acid*. The uronic acids are biologically important, being components of many polysaccharides, for example glucuronic acid (from glucose) is a major component of gums and cell walls, while galacturonic acid (from galactose) makes up pectin. Ascorbic acid or vitamin C is an important aldonic acid found universally in plant tissues, particularly in citrus fruits. Another aldonic acid, gluconic acid, is an intermediate in the pentose phosphate pathway.

sugar alcohol (**alditol**) An alcohol derived from a monosaccharide by reduction of its carbonyl group (CO) so that each carbon atom of the sugar has an alcohol group (OH). For example, glucose yields sorbitol, common in fruits, and mannose yields MANNITOL. GLYCEROL is important in lipid metabolism, and INOSITOL is an intermediate in the synthesis of cell wall polysaccharides.

sugar phosphate A phosphate derivative of a MONOSACCHARIDE.

sulfur An essential element in living tissues, being contained in the amino acids cysteine and methionine and hence in nearly all proteins. Sulfur atoms are also found bound with iron in ferredoxin, one of the components of the electron-transport chain in photosynthesis. It is a component of coenzyme A. Plants take up sulfur from the soil as the sulfate ion SO_4^{2-}. The sulfides released by decay of organic matter are oxidized to sulfur by SULFUR BACTERIA of the genera *Chromatium* and *Chlorobium*, and further oxidized to sulfates by bacteria of the genus *Thiobacillus*. There is thus a cycling of sulfur in nature.

sulfur bacteria Filamentous autotrophic chemosynthetic bacteria that derive energy by oxidizing sulfides to elemental sulfur and build up carbohydrates from carbon dioxide. They use sulfides instead of water as a source of electrons in photosynthesis, releasing sulfur instead of oxygen. An example is *Beggiatoa*. They are found mainly in sulfur-rich muds and springs, including hydrothermal vents. There are two main groups: the green sulfur bacteria (Chlorobia) and the purple sulfur bacteria (certain members of the Proteobacteria). A few Archaea, e.g. *Sulfolobus*, can oxidize elemental sulfur. As well as sulfides, some bacteria oxidize thiosulfates, polythionates, and sulfites. Sulfur bacteria play an important role in the cycling of sulfur in the ecosystem.

supergene A collection of closely linked genes that determine a particular trait or series of inter-related traits and behave as a single unit because crossing over between them is very rare. An example is the S/s gene, which determines pin (s) and thrum

(S) style length in certain species of *Primula*.

superior In plant science, the term is used with reference to the position of the ovary in relation to the other parts of the flower. When the ovary is superior, the petals, sepals, and stamens are inserted at the base of the ovary where it joins the flowerstalk, as in buttercup (*Ranunculus*) flowers. In a floral formula a superior ovary is denoted by a line below the carpel number. *Compare* inferior. *See also* hypogyny; perigyny.

suspension culture A method of growing free-living single cells or small clumps of cells in a liquid medium; the liquid medium is agitated to keep the cells in suspension. Microorganisms or cells of plant callus tissue may be grown in this way.

suspensor A temporary stalklike row of cells, found in angiosperms, that differentiates from the proembryo by mitosis and pushes the embryo into the nutritive endosperm after fertilization. In certain Lycophyta (e.g. *Selaginella*), a few ferns (e.g. *Botrychium*), and in gymnosperms it pushes the embryo into the female gametophyte tissue. It also serves as a passage for nutrients to pass from the parent tissue to the embryo.

swamp An area of vegetation dominated by trees that develops on ground that is normally waterlogged or covered by water all year round, such as the margin of a lake, a river floodplain, or an area of water-retentive clay in an arid region. Swamp vegetation represents the early stages of a HYDROSERE or HALOSERE. The presence of trees distinguishes it from a *marsh*. Coastal and estuarine swamps in warm climates are dominated by MANGROVES. The swamp vegetation slows the flow of water, and dead plant matter builds up.

syconium (**syconus**) A composite inflorescence or fruit in which the small individual flowers are borne on the inside of a hollow receptacle, as in figs (*Ficus*). These develop into achenes (the 'pips' of the fruit). *See* pseudocarp.

symbiosis (*pl.* **symbioses**) Any close association between two or more different organisms, as seen in parasitism, mutualism, and commensalism. Often one or both organisms is dependent on the other. The term is usually used more narrowly to mean MUTUALISM, but many mutualistic associations may have once been parasitic or may become so at some stage in their life cycle, e.g. plants and their mycorrhizal fungi.

sympatric species *See* species.

symplast The living system of interconnected protoplasts extending through a plant body. Cytoplasmic connections between cells are made possible by the plasmodesmata. The symplast pathway is an important transport route through the plant. *Compare* apoplast.

symplastic growth A form of plant growth in which neighboring cell walls stay in contact and grow at the same rate.

sympodial Describing the system of branching in plants in which the terminal bud of the main stem axis stops growing and growth is taken over by lateral buds. These in turn lose their dominance and lateral buds take over their role. *Compare* monopodial.

symport A process involving a membrane protein that transports two kinds of substance (substrates) across a membrane, one of which is usually an ion. *Compare* antiport; proton pump.

synapsis (**pairing**) The association of homologous chromosomes during the prophase stage of meiosis that leads to the production of a haploid number of bivalents.

syncarpous *See* carpel.

syncarpy The condition in which an ovary is made up of fused carpels, as in the primrose (*Primula vulgaris*). *Compare* apocarpy.

synchronous culture A culture of cells in which all the individuals are at approximately the same point in the cell cycle. Cells can be synchronized by a variety of means, e.g. temperature, shock, or drugs. Such cultures are of great value in physiological and biochemical investigations.

synecology The study of all the interactions between the living organisms in a natural community and the effects of the nonliving components of the environment upon them and on their relationships with each other. *Compare* autecology.

synergid cells Two haploid cells located near the egg cell at the micropylar end of the embryo sac in flowering plants. They may represent what remains of the archegonium in angiosperms and the pollen tube may penetrate a synergid first, but they do not participate in the fertilization process and abort soon afterwards. Together with the egg nucleus they constitute the egg apparatus.

syngamy *See* fertilization.

systematics The area of biology that deals with the diversity of living organisms, their relationships to each other, and their classification. The term may be used synonymously with *taxonomy*.

T

tactic movement *See* taxis.

taiga *See* forest.

tandem DNA *See* repetitive DNA.

tangential Orientated at a tangent to a circle, the circle being, for example, the surface of a plant stem.

tannin One of a mixed group of substances that, as defined by industry, precipitate the gelatin of animal hides to form leather. Tannins are also used in dyeing and ink manufacture. They are polymers derived either from carbohydrates and phenolic acids by condensation reactions, or from flavonoids. Many plants accumulate tannins, particularly in leaves, fruits, seed coats, bark, and heartwood. Their astringent taste may deter animals from eating the plant and they may discourage infection. Species that produce large amounts of tannin include tea (*Camellia sinensis*) and the bark of oak (*Quercus*).

tapetum (*pl.* **tapeta**) A food-rich layer surrounding the spore mother cells in the anthers of vascular plants. These cells usually disintegrate, liberating food substances that are subsequently absorbed by the spore mother cells and the developing spores.

taproot A persistent primary root that grows vertically downward. Swollen taproots used for food storage are common in many biennial plants, such as carrot (*Daucus carota*).

Taxaceae A family of evergreen coniferous trees and shrubs, the yews, with needlelike or linear leaves.

taxis (*pl.* **taxes**) (**tactic movement**) Movement of an entire cell or organism (i.e. locomotion) in response to an external stimulus, in which the direction of movement is dictated by the direction of the stimulus. Movement toward the stimulus is positive taxis and away from the stimulus is negative taxis. It is achieved by protoplasmic streaming, extrusion of cell substances, or by locomotory appendages, such as flagella and undulipodia. *See* aerotaxis; chemotaxis; phototaxis. *Compare* nastic movements; tropism.

taxol A diterpene originally isolated from the bark of the Pacific yew (*Taxus brevifolia*), that has proved an effective treatment for solid tumors, especially breast cancer. It blocks tumor cells by stabilizing and polymerizing microtubules.

taxon (*pl.* **taxa**) A group of any rank in taxonomy. Ranunculaceae (a family) and *Triticum* (a genus) are examples.

taxonomy The area of systematics that covers the principles and procedures of classification, specifically the classification of variation in living organisms. *See* classification; systematics.

TCA cycle *See* Krebs cycle.

telophase The final stage in mitosis and meiosis, in which the separated chromatids (mitosis) or homologous chromosomes of bivalents (meiosis) collect at the poles of the spindle and the nuclei of the daughter cells are formed. During this stage in mitosis and the second division of meiosis the chromosomes uncoil and disperse, the nuclear spindle degenerates, nucleoli reappear, and a new nuclear membrane forms.

The cytoplasm may also divide during this phase.

temperate phage A DNA bacteriophage that becomes integrated into the bacterial DNA and multiplies with it, rather than replicating independently and causing lysis of the bacterium. A bacterium thus infected is said to be lysogenic. *See* lysogeny. *Compare* virulent phage.

temporary stain *See* staining.

tendril A slender structure found in plants, used for twining and support. It may be a modified terminal bud, as in the grapevine (*Vitis*); a modified lateral branch, as in the passion flower (*Passiflora*); a modified leaf, as in *Clematis montana*; a modified leaflet, as seen in many of the pea family (Fabaceae); or a modified inflorescence, as in Virginia creeper (*Parthenocissus*). Tendrils may be branched, or unbranched, and may have terminal adhesive disks, as in the Virginia creeper.

terminal bud *See* bud.

terminalization The movement of chiasmata to the end of the bivalent arms, a process that may occur during late prophase I of meiosis. The chiasmata can slip off the ends of the bivalents, and thus chiasma frequency may be reduced by terminalization.

terpene One of a complex group of lipids based on the hydrocarbon skeleton C_5H_8 (isoprene). *Monoterpenes* are built from two C_5 residues ($C_{10}H_{16}$), *diterpenes* from four, etc. The C_{10} to C_{20} terpenes are present in essential oils, giving the characteristic scent of some plants (e.g. mint, lemon). Some terpenoid substances are physiologically active, e.g. vitamins A, E, and K.

Tertiary The larger and older period of the Cenozoic, being composed of the Paleocene, Eocene, Oligocene, Miocene, and Pliocene epochs (65–2 million years ago). Literally the 'third age', it is characterized by the rapid evolution and expansion of angiosperms and the emergence of mammals. The ferns underwent adaptive radiation, and the angiosperms underwent an even greater explosion of adaptive radiation, in which flowers became more specialized. This went hand in hand with the evolution of specialist insect pollinators, especially bees, butterflies, and moths. The climate was warm and wet at the start of the period, but became colder and drier later, and this was accompanied by the expansion in range of grasses. Many modern genera of conifers arose in the Tertiary, but other gymnosperms declined. *See also* geological time scale.

tertiary structure The mode of folding of a polypeptide (α-helix or β-pleated sheet) in a PROTEIN. The shape may be stabilized by HYDROGEN BONDS, disulfide bridges, and hydrophobic interactions, whereby hydrophobic parts of the polypeptide are protected from the surrounding aqueous solution by hydrophilic parts. Such folding determines the shape of the active site of enzymes. *See* conformation; primary structure; quaternary structure; secondary structure.

testa The hard dry protective covering of a seed, formed from the integuments of the ovule. After fertilization the layers of the integuments fuse and become thickened and pigmented.

test cross *See* backcross.

tetrad **1.** A group of four cells formed as a result of meiosis in a spore mother cell. The four cells may develop into spores, as in the formation of pollen grains.
2. In meiosis, the association of four homologous chromatids seen during the pachytene stage of prophase.

tetraploid A cell or organism containing four times the haploid number of chromosomes. Tetraploid organisms may arise by the fusion of two diploid gametes that have resulted from the nondisjunction of chromosomes at meiosis, or by nondisjunction of the chromatids during the mitotic division of a zygote. Many stable tetraploids arise initially from hybridization. *See also* polyploid; allotetraploid; autotetraploid.

thallus (*pl.* **thalli**) A simple plant body showing no differentiation into root, leaf, and stem and lacking a true vascular system. It may be uni- or multicellular, and is found in the algae, lichens, bryophytes, and the gametophyte generation of the Filicinophyta, Lycophyta, and Sphenophyta.

Theophrastus of Eresus (372–287 BC) Ancient Greek botanist and philosopher. Theophrastus was a pupil of Plato and assisted Aristotle at the Lyceum in Athens after Plato's death. He became head of the school when Aristotle died. He is often considered to be the father of scientific botany because of his work on classifying and naming plants, which he recorded in his prolific writings. His pupils were encouraged to observe plants growing near their homes, which were often far away from Athens, and possibly helped him to make the connection between plants and the environment, especially their adaptations to soil and climate. His *Enquiry into Plants* (nine volumes) is important even today. It is evident from his writings that he had distinguished between monocotyledons and dicotyledons and had discovered that seeds and fruits were derived from flowers.

thermocline *See* stratification.

thermonasty (**thermonastic movements**) A NASTIC MOVEMENT in response to change in temperature. For example, *Crocus* flowers open rapidly if the temperature increases by 5 to 10°C, as the inner side of the petals grows faster than the outer side.

thermoperiodism The phenomenon shown by certain plants, e.g. *Chrysanthemum* and tomato (*Lycopersicon esculentum*) in which there is a response to daily alternations of low and high temperature. Such plants flower earlier and more profusely if subjected to low night and high day temperatures. *See also* photoperiodism; vernalization.

thermophilic Describing microorganisms that require high temperatures (around 60°C) for growth. It is exhibited by certain bacteria that grow in hot springs or compost and manure. *Compare* mesophilic; psychrophilic.

therophyte A plant that survives part of the year as a seed and completes its life cycle during the remainder of the year. *See also* Raunkiaer's plant classification.

thiamine (**vitamin B_1**) One of the water-soluble B-group of VITAMINS. It is a pyrimidine compound that is synthesized by plants but not by some microorganisms and most vertebrates. It is therefore required in the diet of mammals. Good sources of thiamine include unrefined cereal grains.

thigmotropism (**haptotropism**) A tropism in which the stimulus is touch. The tendrils of climbing plants are thigmotropic. The part touching the substrate bends toward the stimulated side, so bringing more of the tendril into contact with the substrate, until it encircles it. *See* tropism.

thin-layer chromatography A chromatographic method in which a glass plate is covered with a thin layer of inert absorbent material (e.g. cellulose or silica gel) and the materials to be analyzed are spotted near the lower edge of the plate. The base of the plate is then placed in a solvent, which rises up the plate by capillary action, separating the constituents of the mixtures. The principles involved are similar to those of paper chromatography and, like paper chromatography, two-dimensional methods can also be employed. *See* paper chromatography.

thorn A stiff sharply pointed woody process that may be found on the stems of vascular plants. It is a modified branch and is supplied with vascular tissue, e.g. hawthorn (*Crataegus monogyna*). *Compare* prickle; spine.

thorn forest *See* forest.

threonine An AMINO ACID derived from aspartic acid. It is broken down to form glycine and acetyl CoA. Isoleucine can be synthesized from threonine.

thylakoid An elongated flattened fluid-filled sac that forms the basic unit of the photosynthetic membrane system in chloroplasts and photosynthetic bacteria. *See* chloroplast.

thymidine The nucleoside formed when thymine is linked to D-ribose by a β-glycosidic bond. *See* nucleoside.

thymine A nitrogenous base found in DNA, but not in RNA. It has a PYRIMIDINE ring structure. It base pairs with adenine in the complementary strand of DNA. *See* base pairing; thymidine.

tissue In a multicellular organism, a group of cells that is specialized for a particular function, e.g. palisade mesophyll. Several different tissues may be incorporated into an organ, e.g. a leaf. *Compare* organ.

tissue culture The growth of cells, tissues, or organs in suitable media *in vitro*. Such media must normally be sterile, correctly pH balanced, and contain all the necessary micro and macronutrients and hormones for growth. Studies of such cultures have shed light on physiological processes that would be difficult to follow in the living organism. For example, the cytokinins were discovered through work on tobacco pith tissue culture.

tocopherol (**vitamin E**) A terpenoid-like substance found especially in certain seeds, such as cereals, where it probably prevents oxidation of lipids, thus prolonging seed viability.

tonoplast The membrane that surrounds the large central vacuole of plant cells.

torus (*pl.* **tori**) A disk-shaped structure formed from lignified primary cell-wall material on the middle lamella of a bordered pit. The structure is found mainly in the conifers. It is thought to act like a valve, sealing the pit when the pressure on the two sides is unequal, e.g. when an adjacent tracheid is damaged.

totipotency The ability shown by many living cells to form all the types of tissues that constitute the mature organism. This may be achieved, even if the cells have completely differentiated, provided that the appropriate balance of nutrients and hormones is given. The best example of this phenomenon is the formation of adventitious embryos in carrot tissue cultures.

toxin A chemical produced by a pathogen (e.g. bacteria, fungi) that causes damage to a host cell in very low concentrations. The toxin may be formed within the pathogen and released when it dies (*endotoxin*), or secreted through its cell wall (*exotoxin*). Toxins are often similar to the enzymes of the host and interfere with the appropriate enzyme systems.

trabecula (*pl.* **trabeculae**) An elongated cell wall, cell, or line of cells across a cavity. In plants, trabeculae may be found in *Selaginella* stems, where they suspend the steles in large air spaces.

trace element *See* micronutrient.

tracheid An elongated xylem conducting element with oblique end walls. Tracheids have heavily lignified walls and the only connection between adjacent tracheids is through paired pits, which are mainly concentrated in the end walls. Many tracheids lack protoplasm at maturity and have lignified secondary cell walls. Tracheids form the only xylem conducting tissue of vascular plants other than angiosperms, some Gnetales, and a few ferns such as bracken. *Compare* vessel.

tracheophyte (**Tracheophyta**) Any plant with a differentiated vascular system; i.e.

all plants except the liverworts, mosses, and hornworts.

trama The inner tissue of the gills in basidiomycete fungi that is made up of loosely packed hyphae.

transamination The transfer of an amino group from an AMINO ACID to an α-keto acid, producing a new α-keto acid and a new amino acid. This is catalyzed by a transaminase enzyme in conjunction with the coenzyme pyridoxal phosphate. The amino group becomes attached to the coenzyme to form pyridoxamine phosphate, and is then transferred to the α-keto acid, which is usually pyruvic acid, oxaloacetic acid, or α-ketoglutaric acid. Transamination is a key step in the biosynthesis and breakdown of most amino acids.

transcription The process in living cells whereby RNA is synthesized according to the template embodied in the base sequence of DNA, thereby converting the cell's genetic information into a coded message (messenger RNA, mRNA) for the assembly of proteins, or into the RNA components required for protein synthesis (ribosomal RNA and transfer RNA). The term is also applied to the formation of single-stranded DNA from an RNA template, as performed by the enzyme reverse transcriptase, for example in retrovirus infections. Details of DNA transcription differ between prokaryote and eukaryote cells, but essentially it involves the following steps. Firstly, with the aid of a helicase enzyme, the double helix of the DNA molecule is unwound in the region of the site marking the start of transcription for a particular gene. The enzyme RNA polymerase moves along one of the DNA strands, the transcribed strand (or anticoding strand, since the code is carried by the complementary base sequence of the RNA), and nucleotides are assembled to form a complementary RNA molecule. The polymerase enzyme proceeds until reaching a stop signal, when formation of the RNA strand is terminated. Behind the enzyme, the DNA double helix re-forms, stripping off the newly synthesized RNA strand. In eukaryotes, transcription is initiated and regulated by a host of proteins called *transcription factors*; in prokaryotes an accessory sigma factor is essential for transcription. *See also* translation.

transduction The transfer of part of the DNA of one bacterium to another by a temperate bacteriophage. The process does occur naturally but is mainly known as a technique in RECOMBINANT DNA TECHNOLOGY, and has been used in mapping the bacterial chromosome.

transect A line or belt designed to study changes in species composition across a particular area. The transect sampling technique is most often used in plant ecology to study changes in the composition of vegetation. *See also* quadrat.

transferase An enzyme that catalyzes reactions in which entire groups or radicals are transferred from one molecule to another. For example, hexokinase catalyzes the transfer of a high-energy terminal phosphate group from ATP to glucose to give glucose 6-phosphate and ADP.

transfer cell A specialized type of plant cell in which the cell wall forms protuberances into the cell, thus increasing the surface area of the wall and plasma membrane. Transfer cells are active cells containing many mitochondria, and are concerned with short-distance transport of solutes. They are common in many situations, for example as gland cells and epidermal cells, and in xylem and phloem parenchyma, where they are concerned with active loading and unloading of vessels and sieve tubes.

transfer RNA (tRNA) A type of RNA that participates in protein synthesis in living cells. It attaches to a particular amino acid and imports this to the site of polypeptide assembly at the ribosome when the appropriate codon on the messenger RNA is reached. Each tRNA molecule consists of roughly 80 nucleotides; some regions of the molecule undergo base pairing and form a double helix, while in others the two strands separate to form loops. When flattened out the tRNA molecule has a characteristic 'cloverleaf' shape with three loops; the base of the 'leaf' carries the amino acid binding site, and the middle loop contains the ANTICODON, whose base triplet pairs with the complementary codon in the mRNA molecule. Hence, because there are 64 possible codons in the genetic code, of which about 60 or so code for amino acids, there may be up to 60 or so different tRNAs in a cell, each with a different anticodon, although some of them will bind the same amino acid. The correct amino acid is attached to a tRNA molecule by an enzyme called an *amino-acyl-tRNA transferase*. There are 20 of these, one for each type of amino acid. This attachment also involves the transfer of a high-energy bond from ATP to the amino acid, which provides the energy for peptide bond formation during TRANSLATION.

transformation A permanent genetic recombination in a bacterial cell, in which a DNA fragment is incorporated into the DNA of the cell. This may be demonstrated by growing bacteria in the presence of dead cells, culture filtrates, or extracts of related strains. The bacteria acquire genetic characters of these strains.

transgenic Describing organisms, especially eukaryotes, containing foreign genetic material. Genetic engineering has created a wide range of transgenic animals, plants, and other organisms for both experimental and commercial purposes. Examples include herbicide-resistant crop plants and plants that secrete pharmaceuticals. *See* recombinant DNA.

transition zone The zone in a vascular plant where the root and shoot structures merge and where arrangement of the vascular tissue is intermediate between that of the root and shoot. *See* hypocotyl.

translation The process whereby the genetic code of messenger RNA (mRNA) is deciphered by the machinery of a cell to make proteins. Molecules of mRNA are in effect coded messages based on information in the cell's genes and created by the process of TRANSCRIPTION. They relay this information to the sites of protein synthesis, the RIBOSOMES. In eukaryotes these are located in the cytoplasm, so the mRNA must migrate from the nucleus. The first stage in translation is *initiation*, in which the two subunits of the ribosome assemble and attach to the mRNA molecule near the initiation codon, which signals the beginning of the message. This also involves various proteins called *initiation factors*, and the initiator TRANSFER RNA (tRNA), which always carries the amino acid N-formyl methionine. The next stage is *elongation*, in which the peptide chain is built up from its component amino acids. This involves the participation of several proteins called *elongation factors*, and tRNA molecules that successively occupy two sites on the larger ribosome subunit in a sequence determined by consecutive codons on the mRNA. As each pair of tRNAs occupies the ribosomal sites, their amino acids are joined together by a peptide bond. As the ribosome moves along the mRNA to the next codon, the next tRNA enters the first ribosomal site, and so on, leading to elongation of the peptide chain. Having delivered its amino acid, the depleted tRNA is released from the second ribosomal site, which is then occupied by the tRNA with the growing chain. This process continues until the ribosome encounters a termination codon. A protein called a *release factor* binds directly to the stop codon, causing the addition of a water molecule instead of an amino acid to the polypeptide chain. The polypeptide chain is then released and the ribosome complex dissociates, marking the termination of translation. Following its release, the polypeptide may undergo various changes, such as the removal or addition of chemical groups, or even cleavage into two parts. This post-translational modification produces the fully functional protein. Folding of the protein is assisted by a class of molecules called *chaperones*.

translocation 1. The movement of mineral nutrients, elaborated food materials, and hormones through the plant. In vascu-

lar plants, the xylem and phloem serve to translocate such substances.
2. *See* chromosome mutation.

transpiration The loss of water vapor from the surface of a plant. Most is lost through stomata when they are open for gaseous exchange. It has been shown that transpiration through the stomata amounts to 90% of the transpiration that occurs from a water surface of the same area as a the leaf. Typically, about 5% is lost directly from epidermal cells through the cuticle (*cuticular transpiration*) and a minute proportion through lenticels. A continuous flow of water, the *transpiration stream*, is thus maintained through the plant from the soil via root hairs, root cortex, xylem, and tissues such as leaf mesophyll served by xylem. Evaporation of water from the leaf surface sets up a water potential gradient that extends to the roots, helping to draw up water; water flow is enhanced by the narrow diameter of the xylem vessels, which favor capillary action. ROOT PRESSURE may be a contributing factor. Water evaporates from wet cell walls into intercellular spaces and diffuses out through stomata. Transpiration may be useful in maintaining a flow of solutes through the plant and in helping to cool leaves through evaporation, but is often detrimental under conditions of water shortage, when wilting may occur. It is favored by low humidity, high temperatures, and moving air. A major control is the degree of opening of the stomata. *Compare* guttation. *See* antitranspirant; cohesion theory.

transposon (**transposable genetic element; jumping gene**) A segment of an organism's DNA that can insert at various sites in the genome, either by physically moving from place to place, or by producing a copy that inserts elsewhere. The simplest types are called insertion sequences; these comprise about 700–1500 base pairs. More complex ones, called composite transposons, have a central portion, which may contain functional genes, flanked by insertion sequences. Transposons can affect both the genotype and phenotype of the organism, e.g. by disrupting gene expression, or causing deletions or inversions. They are common in bacteria, where they usually contain only one or two genes. In eukaryotes, transposons account for much of the repetitive DNA in the genome. *Compare* plasmid. *See* repetitive DNA.

tree ferns *See* Filicinophyta.

triacylglyceride (**triglyceride**) An ester of glycerol in which all the –OH groups are esterified; the acyl groups may be the same or different. Plant triacylglycerols are usually unsaturated. They are synthesized from glycerol phosphate and fatty acyl CoAs, usually in the microsomes. Lipases catalyze the breakdown of triacylglycerols to glycerol and free fatty acids. Triacylglycerols are important food reserves, especially in seeds such as rape (*Brassica napus*), linseed (*Linum usitatissimum*), castor bean (*Ricinus communis*) and coconut (*Cocos nucifera*). They are commercial sources of fats (solid triacylglycerols) and oils (liquid triacylglycerols). *See also* carboxylic acid; glyceride; lipid; microsome.

Triassic The oldest period of the Mesozoic era, 250–215 million years ago. During the Triassic, the climate gradually changed from arid to more temperate. Conditions for preservation in the drier period were poor, so there are few fossils. There was a diversification of the gymnosperms, and fossil groups such as the Bennettitales and Caytoniales evolved during this period. *See also* geological time scale.

tribe A taxonomic rank consisting of a group of closely related genera. A number of tribes comprise a subfamily. The tribe is introduced only in classifications of very large families such as the grasses (Poaceae).

tricarboxylic acid cycle *See* Krebs cycle.

trichocyst *See* Dinomastigota.

trichome *See* hair.

triglyceride *See* triacylglyceride.

triplet A sequence of three nucleotide bases on a DNA or messenger RNA molecule. *See* genetic code.

triploid A cell, tissue or organism containing three times the haploid number of chromosomes. In flowering plants the endosperm tissue is usually triploid, resulting from the fusion of one of the pollen nuclei with the two polar nuclei. Triploids are usually sterile.

trisomy *See* aneuploidy.

trophic level In complex natural communities, organisms whose food is obtained from plants by the same number of steps are said to belong to the same trophic or energy level. The first and lowest trophic level contains the producers, green plants that convert solar energy to food by photosynthesis. Herbivores occupy the second trophic level and are primary consumers: they eat the members of the first trophic level. At the third level carnivores eat the herbivores (the secondary consumer level), and at the fourth level secondary carnivores eat the primary carnivores (the tertiary consumer level). These are general categories, as many organisms feed on several trophic levels, for example omnivores eat both plants and animals. Decomposers or transformers occupy a separate trophic level, which consists of organisms such as fungi and bacteria, which break down dead organic matter into nutrients usable by the producers. *See* food chain.

tropism (**tropic movement**) A directional growth movement of part of a plant in response to an external stimulus. Tropisms are named according to the stimulus. The organ is said to exhibit a positive or negative tropic response, depending on whether it grows toward or away from the stimulus respectively, e.g. shoots are positively phototropic but negatively gravitropic. Growth straight toward or away from the stimulus (0° and 180° orientation respectively) is called *orthotropism*. Primary roots and shoots are orthotropic to light and gravity. By contrast, growth at any other angle to the direction of the stimulus as by branches or lateral roots is called *plagiotropism*. The mechanism involved in the latter is poorly understood. In *diatropism* the plant part is orientated at right angles to the direction of the stimulus. For example, rhizomes grow horizontally in response to gravity (diagravitropism), and leaves often grow at right angles to the incident light. Since the receptor for the stimulus is often separate from the region of growth, tropic movements are often mediated by hormones. *See* phototropism; gravitropism. *Compare* nastic movements; taxis.

tryptophan *See* amino acids.

tuber A swollen underground stem or root that contains stored food and acts as an organ of perennation and vegetative propagation. Stem tubers, e.g. potato (*Solanum tuberosum*) develop at the end of underground stems by swelling of nodes and internodes. There is an increase in pith tissue to form a round tuber that bears buds in the axils of greatly reduced scale leaves. The stem connecting the tuber to the parent plant then severs. Root tubers may develop in the same way from adventitious roots, as in *Dahlia*. A root tuber can be distinguished for a stem tuber by the absence of buds or 'eyes'. *See* perennating organ.

tubulin *See* microtubule.

tundra A region or type of vegetation at high latitudes and high altitude. Tundra is widespread in the Arctic north of the tree lines of North America and Eurasia, but occurs in only scattered localities on Antarctic islands, where it consists mainly of lichens and mosses.

tunica–corpus theory A theory of apical organization and development that distinguishes two separate tissue zones, the *tunica* and the *corpus*, in the apex of a flowering plant. The tunica is made up of one or more peripheral layers in which cell division is mostly anticlinal. The corpus is the inner area of tissue in which cell arrangement and division is irregular. The epidermis originates in the tunica region and the other stem tissues arise from either the tunica or the corpus, depending on the species. *Compare* histogen theory.

turgor The state, in a plant or prokaryote cell, in which the protoplast is exerting a pressure on the cell wall owing to the intake of water by osmosis. Being slightly elastic, the cell wall bulges, but is rigid enough to prevent water entering to the point of bursting. The cell is then said to be turgid. Turgidity is the main means of support of herbaceous plants and other unlignified tissues. *See* osmosis; plasmolysis; turgor pressure.

turgor pressure (**pressure potential**) The hydrostatic pressure exerted by the contents of a cell against the cell wall. It results from the uptake of water into the cell by osmosis. It may be alternatively expressed as pressure potential – the hydrostatic pressure to which water in a liquid phase is subjected.

turion **1.** A swollen detached winter bud that contains stored food and is protected by an outer layer of leaf scales and mucilage. It is an organ of perennation or vegetative propagation, and is characteristic of various water plants (e.g. *Sagittaria*). Turions are usually released from the plant and lie dormant on the bed of the pond or river until conditions become favorable for growth.
2. Any vegetative shoot or sucker.

tylose A bladderlike ingrowth from a parenchyma cell into an adjacent tracheid or vessel through a paired pit. Tyloses are often found in injured tissue, older wood, and below an abscission layer, and can completely block the conducting vessel. They often become filled with tannins, resins or pigments, giving the heartwood its dark color, and helping to preserve and strengthen it. Some of these pigments are used commercially as dyes, e.g. hematoxylin.

type **1.** The material used to define a species. It is usually a dried specimen stored in a herbarium but may also be a drawing. Type specimens of microorganisms are often living cultures.
2. The representative species of a genus, the representative genus of a family, etc. For example the genus *Solanum* is the type genus of the family Solanaceae.

tyrosine An aromatic amino acid derived from erythrose 4-phosphate and phosphoenolpyruvate. It is broken down via phenylalanine to give acetyl CoA and fumaric acid. Tyrosine is a precursor of certain alkaloids, and of various phenolic inhibitors such as coumaric acid. *See* amino acids.

U

ubiquinone *See* coenzyme Q.

ultracentrifuge A high-speed centrifuge, operating at up to a million revolutions per second that is used to sediment protein and nucleic acid molecules. Ultracentrifuges operate under refrigeration in a vacuum chamber and forces 50 million times that of gravity may be reached. The rate of sedimentation depends on the molecular weight of the molecule and thus the ultracentrifuge can be used to separate a mixture of large molecules, such as large colloidal proteins, and estimate particle sizes.

ultramicrotome *See* microtome.

ultrastructure (**fine structure**) The detailed structure of biological material as it can be revealed, for example, by electron microscopy, but not by light microscopy.

umbel *See* inflorescence.

undulipodium (*pl.* **undulipodia**) A whiplike organelle that protrudes from a eukaryotic cell and is used chiefly for locomotion (e.g. motile gametes) or feeding (e.g. ciliate protoctists). Undulipodia include all eukaryotic cilia and flagella, which share the same essential structure, and differ markedly from bacterial flagella (*see* flagellum). They are complex structures, containing over 500 different proteins. The shaft comprises a cylindrical array of nine doublet microtubules surrounding a central core of two single microtubules. The outer wall of the shaft is an extension of the cell membrane. Its structure is similar to that of a centriole. Strictly, the term flagella should be used only in relation to bacteria. Long undulipodia (over 10 μm), formerly called flagella, occur singly or in pairs and produce successive waves of bending that are propagated to the tip of the shaft, e.g. *Chlamydomonas*.

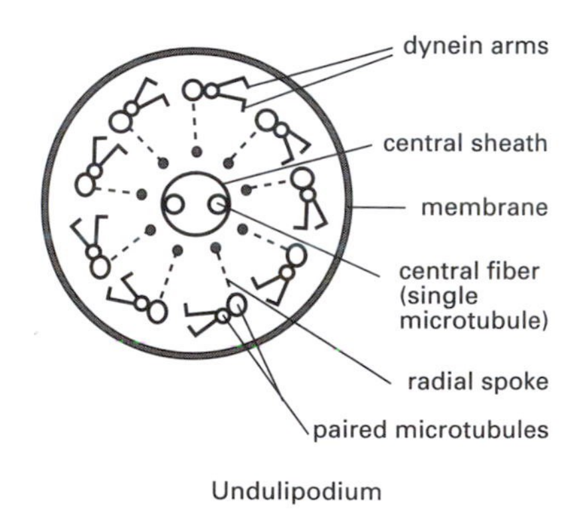

Undulipodium

unicellular Describing organisms that exist as a single cell. Such a state is characteristic of most protoctists and bacteria and is also found in many fungi. *Compare* multicellular; acellular.

unisexual Describing organisms that have either male or female sex organs, but not both. Unisexual plants are said to be dioecious. *Compare* hermaphrodite; monoecious.

universal indicator *See* indicator.

uracil A nitrogenous base that is found in RNA but not in DNA. During TRANSCRIPTION, the thymine of the DNA template is substituted by a uracil in the RNA copy. Uracil has a PYRIMIDINE ring structure, and is derived from sugars and amino acids.

uridine The NUCLEOSIDE formed when uracil is linked to D-ribose by a β-glycosidic bond.

V

vacuole A fluid-filled organelle of variable size found in the cytoplasm and separated from it by a single membrane, the TONOPLAST. Many mature plant cells have a single large central vacuole that confines the cytoplasm to a thin peripheral layer. Newly formed plant cells often have a number of small vacuoles derived from the endoplasmic reticulum. As they fill with sap, they enlarge and coalesce to form a single vacuole. The vacuole is filled with cell sap, which contains substances in solution, e.g. sugars, salts, and organic acids, often in high concentrations resulting in a high osmotic pressure. Water therefore moves into the vacuole by osmosis making the cell turgid. Vacuoles may also contain storage products such as starch grains, oil droplets, crystals, and waste substances that are byproducts of the cell's metabolism.

valine An AMINO ACID synthesized from pyruvate. It is broken down to yield succinyl CoA.

variation The extent to which the characteristics of the individuals of a species can vary. Variation can be caused by environmental and genetic factors. Environmental variation (phenotypic plasticity) results in differences in the appearance of individuals of a species because of differences in nutrition, disease, light intensity, etc. Genetic variation is caused by recombination and occasionally by MUTATION. These differences may be favored or discriminated against by NATURAL SELECTION. *See* discontinuous variation; continuous variation; recombination.

variegation The occurrence of patches of different colors on leaves or flowers. It may be caused by infection with a pathogen, mineral deficiency, or genetic or physiological differences between patches of cells. It occurs naturally in some species, such as the dumb canes (*Dieffenbachia*).

variety The taxonomic group below the subspecies level. The term is often loosely used to describe breeds of livestock or various cultivated forms of agricultural and horticultural species. Varieties are morphological variants, which may differ in color or growth habit. *See also* cultivar.

vascular bundle A strand of conducting tissue found in vascular plants. In plants showing secondary thickening (e.g. many dicotyledons) the vascular bundles join together to form a ring and contain meristematic cambium tissue between the xylem and phloem, but this is generally lacking in plants without secondary thickening (e.g. most monocotyledons). *See also* vascular plants.

vascular cambium The ring of CAMBIUM formed by the activity of the interfascicular cambium and the intrafascicular cambium that gives rise to secondary xylem and secondary phloem. It is a lateral meristem and contains *fusiform initials* giving rise to secondary xylem and phloem and *ray initials* giving rise to *medullary rays*. *See also* secondary growth.

vascular plants Plants containing differentiated cells forming conducting tissue (*vascular tissue*, also known as *fascicular tissue*), which comprises the xylem and phloem. Vascular tissue transports water and nutrients through the plant and also provides strength and support. It consists principally of XYLEM and PHLOEM (water- and food-conducting tissues respectively), and also contains strengthening tissue

(SCLERENCHYMA) and packing tissue (PARENCHYMA). The arrangement of vascular tissue (the *vascular system*) in the stem is very varied, giving a number of different types of STELE. Primary vascular tissue, which is found in all vascular plants, is formed from the procambium. Secondary vascular tissue, found only in plants with secondary thickening, develops from the vascular cambium. The vascular cambium extends to form a complete ring of meristematic tissue around the stem, the separate vascular bundles being linked by interfascicular cambium. Vascular plants are able to achieve considerable vertical growth both upwards and into the soil. They have thus been able to colonize the drier habitats that are inaccessible to the nonvascular bryophytes. The vascular plants may be grouped into a single division, the *Tracheophyta*.

vascular system (**vascular tissue**) *See* vascular plants.

vascular tissue *See* vascular plants.

vector 1. An agent that carries a disease-causing organism to a healthy plant or animal, causing the latter to become infected;. In its widest sense, the term includes agents such as wind, rain, etc., but it is usually restricted to vectors that are animals or microorganisms. Insects such as aphids carry viruses from plant to plant as they feed, and mites and nematodes can also carry disease. Humans, including gardeners and scientists, may carry fungal and bacterial spores on their clothing or skin.
2. (**cloning vector**) An agent used as a vehicle for introducing foreign DNA, for example a new gene, into host cells. Several types of vector are used in GENE CLONING, notably bacterial plasmids and bacteriophages. The segment of DNA is first spliced into the DNA of the vector, such as a plasmid, then the vector is transferred to the host cell (e.g. the bacterium, *E. coli*), where it replicates along with the host cell. The result is a clone of cells, all of which contain the foreign gene, which is expressed by the host cell machinery. A common vector used to transfer cloned DNA to plant cells is the bacterium *Agrobacterium tumefasciens*. The foreign DNA is spliced into a plasmid, which is taken up by the bacterium. The bacterium then infects a plant cell, and the plasmid DNA may become integrated into the plant chromosome.

vegetative cell *See* vegetative nucleus.

vegetative growth (**vegetative propagation; vegetative reproduction**) Growth of any parts of the plant other than flowers. This involves both upwards growth and outwards. Sometimes outward growth involves STOLONS or RHIZOMES or the budding of BULBS or CORMS. These parts may get detached from the parent and become independent, although they are identical genetically. Some plants have specialized organs such as GEMMAE or BULBILS for such spread. *See also* budding; cutting; graft; layering; perennating organ.

vegetative nucleus (**tube nucleus; vegetative cell**) One of the two or three nuclei in a young pollen grain that are formed after division of the haploid nucleus. After the pollen grain germinates on the surface of the stigma, the vegetative nucleus migrates to the tip of the pollen tube. It is thought to regulate the growth and development of the pollen tube. The tube nucleus disintegrates as the pollen tube penetrates the nucellus.

vegetative parts All parts of a plant except for those involved in sexual reproduction.

vein One of the vascular bundles in a leaf. The pattern in which the veins are arranged in a leaf is called the VENATION.

velamen A layer surrounding the aerial roots of epiphytic plants (e.g. orchids) which, due to the spongy nature of the cells, is able to soak up surface water. It is made up of several layers of dead empty cells situated on the epidermis. The cells are spirally thickened, and are translucent to allow light through to the photosynthetic tissue beneath.

velum *See* annulus.

venation The distribution of veins (vascular strands) in a leaf. Dicotyledons usually show a netlike arrangement whereas monocotyledons generally show a parallel distribution of veins. Venation is a character sometimes used in taxonomy.

venter The swollen base of an archegonium that contains the egg cell (oosphere).

ventral **1.** In thallose plants, such as some liverworts, describing the lower surface closest to the substrate.
2. Designating the upper or adaxial surface of the lateral organs of plants, such as leaves. The term is not in common use in this context. *Compare* dorsal.

vernalization The cold treatment of ungerminated or partially germinated seeds. Certain plants will germinate and other flower only if exposed to low temperatures (1–2% C) at an early period of growth, i.e. they have a chilling requirement. Thus winter varieties of cereals will only flower in summer if sown the previous autumn. Spring-sown winter varieties remain vegetative throughout the season unless they have been vernalized.

vernation The arrangement of leaves in relation to each other in the bud, or the arrangement of perianth segments in a flower bud. See illustration overleaf. *See also* ptyxis.

versatile Describing an anther that is attached to the tip of the filament about halfway along its length in such a way that it can turn freely in the wind, thus aiding pollen dispersal. *Compare* basifixed; dorsifixed.

vesicle A small vacuole of variable origin and shape. It may contain secretory products, e.g. a vesicle budded off from the Golgi apparatus, or particles, e.g. endocytotic vesicle. The membrane envelope isolates the contents from the surrounding cytoplasm.

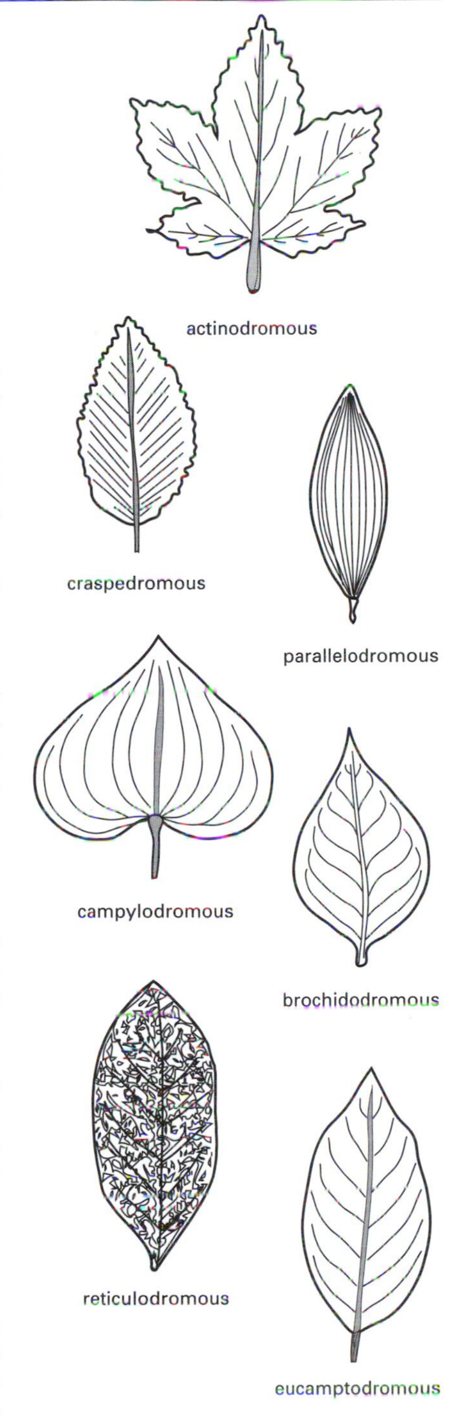

Venation

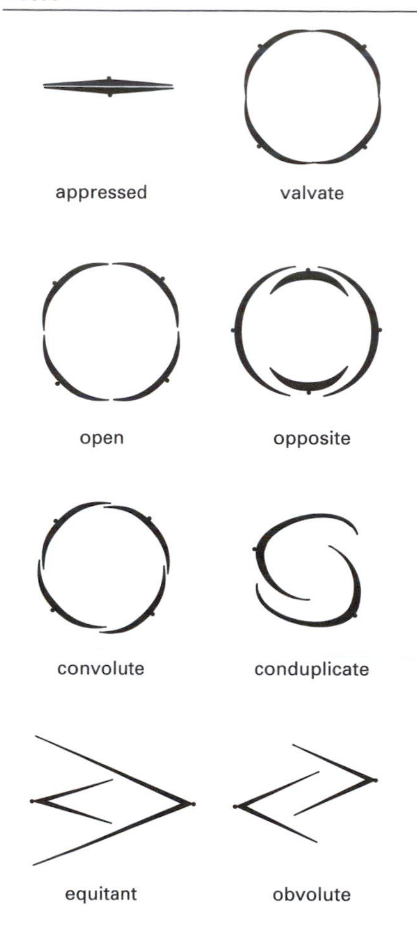

Vernation: types of vernation

vessel A form of xylem conducting tissue composed of vertically arranged *vessel elements* (the individual cells that make up a xylem vessel) joined end to end, generally broader than TRACHEIDS. The end walls of the vessel elements have broken down to leave perforation plates. These are usually horizontal rather than the slanting end walls of tracheids. Vessel elements have lignified secondary cell walls and at maturity lack a living protoplast. Vessels are found in most angiosperms, the gymnosperm order Gnetales, and a few other plants such as bracken (*Pteridium aquilinum*).

vestigial organ An organ that is functionless and generally reduced in size but bears some resemblance to the corresponding fully functioning organs found in related organisms. Examples include the scale leaves of parasitic flowering plants or the reduced stamens in females of dioecious species.

virion The extracellular inert phase of a virus. A virion consists of a protein coat surrounding one or more strands of DNA or RNA. Virions may be polyhedral or helical and vary greatly in size.

viroid A tiny infectious agent found in plants that is similar to a virus but lacks a capsid, consisting simply of a circle of RNA, 300–400 nucleotides long. Viroids replicate within the plant cell and cause characteristic disease symptoms; examples are the potato spindle tuber viroid and the hop stunt viroid.

virulence The relative pathogenicity of an organism: its ability to cause disease. Virulence may vary between strains of the same organism.

virulent phage A bacteriophage that infects a bacterial cell and immediately replicates, causing lysis of the host cell. *Compare* temperate phage.

virus An extremely small infectious agent that causes a variety of diseases in plants, such as tobacco mosaic disease. Viruses can only reproduce in living tissues; outside the living cell they exist as inactive particles consisting of a core of DNA or RNA surrounded by a protein coat (*capsid*). Most plant viruses are single-stranded RNA viruses. The inert extracellular form of the virus, termed a VIRION, penetrates the host membrane and liberates the viral nucleic acid into the cell. Usually, the nucleic acid is translated by the host cell ribosomes to produce enzymes necessary for the reproduction of the virus and the formation of daughter virions. The virions are released by lysis of the host cell. Other viruses remain dormant in the host cell before reproduction and lysis, their nucleic acid becoming integrated with that of the host. Viruses are transmitted by VECTORS such as aphids and nematodes. Some produce visible symptoms, such as leaf mosaics or deformed growth. Others have less obvious effects, but can significantly reduce crop yields. A virus that infects a bacterium is termed a BACTERIOPHAGE (phage).

vital stains Nontoxic coloring materials that can be used in dilute concentrations to stain living material without damaging it. *See also* staining.

vitamins Organic chemical compounds that are essential in small quantities for metabolism but are not synthesized by animals, which therefore need to obtain them from plant or microorganisms. The vitamins have no energy value; most of them seem to act as coenzymes for essential chemical changes in the body, each one influencing a number of vital processes. Vitamins A, D, E, and K are fat-soluble vitamins, while vitamins B and C are water-soluble vitamins.

vivipary The germination of seeds or spores that are still attached to the parent plant. An example is mangroves, in which the seeds germinate into good-sized seedlings before they are released from the plant, enabling them to plunge far enough into the mud to establish themselves.

Volvocales An order of the Chlorophyta (green algae) containing motile unicellular algae (e.g. *Chlamydomonas*), colonial forms (e.g. *Gonium*), and coenobial forms (e.g. *Volvox*). They have a single chloroplast, and often a pyrenoid for the storage of starch and an eyespot.

W

Wallace, Alfred Russel (1823–1913) British naturalist. After a basic education and a short career as a surveyor, Wallace met the entomologist Henry Bates and together they embarked on a scientific expedition as naturalists to the Amazon. Wallace described this expedition in *A Narrative of Travels on the Amazon and the River Negro* (1853). His second expedition, to the Malay Archipelago, took eight years during which he collected over 125 000 specimens. He noticed the marked differences between the Australian and Asian flora and fauna and drew a line on the map separating the regions, still referred to as *Wallace's line*. His observations of the marsupials in Australia led him to conclude that Australian animals were more primitive than Asian ones. He believed that these animals had been cut off from the Asian animals when the land mass of Australia had broken away from Asia. This resulted in less competition for the Australian animals than those on the larger Eurasian land mass. These conclusions led him to formulate a theory of evolution by natural selection, which he communicated to DARWIN; as a result the two men presented joint papers at the famous Linnaean Society meeting of 1858. Unlike Darwin, Wallace never believed that the theory of evolution could be applied to humans. Wallace continued to gather evidence for the theory and published his results in *Geographical Distribution of Animals* (1876) and *Island Life* (1880).

wall pressure *See* turgor pressure.

water culture *See* hydroponics.

water ferns *See* Marsileaceae.

water molds *See* Saprolegniales.

water pore *See* aquaporin.

water potential Symbol: ψ A measure of the energy available in an aqueous solution to cause water to move across a selectively permeable membrane during osmosis. Water potential may be defined as the chemical potential of water in a biological system compared to the chemical potential of pure water at the same temperature and pressure. It is measured in kilopascals (kPa), and pure water has the value 0 kPa; solutions with increasing concentrations of solute have more negative values of ψ, since the solute molecules interfere with the water molecules. This effect is termed the *solute potential* (or osmotic potential), denoted by ψ_s; it is measured in kPa and always has a negative value, with increasing concentrations of solute having increasingly negative values of ψ_s. In turgid plant cells there is also a pressure exerted by the walls of the cell; this is called the *pressure potential*; it is defined by ψ_p, and has a positive value (although in xylem cells it may be negative due to water movement in the transpiration stream). Water potential is also affected by capillary forces, especially in the narrow tubes of the xylem, and in the interstices of cell walls, and by imbibitional forces. These forces are particularly important in cell walls and intercellular spaces, increasing in wilting plants.

wax One of a group of water-insoluble substances with a very high molecular weight; they are esters of long-chain alcohols with fatty acids. Waxes form protective coverings to leaves, stems, fruits, seeds, animal fur, and the cuticles of insects, serving principally as waterproofing.

WCMC *See* World Conservation Monitoring Centre.

WCU *See* World Conservation Union.

Western blotting A technique analogous to SOUTHERN BLOTTING used to separate and identify proteins instead of nucleic acids. The protein mixture is separated by electrophoresis and blotted onto a nitrocellulose filter. Antibodies specific to the protein of interest are applied and bind to their target proteins. A second antibody, specific to the first antibody, is then applied. This carries a radioactive label, so enabling it to be located by autoradiography wherever it binds to the antibody–protein complex.

wild-type The most commonly found form of a given gene in wild populations. Wild-type alleles, often designated +, are usually dominant and produce the 'normal' phenotype.

wilt A plant disease characterized by wilting, usually due to pathogenic fungi or bacteria, or to parasites. It is often a root disease that eventually prevents the plant taking up water. Other wilts occur when the vascular tissues become blocked, e.g. Dutch elm disease, caused by the fungus *Ceratocystis ulmi*, which causes the plant to block off its own infected tissues.

wilting The loss of turgor due to lack of water. The tissues become limp and leaves and flowers droop. Wilting may be due to the rate of transpiration exceeding the rate at which the roots can take up water in the soil during drought periods, or to infection (*see* wilt). The *permanent wilting point* is reached when the plant has wilted to a point where it will not recover, even if more water is added, because the cells have plasmolyzed. *See* turgor.

wood The hard fibrous structure found in woody perennials such as trees and shrubs. It is normally formed from the secondary xylem and thus found in plants that show secondary thickening, namely the conifers, *Gnetum* and dicotyledons. Water and nutrients are only transported in the outermost youngest wood, termed the SAPWOOD. The nonfunctional compacted wood of previous seasons' growth is called the HEARTWOOD and it is this that is important commercially.

woody perennial *See* perennial.

World Conservation Monitoring Centre (WCMC) An information service set up by the World Conservation Union, the Worldwide Fund for Nature and the United Nations Environment Program to compile information on conservation and the sustainable use of the world's natural resources. The WCMC collects data on threatened and endangered species and habitats, national parks and nature reserves, international agreements, and conservation and environment programs. *See* Red Data Book.

World Conservation Union (WCU) Formerly the International Union for the Conservation of Nature (IUCN), founded in 1948, an organization that brings together states, government agencies, and a range of nongovernmental organizations to promote conservation and the sustainable use of natural resources. Over 10 000 expert volunteers collect information on particular species and biodiversity conservation projects, providing a base of expertise that enables the WCU to advise countries on national conservation strategies.

Worldwide Fund (for Nature) (WWF) The largest independent conservation organization in the world, comprising a global network of national organizations, associates, and program offices. It provides conservation services based on global policy, fieldwork, and scientific information. Its goal is to prevent genetic, species, and ecosystem diversity, to ensure the sustainable use of resources, and to promote actions to reduce pollution.

WWF *See* Worldwide Fund (for Nature).

XYZ

xanthophyll One of a class of yellow to orange pigments derived from carotene, the commonest being lutein. In most plants, xanthophylls function as accessory pigments in photosynthesis: they absorb light at wavelengths where chlorophyll has low absorption and pass the absorbed energy to chlorophyll. In the Phaeophyta (brown algae) the xanthophylls fucoxanthin and peridinin act as the primary light-absorbing pigments. *See* carotenoids; photosynthetic pigments.

xeromorphic Structurally adapted to withstand dry conditions. *See* xerophyte.

xerophyte Any plant adapted to growing in dry conditions or in a physiologically dry habitat, such as an acid bog or a salt marsh, or an exposed, very windy situation, by storing available water, reducing water loss, or possessing deep root systems. Succulents, such as cacti and agaves (family Agavaceae), have thick fleshy stems or leaves that store water. Features associated with reducing water loss include: shedding or dieback of leaves, e.g. ocotillo (*Fouquieria splendens*); waxy leaf coatings coupled with closure or plugging of stomata, e.g. *Kalanchoe*; sunken or protected stomata e.g. marram grass (*Ammophila arenaria*); folding, rolling or repositioning of leaves to reduce sunlight absorption, e.g. marram grass; a dense covering of white reflective spines, e.g. cacti; and the development of a dense hairy leaf covering, e.g. *Espeletia*. *Compare* hydrophyte; mesophyte. *See also* halophyte.

xylem The water-conducting tissue in vascular plants. It consists of dead hollow cells (the TRACHEIDS and VESSELS), which are the conducting elements. It also contains additional supporting tissue in the form of fibers and sclereids and some living parenchyma. The proto- and metaxylem, which together constitute the primary xylem, are formed from the procambium of the apical meristems, while the secondary xylem is differentiated from the cambium or lateral meristem. *See* secondary growth.

yeasts *See* Saccharomycetales.

zinc A micronutrient that is needed for the activity of the enzyme alcohol dehydrogenase, which catalyzes the conversion of acetaldehyde to ethanol in anaerobic respiration. *See* micronutrient.

zoosporangium (*pl.* **zoosporangia**) A sporangium that produces ZOOSPORES.

zoospore An asexual motile spore produced by a zoosporangium. It has one or more undulipodia. These motile spores may encyst in adverse conditions or may be the means by which a fungus or parasitic protoctist penetrates a new host.

zwitterion An ion with both a positive and a negative charge. Amino acids can form zwitterions: the amino group has the form $-NH_3^+$ and the acid group is ionized as $-COO^-$.

zygomorphy *See* bilateral symmetry.

zygomycete Any fungus belonging to the phylum Zygomycota.

Zygomycota A phylum of fungi whose members have haploid, nonseptate hyphae (i.e. hyphae that lack cross walls). Their cell walls contain chitin or chitosan. They do not produce motile spores at any stage of the life cycle. Zygomycetes are mostly saprophytic, absorbing nutrients from decaying vegetation and other organic matter; the bread molds *Rhizopus* and *Mucor* are examples.

zygospore A resistant sexual spore formed when a zygote develops a thick wall. Zygospores are characteristic of the ZYGOMYCOTA, especially the Mucorales (which include such fungi as *Mucor* and *Rhizopus*) and the GAMOPHYTA (conjugating green algae, e.g. *Spirogyra* and *Zygnema*).

zygote The diploid cell resulting from the fusion of two haploid gametes. A zygote usually undergoes mitosis immediately. *See* sexual reproduction.

zygotene *See* prophase.

APPENDIXES

Appendix I

SI Units

BASE AND DIMENSIONLESS SI UNITS

Physical quantity	*Name of SI unit*	*Symbol for SI unit*
length	meter	m
mass	kilogram(me)	kg
time	second	s
electric current	ampere	A
thermodynamic temperature	kelvin	K
luminous intensity	candela	cd
amount of substance	mole	mol
*plane angle	radian	rad
*solid angle	steradian	sr
*supplementary units		

DERIVED SI UNITS WITH SPECIAL NAMES

Physical quantity	*Name of SI unit*	*Symbol for SI unit*
frequency	hertz	Hz
energy	joule	J
force	newton	N
power	watt	W
pressure	pascal	Pa
electric charge	coulomb	C
electric potential difference	volt	V
electric resistance	ohm	Ω
electric conductance	siemens	S
electric capacitance	farad	F
magnetic flux	weber	Wb
inductance	henry	H
magnetic flux density	tesla	T
luminous flux	lumen	lm
illuminance (illumination)	lux	lx
absorbed dose	gray	Gy
activity	becquerel	Bq
dose equivalent	sievert	Sv

DECIMAL MULTIPLES AND SUBMULTIPLES USED WITH SI UNITS

Submultiple	*Prefix*	*Symbol*	*Multiple*	*Prefix*	*Symbol*
10^{-1}	deci-	d	10^{1}	deca-	da
10^{-2}	centi-	c	10^{2}	hecto-	h
10^{-3}	milli-	m	10^{3}	kilo-	k
10^{-6}	micro-	μ	10^{6}	mega-	M
10^{-9}	nano-	n	10^{9}	giga-	G
10^{-12}	pico-	p	10^{12}	tera-	T
10^{-15}	femto-	f	10^{15}	peta-	P
10^{-18}	atto-	a	10^{18}	exa-	E
10^{-21}	zepto-	z	10^{21}	zetta-	Z
10^{-24}	yocto-	y	10^{24}	yotta-	Y

Appendix II

Webpages

The following all have useful information:

American Society of Plant Biologists	www.aspb.org/education/
Botanical Society of America	www.botany.org
European Initiative for Biotechnology Education	www.rdg.ac.uk/EIBE/
Internet Directory for Botany	www.botany.net/IDB/info.html
Missouri Botanical Garden	www.mobot.org/
National Museum of Natural History	nmnhwww.si.edu/departments/botany.html
New York Botanical Garden	www.nybg.org
Photosynthesis and the Web	photoscience.la.asu.edu/photosyn/photoweb/default.html
Royal Botanic Gardens, Kew	www.rbgkew.org.uk/
University of California Botanical Garden	www.mip.berkeley.edu/garden/

Bibliography

Alberts, B., Johnson, A., Lewis, J., Raff, M., Roberts, K., & Walter, P. *Molecular Biology of the Cell.* 4th ed., New York: Garland Science, 2002

Ayala, Francisco J., Fitch, Walter M., & Clegg, Michael T. (eds). *Variation and Evolution in Plants and Microorganisms.* Washington: National Academy Press, 2000

Bell, Arnold D. *Plant Form: An Illustrated Guide to Flowering Plant Morphology.* Oxford, U.K.: Oxford University Press, 1991

Bell, Peter R. & Hemsley, Alan R. *Green Plants: Their Origin and Diversity.* 2nd ed. Cambridge, U.K.: Cambridge University Press, 2000

Campbell, Neil A. & Reece, Jane B. *Biology.* 6th ed. New York: Benjamin Cummings, 2002

Hopkins, W. G. *Introduction to Plant Physiology.* 2nd ed. New York: John Wiley and Sons, 1995

Lack, A. J. & Evans, D. E. *Instant Notes: Plant Biology.* New York: Springer-Verlag, 2001

Mabberley, D. J. *The Plant-Book: A portable dictionary of the vascular plants.* 2nd ed. Cambridge, U.K.: Cambridge University Press, 1997

Margulis, L. & Schwartz, K. V. *Five Kingdoms: An Illustrated Guide to the Phyla of Life on Earth.* 3rd ed. New York: W. H. Freeman, 1998

Niklas, Karl J. *The Evolutionary Biology of Plants.* Chicago: University of Chicago Press, 1997

Purves, William K., Sadava, David, Orians, Gordon H., & Heller, H. Craig. *Life: The Science of Biology.* 6th ed. New York: W. H. Freeman, 2001

Raven, P. H., Evert, R. F. & Eichorn, S. E. *Biology of Plants.* 6th ed. New York: W.H. Freeman, 1998

Ridge, Irene (ed). *Plants.* New York: Oxford University Press, 2002

Stern, K. R. *Introductory Plant Biology.* 8th ed. New York: Sinauer Associates, 1998

Westhoff, P., Jeske, H., Jurgens, G., Kloppstech, K., & Link, G. *Molecular Plant Development - from Gene to Plant.* New York: Oxford University Press, 1998

WITHDRAWN
UNIV OF MOUNT UNION LIBRARY